James Clerk Maxwell

JAMES CLERK MAXWELL

Perspectives on his Life and Work

Edited by

RAYMOND FLOOD

MARK McCARTNEY

and

ANDREW WHITAKER

OXFORD
UNIVERSITY PRESS

OXFORD
UNIVERSITY PRESS

Great Clarendon Street, Oxford, OX2 6DP,
United Kingdom

Oxford University Press is a department of the University of Oxford.
It furthers the University's objective of excellence in research, scholarship,
and education by publishing worldwide. Oxford is a registered trade mark of
Oxford University Press in the UK and in certain other countries

First Edition published in 2014

Published in the United States of America by Oxford University Press
198 Madison Avenue, New York, NY 10016, United States of America

British Library Cataloguing in Publication Data
Data available

Library of Congress Control Number: 2013942195

ISBN 978–0–19–966437–5

FOREWORD

The names of Archimedes, Newton, Einstein and Crick have entered into legend because fragments of their achievements and evidence of their personalities have captured the public's imagination. However, another very great scientist, whose fundamental research today pervades all the physical sciences and engineering and enables the technology that makes advances in the life sciences possible, remains virtually unknown outside scientific circles where he is rightly regarded as a giant.

According to everyone who knew him, Maxwell was practical, modest, kind and had a strong sense of humour. For him, science was seemingly effortless, a gentleman landowner's hobby rather than an obsession driven by ambition and riven with rivalry, as was the case with Newton. He was humane and sensitive while possessing extraordinary power and virtuosity in science, and seems to have been devoid of any shred of pomposity. He was very fond of dogs and always had one which he always called Toby.

Maxwell was born in Edinburgh, in 1831, to a landed family, and many of his ancestors on both sides were intellectually accomplished, members of the Royal Society, or the Royal Society of Edinburgh, or both. After a precocious childhood on the family estate and university at Edinburgh and Cambridge, and a university position in Aberdeen, where he explained the rings of Saturn in a prize-winning essay and was made redundant, he moved to King's College, London, in 1860. There followed five *anni mirabiles* during which he consolidated his work on electromagnetism, founded the theory of statistical mechanics and developed his work on colour vision, producing, in 1861, the first colour photograph (appropriately of a tartan ribbon) at the Royal Institution, where he came into regular contact with Michael Faraday, by then a much older man. While at King's he was awarded the Royal Society's 1860 Rumford Medal for his work on colour and elected FRS a year later.

However, in 1865 (the year in which he proposed that radiant heat and light are both examples of electromagnetic waves), he resigned his position at King's to run the family estate while continuing to do research in his own time. There he remained until 1871, when he was prevailed upon to be the first Cavendish Professor in Cambridge. He died there in 1879, aged 48, and was buried in the graveyard of the Old Kirk in Parton, close to the family home.

If he had lived as long as Archimedes or Einstein (76) he would have seen Marconi receive transatlantic radio signals, learned of Einstein's theories of special relativity, of the photoelectric effect and of Brownian motion, on all of which his own work had exerted a profound influence.

If he had lived as long as Crick (88) he would have seen Einstein's theory of general relativity and Max Planck win the Nobel Prize for quantum theory. Maxwell was the equal of these great scientists and his early death was a huge loss.

How can someone with Maxwell's presence and achievement be so uncelebrated in his own time and place and later? When the popular historical novelist, Sir Walter Scott, died in 1832, a 61-metre high tower to his memory was erected on Princess Street, Edinburgh in 1846. When Bobby, a dog much-loved for his fidelity, died in 1872, a statue to his memory was unveiled at Greyfriars Churchyard in Edinburgh a year later. Bobby's legend has survived and his life was made into a Walt Disney film in 1961. Maxwell might have approved.

However, when James Clerk Maxwell, arguably the greatest scientist of the nineteenth century, died in 1879, no statue in Edinburgh was erected to his memory for 129 years. Today there is a statue depicting Maxwell, looking down George Street with his back to St. Andrew's Square, sitting, with Toby loyally at his feet, holding his colour wheel, with his famous equations for the electric and magnetic fields written on the plinth. He might have had misgivings, but one hopes that Maxwell would have been persuaded of the good intentions of his admirers in honouring him for his achievements.

That also is the intention of this book.

John Toland
Isaac Newton Institute for Mathematical Sciences
Cambridge

PREFACE

One hundred and fifty years ago, in the autumn of 1863, James Clerk Maxwell was living at 8 Palace Gardens Terrace, Kensington, London, and was professor at King's College. It was the second of three professorial appointments (the first being at Marischal College, Aberdeen, the last at Cambridge). He was 31, a fellow of the Royal Societies of both London and Edinburgh, and had an established reputation within the Victorian scientific community. However, though he could not know it at the time, he was midway through, rather than early on, in his scientific career. By the autumn of 1879, aged 48, he would be dead. It was a relatively brief, but remarkable, life, lived in his beloved rural home of Glenlair, and variously in Edinburgh, Aberdeen, London and Cambridge. His scholarship also ranged wide – covering all the major aspects of Victorian natural philosophy.

In scientific terms, his immortality is enshrined in electromagnetism and Maxwell's equations but, as this book seeks to show, there was much more to Maxwell than electromagnetism, both in terms of his science and his wider life. Indeed, as editors of this book we are of the view that Maxwell's life and contributions to science are so rich as to demand the expertise of a range of academics – physicists, mathematicians and historians of science and literature – to do him justice. We hope that the various chapters will enable Maxwell to be seen from a range of perspectives. Chapters 1 to 4 deal with wider aspects of his life in time and place, before Chapters 5 to 12 go on to look in more detail at his wide-ranging contributions to science, with concluding chapters on Maxwell's poetry and Christian faith. Each chapter is self-contained and can be read independently of the others, and though this has inevitably led to some repetition and overlap, we trust that this will not be a burden to the reader.

We wish to thank each of the contributors for their chapters, their expertise, and their enthusiasm for both Maxwell and this book. They have submitted to our editorial requests and queries along the way with good humoured compliance. We also wish to thank the staff at OUP (including Clare Charles, Keith Mansfield, Caroline McPherson, Victoria Mortimer and Hemalackshmi Niranjan) for their help, advice and patience on the occasionally winding route which this book took from commissioning to final production.

Raymond Flood
Mark McCartney
Andrew Whitaker
November, 2013

CONTENTS

PART I

Life

CHAPTER 1

Introduction

RAYMOND FLOOD

James Clerk Maxwell's work on electricity and magnetism was such that Einstein enthused:

> Since Maxwell's time, Physical Reality has been thought of as represented by continuous fields . . . and not capable of any mechanical interpretation. This change in the conception of Reality is the most profound and the most fruitful that physics has experienced since the time of Newton.[1]

While the distinguished physicist, Richard Feynman, predicted:

> From a long view of the history of mankind — seen from, say, ten thousand years from now — there can be little doubt that the most significant event of the 19th century will be judged as Maxwell's discovery of the laws of electrodynamics.[2]

I will start by giving an overview of Maxwell's life and will then consider his major areas of interest and his achievements.[3]

Maxwell's Life

James Clerk Maxwell was born in Edinburgh on 13 June 1831. Two years later, the family moved to Glenlair, an estate that his father had inherited, about 90 miles southwest of Edinburgh. The family was financially well off and well established. James's father, John, had trained as a lawyer but seems to have been more interested in what we might now call 'design and technology'. His mother, Frances, was 40 when James was born and she died in December 1839. From all accounts James was a much loved, precocious, curious and intelligent child. After his mother's death his father appointed a tutor for him, which was not a success – to put it mildly – and as a result James was sent to Edinburgh Academy and to live with relatives just under two years after his mother's death.

P. G. Tait was another pupil at the Academy and they became lifelong friends, even though as adults they competed for the same job on more than one occasion. It was while James was at

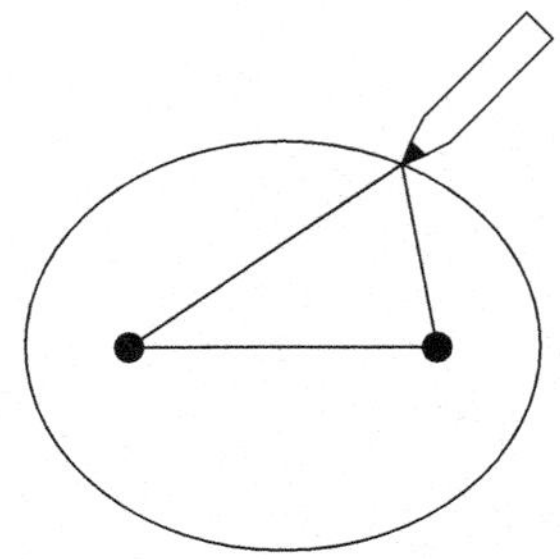

Fig. 1.1 Method of drawing an ellipse with a piece of string.

Edinburgh Academy that he published his first academic paper, at the age of 14, in the *Proceedings of the Royal Society of Edinburgh*. It arose out of his attempts to generalize the well-known process for drawing an ellipse by attaching a string to two pins and keeping the string stretched with the point of your pencil and then moving the pencil around to trace an ellipse (see Fig. 1.1).

James investigated what happens when the string is folded back on itself towards each pin. Figure 1.2 shows the opening page of his paper, as it was later reprinted in his collected works. In the bottom left picture, the string starts at the left pin, goes round the pencil and then round the right pin back to the pencil where it is attached. So the length of the string is the sum of the distance to the left pin plus twice the distance to the right. In the bottom right picture, the string starts attached to the pencil, goes round the right pin, back to the pencil, then round the left pin, back to the pencil and is then attached to the right pin. So the length of the string is the sum of twice the distance to the left pin plus three times the distance to the right. The page shows some of the ovals that can be generated. And Maxwell did not stop with just two pins!

Although Descartes had described ways of generating these curves, Maxwell's method was new. Maxwell's father, John, showed the work to a friend, J. D. Forbes, who was Professor of Natural Philosophy at Edinburgh University and who arranged for its publication. Apart from showing his talent at such an early age, it also demonstrates two of Maxwell's lifelong characteristics: the ability to generalize and his ability with, and enthusiasm for, geometrical reasoning.

In 1847, James enrolled at Edinburgh University, where he would study mathematics under Philip Kelland, natural philosophy under James Forbes and logic under Sir William Hamilton. He was 16 at the time and undecided on his future career, in particular whether he should follow his father's wishes and become a lawyer. When he left Edinburgh University three years later he was decided on a scientific career or as he put it, I think most charmingly, to pursue 'another kind of laws'.[4]

I want to pick out one investigation that Maxwell undertook at Edinburgh University because it illustrates that he was a gifted experimentalist as well as a theoretician – he was like Newton in this regard. This work was on the 'Equilibrium of Elastic Solids' and was published in 1850. Elasticity will come into our story again when we turn to electromagnetism later. In this paper Maxwell axiomatized the equations of elasticity and applied the results to a number of problems (see Fig. 1.3).

Maxwell went up to Cambridge in 1850. He was at Peterhouse College for his first term, but then migrated to Trinity College, possibly because there might have been better career opportunities there. He graduated as Second Wrangler, with E. J. Routh taking the top spot of Senior Wrangler. In the subsequent competition for the Smith's Prize, Maxwell and Routh were declared

[From the *Proceedings of the Royal Society of Edinburgh*, Vol. II. April, 1846.]

I. *On the Description of Oval Curves, and those having a plurality of Foci; with remarks by Professor Forbes.* Communicated by Professor Forbes.

Mr Clerk Maxwell ingeniously suggests the extension of the common theory of the foci of the conic sections to curves of a higher degree of complication in the following manner:—

(1) As in the ellipse and hyperbola, any point in the curve has the *sum* or *difference* of two lines drawn from two points or *foci*=a constant quantity, so the author infers, that curves to a certain degree analogous, may be described and determined by the condition that the simple distance from one focus *plus* a multiple distance from the other, may be=a constant quantity; or more generally, m times the one distance+n times the other=constant.

(2) The author devised a simple mechanical means, by the wrapping of a thread round pins, for producing these curves. See Figs. 1 and 2. He

Fig. 1. Two Foci. Ratios 1, 2.

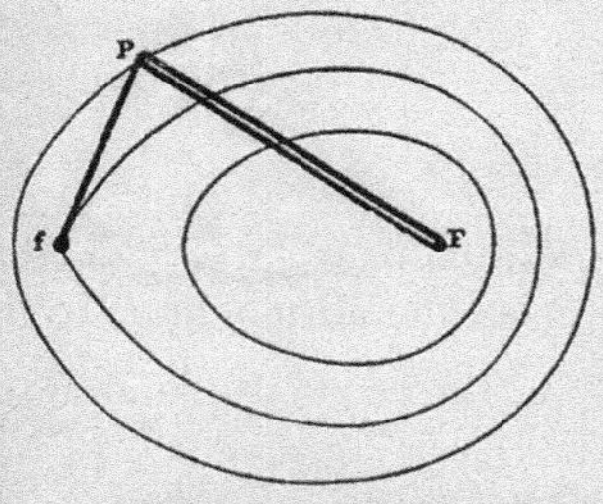

Fig. 2. Two Foci. Ratios 2, 3.

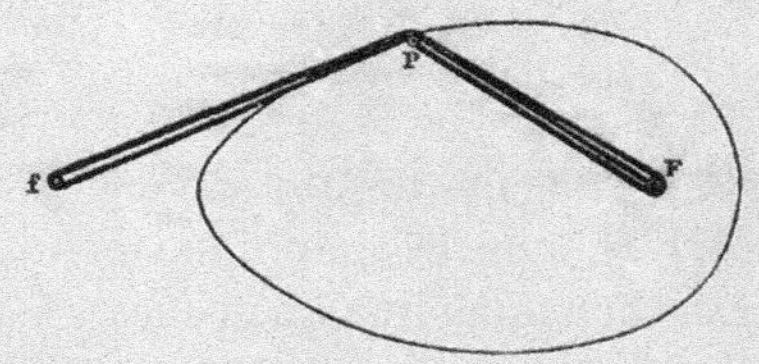

then thought of extending the principle to other curves, whose property should be, that the sum of the simple or multiple distances of any point of

Fig. 1.2 'On the Description of Oval Curves and those having a Plurality of Foci; with Remarks by Professor Forbes', published by James Clerk Maxwell at the age of 14.

joint winners. In the Smith's Prize examination paper that year, one of the questions was on Stokes' Theorem, connecting surface and line integrals (see Fig. 1.4). Maxwell later made great use of this theorem in his work on electromagnetism.

Maxwell had done well, not as well perhaps as Tait, who two years earlier had been Senior Wrangler *and* Smith's Prizeman, but then Tait was not up against Routh. All of them had studied under the private tutor William Hopkins, as had Stokes, Cayley and William Thomson.

In 1855, Maxwell became a fellow of Trinity. This year also saw him publish the paper 'Experiments on colour as perceived by the eye', as well as the first part of his paper on Faraday's lines of force, which showed the first evidence of true genius. The second part of the paper was published the following year.

In 1856, Maxwell received a letter from Forbes telling him of a vacancy in the Chair of Natural Philosophy at Marischal College, one of the two university level colleges in Aberdeen at that time, and suggesting that he should apply. At the age of 24, Maxwell was young for a professorship and relatively inexperienced, but others had been appointed at similar ages: William Thomson

I have therefore substituted for the assumption of Navier the following axioms as the results of experiments.

If three pressures in three rectangular axes be applied at a point in an elastic solid,—

1. *The sum of the three pressures is proportional to the sum of the compressions which they produce.*

2. *The difference between two of the pressures is proportional to the difference of the compressions which they produce.*

The equations deduced from these axioms contain two coefficients, and differ from those of Navier only in not assuming any invariable ratio between the cubical and linear elasticity. They are the same as those obtained by Professor Stokes from his equations of fluid motion, and they agree with all the laws of elasticity which have been deduced from experiments.

Fig. 1.3 The axioms from Maxwell's 'On the Equilibrium of Elastic Solids', 1850.

8. If X, Y, Z be functions of the rectangular co-ordinates x, y, z, dS an element of any limited surface, l, m, n the cosines of the inclinations of the normal at dS to the axes, ds an element of the bounding line, shew that

$$\iint \left\{ l\left(\frac{dZ}{dy}-\frac{dY}{dz}\right)+m\left(\frac{dX}{dz}-\frac{dZ}{dx}\right)+n\left(\frac{dY}{dx}-\frac{dX}{dy}\right)\right\} dS$$
$$=\int\left(X\frac{dx}{ds}+Y\frac{dy}{ds}+Z\frac{dz}{ds}\right)ds,$$

the differential coefficients of X, Y, Z being partial, and the single integral being taken all round the perimeter of the surface.

9. Explain the geometrical relation between the curves, referred to the rectangular co-ordinates x, y, z, whose differential equations are

$$\frac{dx}{P}=\frac{dy}{Q}=\frac{dz}{R},$$

and the family of surfaces represented by the partial differential equation

$$P\frac{dz}{dx}+Q\frac{dz}{dy}=R.$$

Fig. 1.4 Question 8, on Stokes' Theorem, set in the Smith's Prize examination paper taken by James Clerk Maxwell in 1854.

to his chair at Glasgow at the age of 22, and Tait to his professorship of mathematics at Queen's College, Belfast, at the age of 23. Opportunities did not come up very often, and a position at Aberdeen would also allow Maxwell to be closer to his father. However, his father died at Easter in 1856 before James heard that he had been appointed. His appointment to the chair and his time at Marischal College is described in John Reid's Chapter 2, *Maxwell at Aberdeen.*

The university year ran from November to April, when the students went home to help with their parents' farms, trades or professions, so Maxwell was able to spend half the year at Aberdeen and the rest at Glenlair. In 1858, he married Katherine Mary Dewar, daughter of the Principal of Marischal College. She helped him in his scientific experiments, both in Aberdeen with his colour experiments and later in London on kinetic theory, in particular on the viscosity of air (see Fig. 1.5).

Fig. 1.5 Maxwell and his wife Katherine were devout Christians. Maxwell's faith is discussed in Philip Marston's Chapter 14, *Maxwell, Faith and Physics.*

The same year that he married, Maxwell was awarded the Adams Prize of Cambridge University for his work on the stability of the Rings of Saturn. He had also been thinking about the structure of gases, and in an 1860 paper introduced statistical methods into physics.

When, in 1860, Aberdeen merged its two colleges into one university, Maxwell lost out to the other Professor of Natural Philosophy, David Thomson. Maxwell had already applied for the Edinburgh Chair of Natural Philosophy that Forbes had left to become Principal of the United College of St. Salvator and St. Leonard at St. Andrews. But he was up against Tait and Routh, and this time it was Tait who succeeded (Maxwell had previously beaten Tait to the Marischal College post). Maxwell then applied for a vacant chair at King's College, London, and was successful, as described in John Reid's Chapter 3, *Maxwell at King's College, London.* For the next six very productive and fruitful years, the Maxwells divided their time between London and Glenlair.

In his inaugural lecture, as Professor of Physics and Astronomy at King's College, London, he said:

In this class I hope you will learn not merely results, or formulae applicable to cases that may possibly occur in our practice afterwards, but the principles on which those formulae depend, and without which the formulae are mere mental rubbish. I know the tendency of the human mind is to do anything rather than think. But mental labour is not thought, and those who have with labour acquired the habit of application, often find it much easier to get up a formula than to master a principle.[5]

And he continued with what turned out to be an amazingly prophetic statement:

. . . last of all we have the Electric and Magnetic sciences, which treat of certain phenomena of attraction, heat, light and chemical action, depending on conditions of matter, of which we have as yet only a partial and provisional knowledge. An immense mass of facts have been collected and these have been reduced to order, and expressed as the results of a number of experimental laws, but the form under which these laws are ultimately to appear, as deduced from central principles, is as yet uncertain. The present generation has no right to complain of the great discoveries already made, as if they left no room for further enterprise. They have only given science a wider boundary, and we have not only to reduce the regions already conquered but to keep up constant operations on a continually increasing scale.[6]

The most important achievement of Maxwell's London years was undoubtedly the series of articles on electromagnetic theory in which the celebrated Maxwell's equations first appeared, though not in their modern form.

In January 1865, Maxwell resigned, mainly, it appears, to return to Glenlair to occupy his time with experiments and speculations of a physical kind which, as he said, 'I could not undertake as long as I had public duties'.[7] The time he spent at Glenlair was not a period of retirement. As well as preparing his great *Treatise on Electricity and Magnetism*, he published a book on the theory of heat, in which he introduced Maxwell's demon, discussed in Andrew Whitaker's Chapter 9, *Maxwell's Famous (or Infamous) Demon.* He was also developing his ideas on the theory of gases and continued to be involved with the British Association for the Advancement of Science.

Then, in 1871, he was asked to go back to Cambridge to set up and direct the proposed Cavendish Laboratory of Experimental Physics. (Maxwell was not the first to be approached to

A LECTURE ON THOMSON'S GALVANO-METER

DELIVERED TO A SINGLE PUPIL IN AN ALCOVE WITH DRAWN CURTAINS

THE lamp-light falls on blackened walls,
  And streams through narrow perforations;
The long beam trails o'er pasteboard scales,
  With slow-decaying oscillations.
Flow, current! flow! set the quick light-spot flying!
Flow, current! answer, light-spot! flashing, quivering,
  dying.

O look! how queer! how thin and clear,
  And thinner, clearer, sharper growing,
This gliding fire, with central wire
  The fine degrees distinctly showing.
Swing, magnet! swing! advancing and receding;
Swing, magnet! answer, dearest, what's your final reading?

O love! you fail to read the scale
  Correct to tenths of a division;
To mirror heaven those eyes were given,
  And not for methods of precision.
Break, contact! break! set the free light-spot flying!
Break, contact! rest thee, magnet! swinging, creeping,
  dying.

$\frac{dp}{dt}$.

Fig. 1.6 Maxwell published this poem in the 16 May issue of *Nature* in 1872. His poetry is discussed in Stella Pratt-Smith's Chapter 13, *Boundaries of Perception: James Clerk Maxwell's Poetry of Self, Senses and Science.*

direct the new laboratory. Thomson had been approached first and then Helmholtz.) Maxwell's emphasis for the laboratory was to achieve measurements of high precision, sometimes to several orders of magnitude better than any previous attempts. Isobel Falconer describes this in her Chapter 4, *Cambridge and the Building of the Cavendish Laboratory.* Figure 1.6 shows a poem of Maxwell's about measurement, published in 1872.

In 1877, Maxwell's health started to fail and over the following two years there were periods when he was unable to work. He died in November 1879, and was buried in Parton cemetery, near Castle Douglas in Galloway (Fig. 1.7).

Maxwell's Work

The distinguished Charles Coulson, a twentieth-century successor to Maxwell at King's College, London, commented on Maxwell: 'There is scarcely a single topic that he touched upon that he did not change almost beyond recognition'.[8] Maxwell tended to jump between areas of investigation, so I want to collect and discuss his work in the areas of (i) Saturn's rings; (ii) colour vision; (iii) kinetic theory of gases; (iv) electromagnetism.

Fig. 1.7 The grave of James Clerk Maxwell, whose name is preceded by those of his father and mother and followed by that of his wife. Mark McCartney, in the concluding Chapter 15, *I Remember Years and Labours as a Tale That I have Read*, discusses the development of Maxwell's reputation since his death.

Saturn's Rings

It was Maxwell's investigation of the dynamics and stability of Saturn's rings that first gave him a national reputation and earned him the recognition of Stokes and Thomson, for example. The question of the rings' stability was the subject of the 1855 Adams Prize, one of whose setters was Thomson. Entries had to be submitted by December 1856. The problem was difficult, and even the great Laplace, author of the standard work on celestial mechanics, had only obtained partial results in the case where the ring was assumed solid. Laplace had shown that a uniform solid ring would be unstable, but conjectured that a solid ring could be stable if its mass was unevenly distributed.

Maxwell first showed that Laplace's intuition was correct and that a solid ring could not be stable except in a strange configuration where about 80% of the mass was at one point on the ring and the rest uniformly distributed. This was ruled out by observation of Saturn.

The next step was to look at fluid rings and their stability. This was to depend crucially on how internal wave motions behaved, and Maxwell showed that these waves would cause a fluid ring to break up into separate blobs, so would need to consist of separate parts. Next to consider was the stability of a ring of different separate parts. The situation of a ring with a large number of objects of different sizes and masses was too difficult, and Maxwell simplified things by considering a single ring with identical, equally spaced particles. He showed that such a system would show four modes of vibration but, as long as the average density was low enough, it would be stable. When he considered two concentric rings he found that stability imposed conditions on the ratio of their radii. The analytic techniques he used in his essay owed a great deal to his mathematical studies at Cambridge.

More detail and background is in Andrew Whitaker's Chapter 6, *Maxwell and the Rings of Saturn* while Keith Moffatt, in Chapter 12, discusses *The Fluid Dynamics of James Clerk Maxwell.*

Colour Vision

Similarly, I would suggest that his work on colour owed a great deal to his experimental training at Edinburgh.

When at Edinburgh, both Maxwell and Forbes were interested in colour vision. It had been suggested by Thomas Young, in 1800, that there were three sets of colour receptors in the eye. Forbes and Maxwell identified the three primary colours as red, green and blue. They used a wooden disc to which three coloured discs could be attached so that different amounts of each colour could be displayed. When the wooden disc is spun these primary colours are in some sense mixed and the brain perceives a single colour. Persistence of vision, which Maxwell was earlier interested in, seems to be fundamental here. The central area of the wooden disc could contain another colour, and his aim was to find the proportions of the primary colours that, on spinning the disc, gave this colour in the centre. The important point is that he quantified the effect of combining his three primary colours of red, green and blue. He represented this geometrically as a colour triangle.

In 1861, Maxwell was invited to give a lecture on colour vision to the Royal Institution. He wanted to illustrate his lecture and demonstrate that any colour could be made by mixing the three primaries of red, green and blue, but the spinning coloured disc was too small to be seen by the audience. A subsequent invention, the colour box, could only be used by one person at a time. However, there was photography. Perhaps he could use photography to show the phenomenon to such a large audience? It was possible to project a black and white photograph onto a screen so that it was large enough for an audience to see. Maxwell wanted to see if he could make a colour photograph and project it in the same way, illustrating that any colour could be made by mixing the three primaries of red, green and blue. For this, he would need to take three photographs of the same object through red, green and blue filters, in turn, and then project them simultaneously on the screen through the same three filters. It worked, though it shouldn't have!

The result could not be duplicated. The problem was that the then available photographic plates were sensitive to light at the blue end of the spectrum but much less to light at the red end. Maxwell had been lucky because of a sensitivity of the emulsion to the ultra-violet. Instead of photographing the ribbon under red light, Maxwell was actually photographing it under ultra-violet light, so the mixture of colours that Maxwell obtained was ultra-violet, blue and blue-green, not the mixture of red, blue and green he believed he had obtained. He had simply been lucky in obtaining the first colour photograph. Malcolm Longair discusses this story of imagination and good fortune in his Chapter 5, *Maxwell and the Science of Colour.*

Kinetic Theory of Gases

Now I want to turn to an area of investigation that shows Maxwell's mathematical as well as experimental abilities. This is the kinetic theory of gases.

In April 1859, Maxwell read a paper by Rudolf Clausius on the speed of diffusion in gases. In the kinetic theory of gases, it is proposed that gases consist of a great number of molecules whizzing around. Their impact with a surface causes pressure, and what we experience as heat is the kinetic energy of their motion. By the mid nineteenth century, kinetic theory could explain most of the experimentally obtained laws connecting pressure, volume and temperature. But for the theory to explain, for example, pressure, the molecules would have to travel very fast, at hundreds of metres a second. But then, why do smells travel relatively slowly?

If I take the stopper out of a bottle of perfume, the fragrance will only gradually permeate the room. Clausius had suggested that each molecule undergoes an enormous number of collisions and so is continually changing direction. As Maxwell put it in a letter to Tait: 'If you travel at 17 miles per minute and take a totally new course 1,700,000,000 times a second, where will you be in an hour?'.[9] Clausius had assumed that all molecules at a given temperature had the same speed. Maxwell, when analysing Saturn's rings, had similarly simplified by assuming identically spaced particles. But his inspiration now was to capture the motion of many particles in a single probabilistic law. It said nothing about an individual molecule but gave the proportion of molecules that had velocities within a given range. Maxwell's brilliant derivation of this distribution comes out in a few lines and is given in Elizabeth Garber's Chapter 7, *Maxwell's Kinetic Theory 1859–70*.

He showed that the distribution of velocities in any direction is given by a normal distribution. This was the first ever statistical law in physics: the Maxwell distribution of molecular velocities. The proportion of velocities between any two velocities, v and w, is just the area under the normal curve between v and w. The average velocity is zero, and as the temperature of the gas increases the curve becomes flatter and wider.

It is relatively simple to derive the distribution of the molecular speeds from knowledge of the distribution of the molecular velocities. Since the speed is the magnitude of the velocity the average speed will be non-zero. The shape of the distribution of the speeds will also vary with temperature.

Maxwell's discovery of the distribution of velocities and speeds was a result of the first order. It led to a new approach in physics. Previously, statistical methods had been used for analysing errors in observations. Social scientists had used statistical methods to describe populations. What Maxwell had done was to use statistical methods to describe physical processes.

This entirely new approach led to statistical mechanics, an enhanced understanding of thermodynamics, and the use of probability in quantum mechanics. It also helped to confirm that gases consist of fast-moving molecules. This was because Maxwell used kinetic theory to predict a new law, that viscosity is independent of pressure. This result on viscosity is surprising, since you might expect a gas at higher pressure to offer more resistance to a body moving through it. However, if the body is surrounded by more molecules they collide into each other more frequently – have a shorter mean free path – and screen the moving body. The screening effect balances the increase in pressure.

Maxwell's work was to provide an inspiration for Ludwig Boltzmann, who spent most of his career developing the subject of statistical mechanics and thermodynamics and putting it on a sound basis. It is very appropriate that their names are now joined in the Maxwell–Boltzmann distribution of molecular energies.

Further work is described in John Rowlinson's Chapter 8, *Maxwell and the Theory of Liquids.*

Electromagnetism

Maxwell began his researches on electricity and magnetism shortly after graduating in 1854, and continued with them until shortly before his death, refereeing a paper by G. F. Fitzgerald on the subject. It is a complicated and fascinating story described by Daniel Siegel in his Chapter 10, *Maxwell's Contribution to Electricity and Magnetism*, and here I will only select some key features.

Oersted's discovery of electromagnetism in 1820 (see Fig. 1.8) was very important because what he observed differed from previous phenomena in two ways. First, he observed that magnetism was caused by electricity in motion and second, that the magnet was not attracted or repelled but was caused to move at right angles to the wire carrying the current. Michael Faraday viewed this result as crucial in his own thinking about electromagnetism and his development of ideas about lines of magnetic and electric force (Fig. 1.9).

Faraday was an important influence on Maxwell, as indeed was Thomson who, for example, developed a formal connection between the equations of electrostatics and the equations for the flow of heat. Maxwell's first paper on electromagnetism, 'On Faraday's Line of Force', was published in 1855 and 1856 in two parts. The first part dealt with an analogy between lines of force and streamlines in an incompressible fluid. In the second part Maxwell started the development of a theory of electromagnetism and made extensive use of Stokes' Theorem.

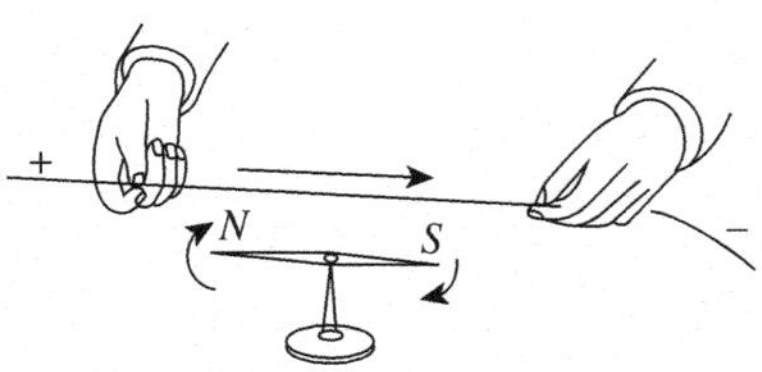

Fig. 1.8 Oersted's experiment.

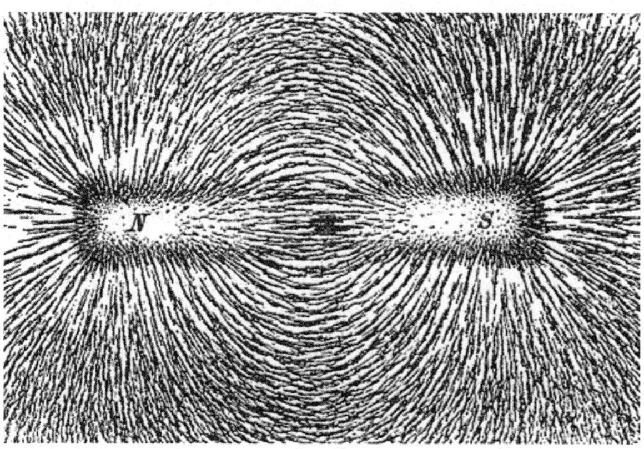

Fig. 1.9 Iron filings scattered on paper over a magnet show the lines of force.

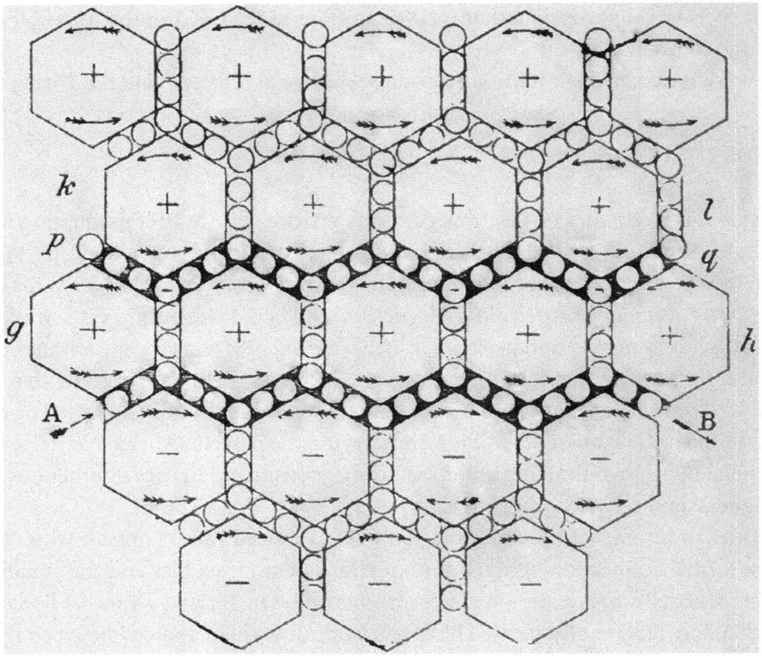

Fig. 1.10 Model of molecular vortices and electric particles, 1861.

Maxwell's next paper, 'On Physical Lines of Force', published in 1861 and 1862, attempted to devise a medium that would fill space and would account for the stresses that Faraday associated with his lines of magnetic force. It finished with the conclusion that vibrations of this medium had properties identical with those of light. Look at the array of hexagonal vortices in an incompressible fluid in Fig. 1.10. Normally the pressure is identical in all directions but if a vortex rotates, centrifugal force will cause it to expand along the middle, in the plane of the page, and contract along the axis of spin, perpendicular to the page – in the same way that the Earth's rotation causes it to bulge at the equator and flatten at the poles. In the plane of the page the rotating cells will push against each other creating a pressure, and along the spin axis, perpendicular to the page, they will try to contract, creating a tension. So the lines along the spin axes will behave just like Faraday's lines of force, exerting an attraction along their length and a repulsion sideways. By making the angular velocity proportional to the local magnetic intensity, Maxwell obtained the existing known connections between magnets and steady currents.

But how did he incorporate electricity into the model? This was done by means of the small spheres between the hexagonal vortices, the electrical particles. These spheres also solve another problem, which is to prevent the vortices rubbing against each other when rotating in the same sense. They act like the 'idle wheels' that engineers put between wheels that need to rotate in the same direction.

The model was to prove very useful as an engine of discovery. For example, when current flows in wire *AB* it makes the adjacent vortices rotate, which do the same to the next layer of particles, and so on, until we have an infinite series of rotating vortex rings giving the magnetic lines of force associated with the current, as seen in Oersted's discovery.

There is one final important feature of the model that Maxwell introduced. This was to make the vortex medium elastic. Now the forces between electrically charged bodies could be attributed to potential energy stored in the medium by electrical distortion, just as magnetic forces are attributed to stored rotational energy. This allowed Maxwell to develop the displacement current, but more spectacularly, the great result was at hand. An elastic substance can transmit waves, and Maxwell calculated that their velocity was the ratio of the electromagnetic and electrostatic units of electric charge, which turned out to be the speed of light. As he wrote:

we can scarcely avoid the inference that *light consists in the transverse undulations of the same medium which is the cause of electric and magnetic phenomena.*[10]

In other words light is propagated as an electromagnetic wave.

Maxwell was unhappy, as were many, with this model of whirling cells and idle wheels and as he wrote:

The conception of a particle having its motion connected with that of a vortex by perfect rolling contact may appear somewhat awkward. I do not bring it forward as a mode of connexion existing in nature, or even as that which I would willingly assent to as an electrical hypothesis. It is, however, a mode of connexion which is mechanically conceivable, and easily investigated, and it serves to bring out the actual mechanical connexions between the known electro-magnetic phenomena; so that I venture to say that anyone who understands the provisional and temporary character of this hypothesis, will find himself rather helped than hindered by it in his search after the true interpretation of the phenomena.[11]

Maxwell's fourth paper, 'A Dynamical Theory of the Electromagnetic Field', published in 1865, resolved this issue as it provided a theoretical framework based on experiment and a few general dynamical principles from which the propagation of electromagnetic waves through space followed, without any recourse to vortices or forces between electric particles. Maxwell wrote about this paper:

I have also a paper afloat, containing an electromagnetic theory of light, which, till I am convinced to the contrary, I hold to be great guns.[12]

In this paper and in his *Treatise on Electricity and Magnetism* (1873), he developed 'Maxwell's equations' which in a modern notation are:

Maxwell's equations

$$\nabla \times \mathbf{H} = \mathbf{j} + \frac{\partial \mathbf{D}}{\partial t}$$

$$\nabla \cdot \mathbf{B} = 0$$

$$\nabla \times \mathbf{E} = -\frac{\partial \mathbf{B}}{\partial t}$$

$$\nabla \cdot \mathbf{D} = \rho$$

H is the magnetic field strength, the magnetic flux density divided by the permeability of the medium.

j is the current density

t is time

B is the magnetic flux density

E is the electric field strength

D is the electric displacement and ρ is the electric charge density.

Maxwell's equations are immensely practically and theoretically important. Chen-Pang Yeang discusses them in his Chapter 11, *The Maxwellians: The Reception and Further Development of Maxwell's Electromagnetic Theory.*

Let me end by repeating the quotation from Einstein given in my opening:

> Since Maxwell's time, Physical Reality has been thought of as represented by continuous fields . . . and not capable of any mechanical interpretation. This change in the conception of Reality is the most profound and the most fruitful that physics has experienced since the time of Newton.

In other words, Maxwell went from a mathematical description of a mechanical model involving vortices separated by rolling spheres to the complete abstraction given in his *Treatise on Electricity and Magnetism*. Maxwell removed all the imagery of the model until finally all that remained was the mathematics. The mathematics *was* the model. This transformation was, possibly, James Clerk Maxwell's greatest achievement.[13]

CHAPTER 2

Maxwell at Aberdeen

JOHN S. REID

Aberdeen

Aberdeen is not a place one happens to pass through. It is a place one deliberately goes to. So it is in the twenty-first century and so it was in the nineteenth century. James Clerk Maxwell chose to go to Aberdeen not only because a vacancy there had appeared at a convenient time in his life but because it was certainly an appropriate place to be. Edinburgh, the city whose people and places Maxwell knew so well, is 120 miles to the south. There is no indication that Maxwell had ever visited Aberdeen before his application for the job, but it was a place he undoubtedly would have known by repute. By way of introduction to Maxwell's Aberdeen years, let us look briefly at the city Maxwell chose to leave Cambridge for. There have been several histories written of Aberdeen that cover its changing character;[1] a briefer view of its situation in the 1850s is revealed in contemporary encyclopaedias.[2]

Aberdeen was a centre for academic life. There were two universities, both of which could boast centuries of serious scholarship and teaching reputation. In the 'New Town' was Marischal College, founded in 1593 by Earl Marischal as a post-reformation university that, within its Arts syllabus, established a reputation for progressive teaching in mathematics and the sciences. In Old Aberdeen, a city that was still separated from commercial Aberdeen by charter, ecclesiastical jurisdiction, administration and a mile of cobbled road that passed through the Spital village, lay King's College. By 1850 King's was the smaller of the two universities, founded by Papal decree in 1495. Old Aberdeen was little more than King's College, the ancient St. Machar's cathedral, the Old Town house and a few streets of housing that served these places; a city of 1000 souls, not many more. The Aberdeen of Marischal College was a thriving, expanding, modernizing city of some 35,000 in Maxwell's day, the most important city in Scotland north of Edinburgh. An academic city it may have been but it was certainly no Cambridge of the north.

Until the beginning of the nineteenth century, mercantile Aberdeen had been a small town of narrow, twisting streets set back on small hills from a usable but not outstanding natural harbour

Fig. 2.1 Undated photograph showing the centre of Aberdeen much as Maxwell would have known it. The street disappearing away from the centre of the picture is Union Street, where Maxwell had his lodgings about 300 m down on the left. Marischal College is off the picture some 200 m to the right of the start of Union Street. Photograph courtesy Aberdeen City Libraries.

at the estuaries of the Denburn, the Holburn and the River Dee. In 1801, the city fathers had initiated a visionary rebuilding of the city, beginning with laying out two streets, Union Street and King Street, approximately at right angles, artificially level and each about a mile long. Union Street, the main street, was 70 feet (21 m) wide and over the next few decades would be lined with fine granite buildings that set the scene for the future expansion of the whole 'granite city' (Fig. 2.1). A self-sustaining sense of civic pride brought with it a high quality building and development programme utilizing the native stone. New schools, churches of several denominations, banks, a central hospital, asylums for the poor and mentally ill, a large public indoor market, assembly rooms, the seven incorporated trades' building, a permanent building for the Mechanics Institution and other public buildings were erected to designs by architects of stature. Marischal College itself was completely rebuilt in the late 1830s.

Aberdeen was a thriving commercial city, a centre for textile production, fishing, the rapidly expanding granite industry that served both the locality and exported finished and polished granite, and the shipbuilding industry that was producing not only coasters and fishing smacks but the great Aberdeen clippers that opened up trade with the East Indies and China. Aberdeen was also an expanding rural centre, as agricultural reform over the preceding decades had raised both the extent and productivity of northeast Scotland's farming. Not far from the town, several large papermaking mills using the waters of the rivers Dee and Don were making the area into one of the leading paper producers in Britain. Industry on this scale needed good connections with the outside world, and these were provided by the expanding harbour facilities and, from 1850 onwards, the rail link south and to London in particular, a distance of 542 miles (870 km) by the shortest route.

It was to this elegant, bustling, dynamic city that Maxwell came in 1856. It was from this city and its surroundings that most of the pupils in his classes would come. It was with the permanent and temporary residents of Aberdeen that Maxwell would mainly engage during the years he was a resident.

Maxwell was appointed by the Crown as Regent and Professor of Natural Philosophy at Marischal College in April 1856, only weeks before the death of his father. The competition had been strong, with about a dozen candidates, some of whom were highly qualified. He was 24 years old, though he would be 25 when he was formally inducted into the office in August of that year. In April, he had also been elected a Fellow of the Royal Society of Edinburgh. He took lodgings from Mrs Buyers at 129 Union Street, above the premises of J. D. Milne, Advocates, within five minutes walking distance of Marischal College (Fig. 2.2).[3] This was his first professorial post and would be his only Scottish chair. A Regius Chair in an ancient university is a post that is hard to better in academic life. William Thomson and P. G. Tait were both appointed to the

Fig. 2.2a Maxwell's lodgings at 129 Union Street. The entrance between the ground floor shops leads, via a spiral staircase, to the upper floors that were leased. The plaque placed on the side of the building in 1956 is shown in Fig. 2.2b.

Fig. 2.2b The commemorative plaque of 1956 on 129 Union Street, which should read '. . . he was Professor of Natural Philosophy at Marischal College and University'.

corresponding chairs in Glasgow and Edinburgh in their twenties and both stayed in their posts for their academic life. Maxwell might well have been anticipating the same prospect, though he did know in 1856 that proposals were afoot to merge the two universities in Aberdeen. If that issue worried him, it does not come through in any correspondence. He set to, pursuing interests that had already caught his attention in optics, in colourimetry, in electricity, in the challenge of further understanding the dynamics of Saturn's rings and in the broader field of understanding the nature of Nature. One of these topics was probably never far from his thoughts, but what would occupy most of his time while at college would be his teaching, an aspect of the post that he clearly enjoyed.

Maxwell came to Marischal College as a brilliant young man. He would leave four years later as one of the leading scientists of his age, to use the modern description.

Marischal College

Figure 2.3 shows the almost new, sparkling, grey granite Marischal College building that welcomed Maxwell. The college's motto, taken from that of the Earls Marischal, 'Thay haif said, Quhat say thay, Lat thame say' promised a place that valued independence of thought. The clan motto of George Keith, 5th Earl Marischal and college founder was '*Veritas Vincit*' – truth conquers, which was also fitting. Looking at the college from the entrance, his rooms were one floor up, towards the rear on the left. The business of the college was its general MA degree, its substantial medical school and its small post-graduate schools in law and divinity. In 1856, almost

MARISCHAL COLLEGE.

Fig. 2.3 Marischal College, from a contemporary print in possession of the author.

250 students were enrolled in the four years of the MA course.[4] There were subject entrance exams for the regular students (called 'gowned' students), 216 in all, but courses were open to private students as well.[5] Private students of natural philosophy were exempt from any entrance examination but were required to have a 'competent knowledge of mathematics'. Maxwell himself knew the Scottish system very well, for he had been a private student in the University of Edinburgh for three years in his late teens.

Maxwell taught the major component of the degree programme in a student's third year, or tertian year as it was known. Typically he had around 50 students in this class. He also had other teaching, as will be discussed later. He was, of course, only one link in the chain of the four-year MA syllabus. Almost all Maxwell's colleagues were at least twice his age. Four were 'divines', in the sense that they were licentiates of the Church of Scotland, and taught either the religious component of the degree, philosophy or Greek. In addition there was Robert McLure, the Professor of Humanity, whom Maxwell had known in McLure's previous career as a schoolteacher in the Edinburgh Academy. Finally, the other two principal scientific staff besides Maxwell were John Cruickshank, the elderly but very competent Professor of Mathematics who had devoted his life to teaching, and James Nicol FRSE, FGS – at 46 Maxwell's youngest colleague, formerly Professor of Geology and Mineralogy at Cork, a geologist who can justifiably be placed in the same room as his great Scottish contemporaries Murchison, Lyell and Archibald Geikie for his perceptive interpretation of the geological record.[6] There were also professors for the optional classes, such as chemistry and agriculture. The staff in total comprised 14 professors and 6 lecturers, covering the MA degree and the degrees in law, divinity and medicine.

Marischal College was serious about its MA teaching. It had introduced both an entrance exam and final exams long before they became general university requirements in Scotland.[7] The MA degree was available only to enrolled students who had successfully completed their four years' study of the traditional subjects of latin, greek, natural history, mathematics, natural philosophy, moral philosophy, logic and evidences of christianity. In addition, they had to pay a significant fee to graduate.[8] Some took the subjects but did not pay to collect their degree. There were also optional classes in chemistry, agriculture, Roman history, rhetoric and English literature and in advanced Greek, Latin, natural philosophy (introduced by Maxwell) and mathematics. In Maxwell's time, the degree was known by its Latin initials of AM (*Artium Magister*).

Marischal College shared the Scottish reputation for providing good and accessible education. We think of nineteenth-century Britain as being highly structured by the largely unwritten rules of social status. This was not the case within the Scottish universities. Fees were low (about £5 per annum at Marischal College and King's College, Aberdeen) and many entrance bursaries were available. More than half of Maxwell's pupils came from farms or the trades. The poor rubbed shoulders with the well off and all were judged on how they performed. Little else mattered. The motivation for most students was, of course, not only to use one's brains but to escape from following one's father's occupation of long hours, low pay and little status. As it transpired, some exceptionally talented students came from poor backgrounds.

The session ran continuously from early November until early April. There followed a break of seven months, during which pupils went home to their farms or their parents' trades and professions. Non-graduating students were, by and large, boys in the age range 17 to 19 by the end of a session, well capable of earning their keep and contributing in labour towards next year's university expenses. Maxwell followed their example of decamping from the College at the end of session to base himself at his estate in Glenlair,[9] now his own responsibility having inherited it in June 1856.

Maxwell's Undergraduates

We would not be writing about Maxwell in the twenty-first century if he had been simply an exceptional teacher in his day. Few people earn a reputation for their teaching that lasts much beyond their own retirement and the dispersal of their audience to their separate lives. Maxwell, though, seems to have earned himself a 'bad press' in relation to his teaching, a press based on very little contemporary evidence and that only in the form of a few anecdotal comments. His friends, such as P. G. Tait and William Garnet, described him as a lecturer who would lose most of his audience,[10] though they had never been students in Maxwell's class. Comments quoted in the biography of David Gill, who was a pupil of Maxwell's at Marischal College, cast some light on the enigma that was Maxwell's teaching. 'He was not a schoolmaster at all' said Gill,[11] 'his lectures were terrible' but 'his teaching influenced the whole of my life'. Gill never lost an opportunity throughout his life to praise Maxwell's teaching.

Maxwell may have been no schoolmaster but the bad press that labelled him a poor teacher has been misguided. Maxwell was well aware that his insight was both hard-won and, set against the knowledge of society at large, exceptional. He shared with Faraday, and others now remembered particularly for their research, the ethos that searching for the truth is only one part of the task; passing on one's knowledge is equally important. Education in its broadest sweep was a motivation for Maxwell throughout his life. When he came to Aberdeen he threw himself into his teaching, conscientiously prepared his coursework, discoursed with his students long after the lecture hour was up, inspired some and impressed many. Gill related how Mrs Clerk Maxwell would come into the College wondering why he hadn't come home for dinner and find him discoursing with his students. He maintained more private students than his predecessor, larger natural philosophy classes than at King's College and a thriving optional fourth year class that he introduced. In October 1857, after a year in post, he was asked to present evidence to the Royal Commission of enquiry into the possible union of Aberdeen's two universities. In his opening remarks he commented 'I have not so much studied the interests of [the] Colleges as those of the classes and students'.[12] It was plain where his own interests lay.

Teaching hones clarity of thought, at least in the conscientious teacher, and there is a good case to be made that the exceptional clarity of thought that Maxwell brought to all aspects of his work and writing was in part developed through his teaching experience, particularly that of his Marischal College years. In addition, teaching provides an opportunity both for breadth of coverage of one's expertise and for general reflection. For example, one can read in such places as Maxwell's inaugural lecture[13, 14] his general thoughts on 'What is Natural Philosophy?'.[15]

Even today, with specialist honours degrees occupying most of the undergraduate work in a university, graduates often take up careers in areas other than the narrow discipline of their degree. Maxwell was teaching a course that all MA students took, whether they would become future doctors, lawyers, teachers, preachers, engineers, traders, explorers, academics or professionals in some other way. Thanks to the efforts of graduates to keep in touch with each other and with the college, we know much more about many of Maxwell's students than simply their names.[16] The class memoirs contributed to by graduates later in their lives tell of their careers; the college records give details of their background as students.

Reid[17] has presented some statistics relating to Maxwell's pupils. By birth, the majority of the students were from northeast Scotland. The greatest proportion of students were sons of tradesmen: masons, wrights, millers, saddlers, shoemakers, merchants, booksellers, smiths, ironmongers, druggists and so on. These trades, however, got little direct return on their investment, for over two-thirds of his students entered one profession or another, in medicine, religion, teaching or the law. In the 1850s, 'scientist' was not a recognized profession within the educational system. The word had barely been coined. Only a single Aberdeen student became a recognized scientist – David Gill, mentioned above, one of the leading observational astronomers of the century who would become the Queen's Astronomer at the Cape of Good Hope, FRS and President of the Royal Astronomical Society. David Gill was a private student in Maxwell's final year at Marischal College and wrote briefly about his experience.[18]

The fact is that very few men out of the total number who obtained degrees anywhere became scientists of note. David Thomson, Maxwell's 'rival' who taught natural philosophy at neighbouring King's College for 35 years, had four students who went to Cambridge and emerged Senior Wranglers. None became physicists[19] but their achievements attest to the standard of teaching and students in Aberdeen. However, in contrast to Maxwell's experience at Aberdeen, when he himself was to finish his professorial career at Cambridge, he would have very few students in his class but a significant number of them *would* make a name for themselves in the physical sciences. That era of his career is the subject of a later chapter.

Maxwell's Teaching

At Marischal College, Maxwell had a pretty free hand in constructing his undergraduate course; he could choose his own syllabus and his own recommended textbooks. He settled on the monographs of Galbraith and Haughton as supplementary reading. There were no external examiners, and no quality assurance schemes – not that he would have feared either. He inherited a common level of preparation of his students, private ones excepted, who had already received lecture courses in natural history and elementary mathematics but no previous natural philosophy. He also inherited the use of a very large amount of demonstration equipment, some of it made to a high standard over 50 years previously by a predecessor, Professor Copland.[20] He had his own lecture theatre, one of 16 in the college, with a private room attached.

Maxwell's inaugural lecture at Aberdeen, outlining his philosophy of knowledge, was summarized in the press at the time and more recently has been printed. Maxwell writes quite frequently and positively in his correspondence about his teaching (a sure sign of his commitment) but we can be even more precise about what he did, for two sets of student notes have come to light, one from his first year and one from his third year of teaching. The first year notes were taken by Angus Fraser,[21] son of an Aberdeen grocer who went on to become a doctor and later a representative for the University of Aberdeen on the General Medical Council.[22] The second set of notes was taken by Alexander Davidson,[23] a man who continued his father's papermaking business on the River Don that flows through Aberdeenshire, and whose mill survived into this century under the family name but closed in 2005.

Both sets of notes refer to Maxwell's tertian course, the course that was the main reason for his college employment. He chose to include a modest range of subjects, unlike his predecessors. Also unlike his predecessors, he emphasized the numerical and mathematical side of his selected subjects.

Angus Fraser's notes cover 156 pages of a small notebook significantly less than A5 in page size. He dates many of his entries, showing that in this year Maxwell typically gave four lectures per week at 11 am and also two to three 9 am sessions per week that were either problem sessions or in-course examinations. In all there were 14 examinations, each containing 10 questions, written out within the notes and each followed by brief answers. Some days when one might have expected lectures they are missing, with no indication whether the absence is due to the student or to Maxwell. On the whole the topics flow.

The notes are indeed 'notes', an hour's lecture being condensed into typically one to two small pages. (Harman has printed a manuscript version of an introductory lecture by Maxwell in November 1856 on 'Properties of Bodies' that is contained in the University Library, Cambridge.[24] It runs to six printed pages.) Fraser's notes read as if they are a condensed version, in the student's own words, of the ideas Maxwell was trying to get across, written up as a fair copy from rough draft made during the lecture. Nonetheless they start out with two sentences that sound like Maxwell's. He begins: 'The most simple things are the most abstract. There is nothing simple in Nature.' A brilliantly succinct message for those bright enough to read it that natural philosophy is going to be a challenging subject that will get you thinking.

Fraser appends three pages of 'Contents' to the notes that show at a glance the balance of the course. Pages 1–106 cover statics, dynamics and related concepts; pages 107–27 hydrostatics and pneumatics; pages 128–45 heat and a section on determining altitude with a barometer; pages 149–56 a little on basic optics, namely reflection from plane and curved mirrors. The final date is the 10 March. It has to be said that the notebook represents simply Fraser's notes of Maxwell's course and not necessarily the complete course. In fact, as the course progresses, the emphasis in the notes is increasingly on the numerical work. Some of the in-class exam questions ask for facts not recorded, confirming that Fraser's emphasis is on recording the numeric aspects.

The lectures include flashes of nice visual imagery. For example his picture of compounding velocities is: 'This is best seen by a fly walking along one side of a sheet of paper and the paper moved in a direction perpendicular to its motion.'[25] However, unlike the notes of his predecessors, Fraser's notes do not contain multiple examples of practical applications of concepts. Some practical applications come though in worked numerical examples and in the examination questions. The mathematical skills expected of the students are manipulation of algebraic equations, reading the geometry of diagrams, resolving forces into components at right angles (sine and cosine), manipulating expressions involving trigonometrical quantities, properties of a parabola (in connection with trajectories) and manipulation of logarithms (in connection with the relation between barometric pressure and height).

One needs to compare the course not with today's third year of a university degree, but with what one might teach a bright 19 year old who had not done any physics before, for that was the average age of Maxwell's students. It is also worth reflecting that every undergraduate in the university took this course in his third year, not just those intending a career with some technical background. Look very briefly at a few examples from the notes. The first examination after a fortnight's work includes the question 'ABCD is a horizontal rod, 1 yd [yard = 3 feet] long. A weight of 1 lb [pound] is hung at A, 2 lbs at B one foot from A, 3 lbs at D. What is the moment about C of a point 2 feet from A and which way does it act?'. This sets the tenor of the course. In the problem class six days later the following problem is discussed (see Fig. 2.4).[26] 'AB is a rod placed between a smooth wall and a smooth floor. Put a [smooth] peg C to prevent it sliding' and there follows a worked exampled showing how to find the force *S* exerted by the peg as $Wa\cos\alpha/(b\text{-}a\cos 2\alpha)$; '*a*' is the half-length of the rod and '*b*' the distance from the peg to the centre of the rod. Such an example shows the power of the formalism to be precise and tacitly assumes that boys who have passed exams in Latin, Greek and maths are possessed of significant powers of deduction and logic.

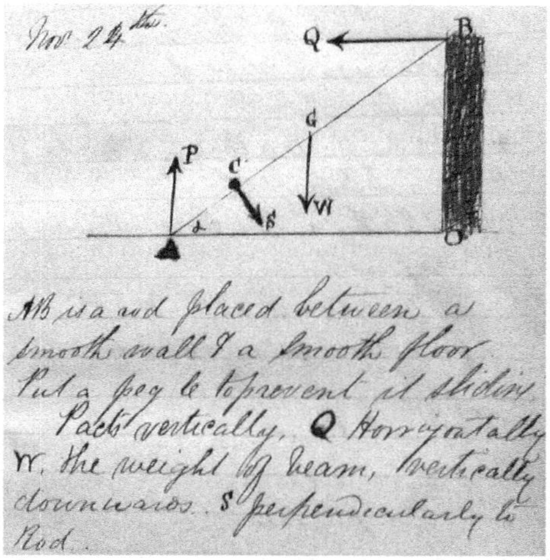

Fig. 2.4 Illustration from Fraser's notes[19] showing the problem of a rod leaning against a wall and held by a peg, mentioned in the text. Courtesy of the University of Aberdeen Library.

Not long before Christmas, having dealt with motion under constant acceleration, the subject of trajectories is discussed. It is at this point that the majority of physics texts launch into a discussion of guns firing at targets, the maximum range attainable and so on. Maxwell, avoiding this aspect, discusses balls being projected but his worked example is 'In a certain waterfall the stream shoots horizontally over a rock and falls into a pool below clearing a horizontal distance equal to one fourth of the height of [the] fall. The time of descent is 3" [seconds]. Find the height of the fall in feet and the velocity of the stream above the fall in miles per hour.' The acceleration due to gravity is taken as 32 ft s^{-2} and the answers are shown to be 144 feet and $8\frac{2}{11}$ mph. With a sporting slant, a later example discusses what conditions need to be satisfied for a partially inelastic ball to rebound to its initial point when it hits a vertical wall. The reader is invited to answer this question.

Alexander Davidson's notes made two years later are substantially fuller and could have been made 'on-the-fly' in the class, or perhaps copied to some extent from Maxwell's own manuscript. He records 109 lectures altogether, excluding exams and problems sessions. Optics is dealt with much more fully, 19 lectures on the subject running to the end of March 1859. Reid[27] discusses these notes in some more detail.

There can be no doubt that Maxwell developed his undergraduate courses himself,[28] designing them to put across his own views on the relevance, desirable content and purpose of natural philosophy. The great majority of his extant letters written during the first three sessions he was in college mention some aspect of his teaching. None acknowledge any specific input into it from other sources. In his fourth session he was applying for the equivalent Edinburgh post and had other things on his mind.

Maxwell was now delivering a parallel course to that of James D. Forbes in Edinburgh, whose lectures had benefitted him over the years 1847 to 1850 and whose friendship had continued beyond Maxwell's departure from Edinburgh. Forbes has been credited with introducing, in the 1830s, a more numerate course than that of his predecessors John Leslie and John Playfair. Forbes' extant notes attest that his course was a fine example of a broad 'old style' Scottish course: full of facts, not shying from numeracy, well illustrated by practical examples of applications of principles and including lecture demonstrations.[29] Maxwell's course differed from Forbes in many ways but particularly in Maxwell's ethos that it was important to show how principles plus mathematics resulted in quantitative deductions about nature.

Maxwell certainly included demonstrations and went out of his way to make up his own demonstrations for the students, doing more than simply using the college's large stock of equipment. Indeed, Maxwell not only put on demonstrations but also ran experiments, as he put it, 'being done by the men on their feet and the experiments not cooked in any way'.[30] One of the messages he wanted to get across was that simply 'doing an experiment' is not enough: accurate results require accurately performed experiments. Maxwell's introduction of practical physics work for students was an early example of such initiatives, albeit on an occasional basis rather than as a formal requirement. William Thomson had introduced practical work for students at Glasgow[31] even earlier. The Scottish universities were well placed for such a development. The popularity of teaching by demonstration ensured that the professors had access to a large cabinet of equipment and were well practised in its use, and the egalitarian atmosphere of the university in which social divisions were largely ignored contributed to the recognition that students were participants in the educational process and not mere recipients of wisdom handed down '*ex cathedra*'.

In his first year of teaching Maxwell inaugurated a voluntary 'advanced class' for final year students. In his second year he formalized the arrangement and commented to his aunt Jane Cay: 'I have a large attendance of my old pupils . . . This is not part of the College course, so they come here merely from choice.'[32] To P. G. Tait: 'My regular class is small this year as I expected from the character of the men below, last year. My advanced class (an institution of mine) is large and evidently jealous. With them I expect to do Newton 1, 2, 3 with a sketch of Physical Astronomy, Magnetism and Electricity in the poor weather and Undulatory theory when the Sun appears in Spring.'[33] His 'ordinary men', as he called them, covered statics, dynamics, hydrostatics, heat and some geometrical optics. This comment agrees with Fraser's notes. He related to Tait in a tone of some surprise and pleasure that when his first class returned after the summer break of almost seven months (early April to November) he set them an exam and four men got full marks and only four less than half marks. Try that experiment today and even the best in the lecturing profession would be very pleased with a comparable performance.

Maxwell was not alone in Aberdeen in offering advanced classes and the system was essentially the pre-cursor to the introduction of an Honours degree, which did not formally exist in Aberdeen in Maxwell's time, though they had the next best thing. Since Cruickshank introduced the degree exam, after 1828 the best students were declared to have attained their degree 'with honourable distinction'. Formal Honours appeared in 1860 in the new, merged, University of Aberdeen.[34]

With Maxwell in particular, it is easy to fall into the trap that snares scientific biographers of raising their subject to saintly status. It is true that one can be knowledgeable, enthusiastic and have well-organized course material, but if one's oral delivery doesn't match the expectation of the audience then the lecture has fallen at the last hurdle, as it were. It seems from some contemporary comments that Maxwell fell down here.[35] Most in the audience saw past any difficulty; some badged him inappropriately as a poor teacher. His biographers Campbell and Garnett devote two pages to discussing Maxwell as a teacher, but the fact remains that if he wrote a better lecture than he sometimes delivered, which seems to be the case, a far wider audience reaped the benefit than had the reverse been true.

In many ways both Maxwell's courses and his methods of delivery look thoroughly modern, and certainly much more closely aligned with twentieth-century than with eighteenth-century practice. With his intimate knowledge of both the Scottish and Cambridge educational systems, he shared with Thomson and Tait the ability to blend the best of both systems: the breadth of vision of the Scottish, which encouraged intellectual versatility, with the attention to detail cultivated at Cambridge. Thomson and Tait's textbooks (their *Treatise* and its simplified version) and Tait's shorter, more specific, textbooks used this expertise to fill a conspicuous need for good English language textbooks.[36] The Aberdeen years grounded Maxwell's teaching expertise. It is often forgotten that Maxwell wrote teaching textbooks that were not only highly successful in the nineteenth century but were valued well into the twentieth century. His *Theory of Heat* first appeared in 1871, where he introduced the public to *Maxwell's demon*. 2000 copies were printed.[37] Before 1872 had passed another two editions each of 2000 copies had been printed. By the time of the posthumous 10th edition in 1891, Lord Rayleigh had taken on the rôle of providing updates and 1921 saw the 18th 'edition', albeit a reprint of the 1897 version. At least five 'modern' reprints have appeared since 1968. Maxwell's *Matter and Motion* first appeared in 1876 and was reprinted before the year was out. The first American edition was printed in 1878. Following several reprints on both sides of the Atlantic, Sir Joseph Larmor added notes and appendices to produce a new edition in 1920. This edition was reprinted in 1925 and at least half-a dozen times since 1952. Maxwell's *An Elementary Treatise on Electricity* was conceived of in 1873 as containing simplified material from the now famous *A Treatise on Electricity and Magnetism*. It was not published, however, until 1881, under the editorship of William Garnett. It ran to a second edition in 1888 and two modern reprints.

One reason for the success of Maxwell's teaching texts, and those of Thomson and Tait, is that they were not simply recycling long-established results but were expounding in clear and persuasive language the new understanding in basic physics that their authors had been responsible for developing – concepts such as energy, the molecular picture of thermodynamics, the new understanding of electricity and more. In a way, Maxwell's contributions to the great ninth edition of the *Encyclopaedia Britannica* were a natural development of his zeal for teaching. They covered *Atom*, *Attraction*, *Capillary action*, *Constitution of bodies*, *Diagrams*, *Diffusion*, *Ether*, *Faraday*, *Harmonic analysis*, and *Physical sciences*. All this was long after his Aberdeen years but these years laid the foundation of his proficiency. Maxwell deserves to be remembered as one of the nineteenth century's notable pedagogues. From his second year onwards at Marischal College he was, as he remarked in a letter to his friend Henry Droop,[38] talking to classes for 15 hours a week. His expertise was built upon understanding, enthusiasm, hard work, and experience.

Maxwell's Evening Class

In October 1857, Maxwell undertook to give the natural philosophy evening course of the 'Aberdeen School of Science'. This course of some two dozen lectures ran for one evening per week (Monday, at 8.10 pm) for exactly the same span as the college classes, namely five months from early November until early April.[39] The fee was 8 shillings per student, per annum. Maxwell had already had the experience of lecturing to 'working men' in Cambridge and had clearly found the experience rewarding. In Aberdeen, he had been present at the meeting to inaugurate the Schools of Navigation and Trade earlier in 1857 and was appointed to the Committee of the School of Science.

The Aberdeen School of Science and Art, which was already established when Maxwell arrived, was aimed at mechanics, tradesmen, and others who could not attend daytime lectures. Indeed, most local tradesmen of the day worked from 6 am to 6 pm and then went home to clean up and be fed. The school was also open to ladies, though it had 'the special object of affording to young Mechanics, Tradesmen, &c., instruction in those branches of Science which are of the greatest practical utility in the chief, trades, arts, or professions, followed in this locality . . . '.[40] It was operated in conjunction with the Aberdeen Mechanics Institution and, in a pioneering venture that was later to be adopted in other cities, with the Committee of the Privy Council on Education. The 34th report of the Mechanics Institution incorporates the report on the school for the last year that Maxwell taught there (1859/60). Appendix II of this report is a table listing the professions and age groups of all the male attendees at all the evening classes. Maxwell's class isn't singled out, but the table shows clearly the ages of the men who attended these classes.[41] The numbers are: ages 11–15 (55 pupils), 16–20 (145 pupils), 21–25 (42 pupils), 26–30 (10 pupils); 55 named professions are listed for the 252 pupils, covering a wide range of the service and manufacturing industries. Apart from the obvious blacksmiths, brassfinishers, brassmakers, engineers, mechanics, shipwrights, tinsmiths and others on the mechanical side, there were basketmakers, carvers and gilders, combmakers, ropemakers, stonecutters, stonepolishers, a tobacco-spinner and a wirefence maker, wrights and 39 other professions, an eye-opening testament that self-improvement had percolated into the pores of pretty well every trade.

Students of the school were given access to the extensive library of the Aberdeen Mechanics' Institution,[42] and could expect the lectures in natural philosophy and in chemistry to be illustrated by Marischal College apparatus, courtesy of the College Senate. The classes took place in the college. Although it was not stated in the advertisements for the class, it would seem that the natural philosophy class was seen as a sequel for those who had acquired the accomplishments of both the junior and senior mathematics classes. It was intended as a serious professional class, in a way the pinnacle of what the school offered, and hence expected to attract only modest numbers. The natural philosophy course had been started by Maxwell's predecessor at Marischal College, Professor Gray, at the opening of the school in 1854. It involved not only preparing and delivering the coursework, but marking the attendees' weekly homework exercises.

Gray had given a wide-ranging course, but Maxwell's course concentrated on mechanics, with perhaps some electrical theory relevant to the telegraph. In one letter of 1857 to C. J. Munro, Maxwell wrote that he had 'also a class of operatives on Monday evening who do better exercises than the University men about false balances, Quantity of Work etc.'.[43] His choice of topics may have been influenced by outside considerations, since participants could present themselves for an exam in statics and dynamics at the end of the course, overseen by the Department of Science and Art in London. Few did. The *Aberdeen Journal* reported in April 1858 just four natural philosophy results,[44] by a draughtsman, a watchmaker, a blacksmith and a clerk.

Marischal College had an enviable track record of supporting extramural lectures to tradesmen, professionals and a social audience that went back more than 70 years before Maxwell's time. A predecessor in Maxwell's chair, the same Professor Copland who built up such a tremendous stock of demonstration equipment, had given a much longer extra-mural course on natural philosophy over a period of almost 30 years from the mid 1780s.[45] Copland's successor at Marischal College, Professor Knight, was the founding lecturer in natural philosophy at the Aberdeen Mechanics' Institution, in 1824, when he lectured to an audience of over 500. In many towns the Mechanics Institutions of the early 1820s foundered as enthusiasm for them waned, but in Aberdeen the local institution survived this dip in interest and secured an established place in the educational framework of the city. Edinburgh's Watt Institution and the Glasgow Mechanics' Institution were the only two other Scottish examples to conspicuously survive. In fact, The Robert Gordon University in Aberdeen can trace its nineteenth century roots to the Aberdeen Mechanics Institution.

Maxwell's Research

This chapter has spent some time on Maxwell as teacher, for it is an aspect of his life and character that has been underplayed in his other biographies. However, what still astonishes modern scientists is Maxwell's penetrating insight into a wide range of disciplines. The eminent Charles Coulson, a twentieth-century successor to Maxwell at King's College, London, has rightly commented that 'There is scarcely a single topic that he touched upon which he did not change almost beyond recognition'.[46] The diversity of Maxwell's interests come out both in the papers he published in his Aberdeen years and in the topics he discussed in his personal correspondence.

Glenlair beckoned for a good half year in the closed season of teaching, but in the 1857 season Maxwell had to establish himself as a worthy successor to his father as landlord of the estate, overseeing the building of houses and dykes (field walls), cutting drains, sowing seed, attending to the welfare of his tenants and generally running Glenlair house as owner. Maxwell had been made a JP (Justice of the Peace) in Edinburgh in 1854, though he does not mention this in any of his letters. 1858 was the year of his marriage, a time for relaxation and enjoyment and introducing his wife to the locality and the estate. 1859 was the year of the Aberdeen meeting of the British Association for the Advancement of Science (BAAS), discussed later. In short, the Glenlair months were not half a year of free time to concentrate on research interests. Nonetheless, Maxwell wrote seven short papers for the BAAS and seven substantial papers while in post at Marischal College, spanning some 200 published pages.[47] His great achievements in this period were his continuing analysis of the dynamics of Saturn's rings, his quantitative colour-matching experiments that laid the foundations for modern colourimetry and his fundamental work on the kinetic theory of gases. The following brief sections on Maxwell's research while at Aberdeen introduce topics mainly dealt with in depth in other chapters of this book and in books devoted to particular aspects of Maxwell's work[48, 49] and in more detail in the well-researched biographies by Everitt,[50] Tolstoy[51] and Goldman.[52]

Technically the first papers Maxwell presented as Professor of Natural Philosophy at Marischal College were the four short papers he delivered at the August 1856 Cheltenham meeting of the British Association for the Advancement of Science. These are summarized in the meeting report and in Maxwell's collected papers.[53] The first three were on what is now known as tritanopia of the yellow spot of the eye, on a method of drawing Faraday's lines of force without calculation and on colour mixing. His final paper in the quartet was 'On an Instrument to Illustrate Poinsôt's Theory of Rotation', an early version of his dynamical top.

The Dynamical Top

Maxwell's main paper describing his dynamical top was his first substantial production while at Marischal College.[54] Nowadays, the vector concept of angular momentum is introduced at a fundamental level in all mechanics textbooks as a natural application of Newton's laws to the motion of rigid bodies. What is generally not made clear to students is that the application of vector calculus to mechanics (and electricity and magnetism) was a development of the second-half of the nineteenth century and the rôle of angular momentum, far from being an obvious development from Newton's laws, was really only made clear by Maxwell and his contemporaries.

'The complexity of the motion of the particles of a body freely rotating renders the subject so intricate, that it has never been thoroughly understood by any but the most expert mathematicians', as Maxwell remarked early in his paper. Euler, D'Alembert, Lagrange and others (to cite Maxwell's examples) were able to tackle problems of rotational motion only through the use of involved mathematics. What they lacked were the 'appropriate concepts'. The work of Louis Poinsôt in the second quarter of the century on the theory of rotation had attracted the attention of both Maxwell and Robert Baldwin Hayward at Cambridge. Hayward developed Poinsôt's approach, and may well have coined the term *angular momentum* in 1856.[55] The theme of finding 'appropriate concepts' runs though much of Maxwell's work. He recognized in Hayward's angular momentum the key appropriate concept for this kind of problem. The use of vector notation was still some years off, but Maxwell's completely adjustable dynamical top was designed to clarify the use of the concept of angular momentum and the complex rotational motion of rigid bodies in the most general circumstances. It showed, in a way, Maxwell the pedagogue at work, not teaching students now but teaching himself, first and foremost, and his peers.

Other tops had been made earlier to illustrate, for example, the precession of the Earth's axis but Maxwell's version allowed for the variation of all the relevant parameters (the moments and products of inertia and the rotation point relative to the spinning body). His original model presented to the British Association was wooden. He continued to develop his top at Aberdeen, substantially refining the design and commissioning John Ramage of the instrument-making firm Smith and Ramage at Marischal Quay (appropriately enough) to make several copies. Another version of the top, even more polished than Ramage's, was produced later in the nineteenth century by the instrument firm of Harvey and Peake and is shown in Fig. 2.5.

It's fair to say that the top impressed his contemporaries and did much at the time to publicize the relevance of the concept of angular momentum to the treatment of rotating bodies. Notwithstanding, his paper is seldom cited these days, mainly because the dynamics of general rotary motion is now so embedded in standard physics that references are considered unnecessary. Maxwell drew conclusions about the precession and nutation of the Earth's rotational motion from considerations of his dynamical top. He had tops made for Trinity College, Cambridge, one for R. B. Hayward, by then Reader in Natural Philosophy at the quite recently founded Durham University but soon to move to Harrow School, one for J. D. Forbes in

Fig. 2.5 Maxwell's dynamical top by Harvey and Peake. Courtesy of the Natural Philosophy Collection of the University of Aberdeen.

Edinburgh, one for Christian Peters, Professor of Astronomy at Königsberg, and he kept one for himself. His own top survives at the Cavendish Laboratory and J. D. Forbes' is in the Royal Museum of Scotland.

Saturn's Rings

The work that made Maxwell's reputation in his day, amongst his contemporaries in Britain at least, was his Adams' Prize essay 'On the Stability of the Motion of Saturn's Rings'. Perhaps spinning tops engendered thoughts on the spinning rings of Saturn. We don't know. The prize topic was announced by the University of Cambridge in 1855 for submission in 1857, following correspondence between James Challis, Director of the Cambridge Observatory, and William Thomson. It was a subject that was particularly topical on both sides of the Atlantic.[56] Maxwell's submission, one third of a kilogram in weight of closely argued mathematical physics of the highest calibre, was received in December 1856 and he was awarded the prize in 1857, barely three years after he was an undergraduate. Maxwell did not publish his submitted manuscript, but spent more time on developing it while at Aberdeen than on any other research topic. He had some lengthy correspondence with William Thomson in particular, notably on issues connected with the stability of the rings. The final version was published as a free-standing work in 1859. His study was a theoretical tour-de-force, effectively Maxwell's trial piece submitted to the guild of the elect professoriate in Britain as an entrance test. Challis, Stokes, Airy, Thomson and others absorbed the argument.

The essay is largely an exploration of the stability of a number of physical models for the constitution of the rings. Maxwell makes significant deductions on points of detail, and overall his analysis proved beyond doubt to his contemporaries that 'The final result, therefore, of the mechanical theory is, that the only system of rings which can exist is one composed of an indefinite number of unconnected particles, revolving around the planet with different velocities according to their respective distances'.[57] This before a single clear photograph of Saturn's rings had been taken. Smith and Ramage also made a mechanical model that illustrated some of the wave motion that Maxwell predicted could exist within the ring system.

Maxwell's work has spawned the modern theory of planetary discs and astronomical accretion discs that are found around dwarf stars orbiting close to giant stars and in matter orbiting black holes.[58] The rings of Saturn and other discs are now recognized to be substantially more complicated than Maxwell could model with his analytical mechanics, and are most frequently treated by a combination of theoretical insight and computer modelling.

On Colour Perception and Optics

Maxwell wrote at least 10 articles on colour perception. It was a subject that fascinated him from his youth. Most of his papers were written just before, during and just after the time he was at Aberdeen. In the 1850s, colour awareness in society was not what it is today and colour science scarcely existed as a topic. The impact of colour science today is almost as pervasive as the impact of electricity and magnetism yet, judging by the minor place it seems to have in university syllabi and undergraduate optics texts, Maxwell seldom gets the credit for the major steps that he took to establish this now essential branch of applied science.

A well-known photograph shows a youthful Clerk Maxwell holding an early version of his spinning colour disc.[59] A variant of this apparatus used a spinning top. Both devices were used

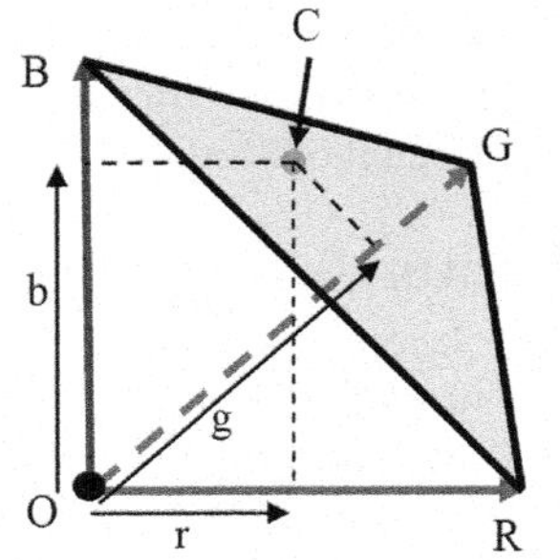

Fig. 2.6 Maxwell's colour triangle (shaded grey) is actually a figure in three dimensions. It is formed to span three axes **R**, **G** and **B** (representing the full amounts of the colours red, green, and blue) drawn from an origin O. Any colour C has coordinates r, g, b that represent the fractions of the primaries necessary to produce the same colour sensation. This is represented mathematically as $C \equiv r\mathbf{R} + g\mathbf{G} + b\mathbf{B}$. All points on the triangle are colours of the same total intensity. Normally the triangle is drawn flat on a page, not showing the colour space it really occupies.

by the young Maxwell in his colour-matching experiments. The fact that three colours are generally enough to match another colour had been postulated by Thomas Young early in the century, but the theory and practice had not advanced much since. Maxwell brought his mathematical skills to bear. 'It is easy to see' he wrote in talking of the number of primary colours needed 'that the number of these sensations corresponds to what may be called in mathematical language the number of independent variables, of which sensible colour is a function'. Moreover, the experimental evidence Maxwell accumulated demonstrated beyond doubt that when mixing colours the variables are combined linearly. Maxwell mapped these variables onto a plane triangle, using the constraint that the sum of the variables always adds to a constant (Fig. 2.6). So was born the *Maxwell colour triangle*, used for quantifying the measurement of colour in terms of a mixture of three primaries. The CIE chromaticity diagram, the basis of all modern colourimetry work and colour standards, is essentially this same diagram using transformed coordinates.

Maxwell's first paper on colour that he wrote at Aberdeen described extensive experimental testing of the linear model of the three-colour mixing hypothesis, going as far as he could. He highlighted the importance of the illuminating conditions in colour matching, adding the observation 'that the results would differ far less if observed by different persons, than if observed by different lights; for the apparatus of vision is wonderfully similar in different eyes'.[60] Although he hadn't fully explored it yet, Maxwell was setting the academic basis for colour photography. He would come back to the topic, but colour wasn't his only optical interest.

Before he came to Aberdeen, at Cambridge Maxwell had been struck by the unusual structure of the crystalline lens of a fish and set off exploring the imaging properties of a spherical lens with a radial refractive index gradient. In 1854 he published the conditions under which such a lens can give a perfect image. The result is known as the *Maxwell fish-eye lens*. Until recently it has been mainly a theoretical curiosity,[61] though the idea is now of interest to microwave engineers. Perhaps spurred by this investigation, Maxwell went on while at Aberdeen to write a general paper on the conditions imposed on optical devices that formed perfect images.[62] He demonstrated that useful results could be obtained by considering the whole instrument as a single imaging device rather than as a sequence of image forming surfaces. However, in retrospect, we can see that the cutting edge of theoretical optics in 1856 was the exposition of lens aberrations by von Seidel in Munich, though it took several decades for this work to be fully appreciated. In addition, Maxwell's paper was found some 90 years later to be less general than he had hoped. It was actually Maxwell's practical optics that would prove more influential than his theoretical excursions.

The most innovative instrument of the decade was Maxwell's own colour box, conceived of before he came to Aberdeen but developed into an accurate tool there. Maxwell realized that the purest primary colours that he could obtain for his colour matching experiments were spectral colours, not the colours of paint or card he had been using on his discs, which each necessarily reflected a range of spectral colours. Maxwell's 'colour box' was the first modern colourimeter. Maxwell thought hard about how to obtain uniformly illuminated fields of view to match his test field (daylight) with a superposition of fields of red, green and blue. The resulting novel arrangement of lenses is called the *Maxwellian view*[63] and as a sideline Maxwell thus introduced a new optical design principle that is to this day a useful tool for the optical instrument designer. The Maxwellian view imaged the slit onto the entrance pupil of the detector (the eye's iris here), unlike the standard spectrometer in which the slit is imaged onto the image plane of the detector (the retina in the case of direct observation).

Maxwell made his most developed colour box in 1858, with the help of John Ramage again, and designed a smaller version in 1859. He had been in communication with G. G. Stokes (the Lucasian Professor at Cambridge) on his colour experiments during much of his time at Aberdeen, and through Stokes (who was also Secretary of the Royal Society of London) sent the main account of his results using the 'colour box' to the Royal Society in late 1859. Stokes had been won over by Maxwell from an agnostic position on colour perception to the realization that Maxwell had pretty well single-handedly put colour science on a quantitative basis. The Royal Society Council responded by appointing him as Bakerian Lecturer, an appointment that was advertised in the press but later had to be withdrawn, following a comment by Maxwell that he had not yet been elected a Royal Society Fellow.

The paper was published in full, with an addendum on results obtained from one of Maxwell's students, who was a dichromat.[64] It reports an extensive set of observations by himself 'J' and his wife 'K' designed to determine the position of the spectral curve on the colour triangle. This curve is the key to quantitative colourimetry, as Maxwell realized, 'since all natural colours are compounded of the colours in the spectrum'. Maxwell's method overcame the difficulty that some spectral colours have negative colour coefficients, which means they cannot be matched by real amounts of three given primaries. It was this problem that led the CIE some 70 years later to use imaginary primaries (in the sense of physically unrealizable) when establishing the international standard colour measurement system. Maxwell's paper laid down the academic foundation, and limitations, of the three-colour photographic process, discussed in further detail in Chapter 3 on his King's College, London, years.

Maxwell on Molecular Science and Kinetic Theory

Maxwell's work on the foundations for understanding properties of matter, particularly gases, by treating them as collections of moving molecules, is absolutely fundamental. Maxwell's work on molecular science and kinetic theory has had two books devoted to it.[65] This work was started, but by no means completed, while he was at Marischal College.

With Saturn and its myriad fragment rings at last out of the way, Maxwell had turned his attention early in 1859 to the work of Rudolf Clausius, who was trying to build up a dynamic molecular theory of gases from first principles. Laying the foundations of a subject on strict mechanical principles (essentially Maxwell's words) was just the sort of task Maxwell was drawn to. It is hard to credit nowadays that the default picture of a gas in the mid nineteenth century was as a tenuous medium in which the constituents were held apart metaphorically at arms' length by long-range repulsive forces.[66] This is just about tenable as a hand-waving suggestion if one has no idea about the size of the constituents, how far apart they are, or tries to work out detailed numerical properties of gases. Surprisingly enough in an age when water power and wind power were clearly giving way to steam power, and lighting was increasingly depending on gas, properties of gases were not a popular subject of investigation. Graham in Glasgow, Stokes in Cambridge and the Meyers on the Continent had produced experimental results on diffusion and viscosity that could not be explained by the de facto model. A new viewpoint was clearly needed.

Maxwell announced some key results of his first-principles approach at the British Association meeting in 1859 (discussed later), and published his first formal paper on the subject in January and June of the following year in two parts under the title 'Illustrations of the Dynamical Theory of Gases'.[67] It is an extraordinary paper. Almost every topic relevant to the foundations of the kinetic theory of gases is discussed: the distribution of velocities, the average velocity and root mean square velocity, the pressure exerted by a gas, the equi-partition of energy among molecular constituents, the mean free path between collisions, the viscosity of a gas, the thermal conductivity of a gas, the diffusion of a gas through a porous plug, and the diffusion along a pipe connecting two containers, the specific heat of a simple gas and other topics. The paper is full of intuition, fundamental physics, incomplete arguments and a few numerical mistakes.

In spite of some blemishes, the paper is still one of Maxwell's greatest. He conjured the *Maxwell distribution* of molecular velocities in a gas almost out of thin air by an argument that still astonishes 150 years later, based on little more than the independence of velocity components along perpendicular directions. The Maxwell distribution is the first law of physics discovered that involves probability. His intuition that the equi-partition of energy has to be true exceeded the strength of his argument, which was based upon a consideration of special cases. However, this paper moved on Clausius' work considerably. Maxwell was able to estimate the mean free path of a molecule by two independent means, and hence the number of collisions per second molecules would have on average, since average speed could be found from pressure and density; he showed that Avogadro's hypothesis (the number of molecules of all gases in unit volume at a given temperature and pressure is a constant) followed from kinetic theory. In fact, in broader terms, it became clear from Maxwell's work that the kinetic theory of gases provided a new way into understanding molecules that would complement the insight of chemists gleaned from chemical reactions.

The prediction that both caught people's attention and offered the prospect of experimental verification was the unexpected result that the viscosity of a gas should be independent of its density. The result seems counter-intuitive. If you reduce the pressure of a gas in a container (and hence its density) by taking out more and more of it, how can the viscosity of what is left remain the same? There were no trustworthy experimental results that could be used immediately to test the idea, but the paper had the effect of stimulating an interest in the three transport properties of gases discussed, namely diffusion, thermal conduction, and viscosity. Maxwell had to leave the topic, for by mid 1860 he was leaving Aberdeen and would have other concerns when he got to London. As it turned out, he would provide definitive results himself that were influential in convincing his contemporaries of the validity of the kinetic theory of gases, but not until 1865. By this time his work in yet another field, electricity and magnetism, had borne fruit.

Maxwell's Marriage

Maxwell's years at Marischal College generally have been covered quickly in his biographies, yet in many respects they were more formative on a personal level than almost any other similar period in his life. Maxwell had been a precocious youth, but it was in the Marischal years that he went from youth to manhood. He began 1856 as the son of a respected Edinburgh lawyer with a modest estate in Kirkcudbrightshire.[68] By the middle of that year he had become laird of Glenlair with all the responsibilities for the upkeep of the estate, its tenant farmers, Glenlair house and its small staff. He began 1856 as a young Fellow of Trinity College, with much time for solving problems of his own choosing, rowing, some tutoring and giving a number of lectures mainly on hydrostatics and optics. He ended the year as Regius Professor, responsible for generating and teaching most of the coursework for the third year of all the MA students in his university, and more, as we have seen. He came to Marischal College after five years of life among the privileged and relatively narrow surroundings of Cambridge academic life, to the bustling commercial and manufacturing environment of Aberdeen with its range of students representing a wide cross-section of society. He now had to judge the abilities of men who would never have made it to university in Cambridge, and the experience would have reinforced his own tendencies to judge people by what they could do rather than who they were. He came to Marischal College a serious and thoughtful young man of independent spirit who partly by choice and partly from circumstances had never been part of the social crowd. He left the college a married man with someone closer to him than he had ever had, his parents excepted.

Following on from the example of Maxwell's first biographers, Campbell and Garnett,[69] successive biographers have left us with but a shadowy outline of Maxwell's wife Katherine Mary Dewar, though all mention that she was the daughter of Principal Dewar of Marischal College. Maxwell himself in his letters has given only fleeting glimpses of his family life, just enough to paint the picture of a devoted couple, a picture echoed by some of his friends. This lack of detail has been a spur to biographers to look in the writings and correspondence of contemporaries for further details. Not a lot has been uncovered. From 1858 onwards, Katherine is ever present in the background to Maxwell's scientific life, and this account will add to Maxwell's story a glimpse of the kind of family Katherine grew up in. It is, surely, the key to her character.

The Reverend Donald Sage has left us some biographical detail of Principal Dewar.[70] 'He was a native of Argyllshire [corrected to *Perthshire* in a longer account by another contemporary],[71] and was born in the humblest circumstances. His father was a blind fiddler, who earned his bread by travelling through the country and playing at weddings, etc. He was attended by his son Daniel for the purpose of being the bearer of the fiddle-case, and his father's guide. Some wealthy individual took notice of the musician's son, and thought he could discover, under the guise of his poverty, the germs of future greatness; and so, at his own private expense, sent him to a public school.' Dewar built up his career in steps from these humble beginnings. He went to London to train as an independent preacher and then to Glasgow, found there was insufficient call for his message in Scotland and enrolled in the University of Edinburgh to become a licensed Minister of the Church of Scotland. He had the extraordinary record of being awarded an LℓD by the University of Glasgow in the same year (1815) he graduated from Edinburgh. His first post was in a small community in Argyllshire, but he preached at times in the Gaelic Church in Edinburgh.

Dewar became the minister at Greyfriars Church in Aberdeen, the college church, and on the strength of this and his publications was appointed in 1817, after some professorial wrangling, as Professor of Moral Philosophy in King's College, Aberdeen. Sage says he was 'on a very friendly footing with him' while he was at King's College. Dewar became an extremely popular speaker in Aberdeen, for Sage relates: 'No public meeting could be conducted without Dr. Dewar. No sermon could be preached for any religious or charitable object but by him only. Not any new scheme could be formed, nor recent society established, without his countenance.' This, and Dewar's reluctance to give up his ministry of Greyfriars, annoyed King's College so much that Dewar left Aberdeen in 1819 for Tron Church in Glasgow. In 1826 he published his extensive two-volume work on *Elements of Moral Philosophy and of Christian Ethics*,[72] after which he returned to Aberdeen in 1832 to fill the vacancy as Principal of Marischal College. The College records note that his appointment had 'the unanimous disapproval of the College'.[73] It was a Crown appointment and the Crown overruled the College's choice. Sage concludes unexpectedly harshly, perhaps throwing light on this comment, 'He sat in a vehicle drawn by two horses, Ambition being the name of the one and Avarice that of the other'. In 1832 he was awarded a second doctorate (DD) by the University of Glasgow and, perhaps motivated by ambition or just plain responsibility, in the following years he played a significant part in obtaining government money to support the complete rebuilding of Marischal College in tasteful and practical style. The picture painted by Sage, by other comments, and by Dewar's own works is of a man of considerable ability and independence of thought, evangelical, knowledgeable on philosophy, biblical hermeneutics, and church history who at least until middle age could preach strongly. He continued to publish theological works throughout the 1850s.

What may have brought Maxwell to the Dewar's home was the arrival, in 1856, of a 125-page treatise in Gaelic on *Gaelic Astronomy* sent to the Principal by the author D. M. Connell, a teacher in Aberdeen. Daniel Dewar had co-authored a Gaelic/English dictionary published in 1839. He had been asked to offer a testimonial for Connell's work, which he duly gave in December of 1856. What would be more natural than to invite the new Professor of Natural Philosophy to discus the book's contents? We shall never know, but when he died Clerk Maxwell had in his scientific library of over 400 titles,[74] well populated with French and German texts, just one book in Gaelic: Connell's treatise on *Gaelic Astronomy*. Was it kept as a memento of how he first met Katherine?

Katherine's mother, originally Susan Place, was a grand-daughter of George Gordon, third Earl of Aberdeen.[75] This made her a cousin of George Hamilton-Gordon, Prime Minister from 1852–55.[76] Eleven years younger than Daniel and born in Yorkshire, she was described in a letter by Maxwell to his aunt as 'a first-rate lady, very quiet and discreet, but has stuff in her to go through anything in the way of endurance'.[77] There we have the in-laws. In the 1851 census covering the extensive Old Machar parish of Aberdeen, Daniel is recorded as aged 66, Katherine, who was born in Glasgow, 27, brother Donald 19. However, there had been a younger daughter, Susan Place Dewar, who had died in 1844 in York; a son, John Dewar, who had died in 1849; another son, the Reverend Edward Place Dewar,[78] who died in 1855; and there was also a son, William Gordon Dewar, and another daughter, Ann Gordon Dewar.[79]

According to Campbell and Garnett,[80] in 1857 Maxwell was a frequent visitor to Dewar's house (Fig. 2.7). 'His deep and varied knowledge, not only of his own and kindred subjects, but of history, literature, and theology, his excellence of heart, and the religious earnestness which underlay his humorous "shell," were there appreciated and admired.' In the comfort of one of Dewar's armchairs, Maxwell must have enjoyed articulating and refining his own sentiments on religious and moral topics with the learned principal. He was invited to join the Dewars in September 1857 for their annual visit to Ardhallow near Dunoon, the house in 10 acres of land of John McCunn, the husband of Ann Gordon Dewar. Maxwell was engaged to Katherine in February 1858 and married on Wednesday 2 June (Fig. 2.8) in the Old Machar district of Aberdeen (in which Dewar's house was located at 13 Victoria Street West). The witnesses were John McCunn and Katherine's brother Donald. Katherine's address is given as Craiglochie in the Parish of Kilspindie, near Perth,

Fig. 2.7 Principal Dewar's house at 13 Victoria Street West (now an office called 'Victoria House' with address 13 Victoria Street, photographed in 2012), a relatively new house in the 1850s in the expanding western periphery of Aberdeen, about a mile from Maxwell's downtown lodgings. Maxwell was a frequent visitor during 1857, indulging in varied social discussions that no doubt brought in Dewar's daughter Katherine.

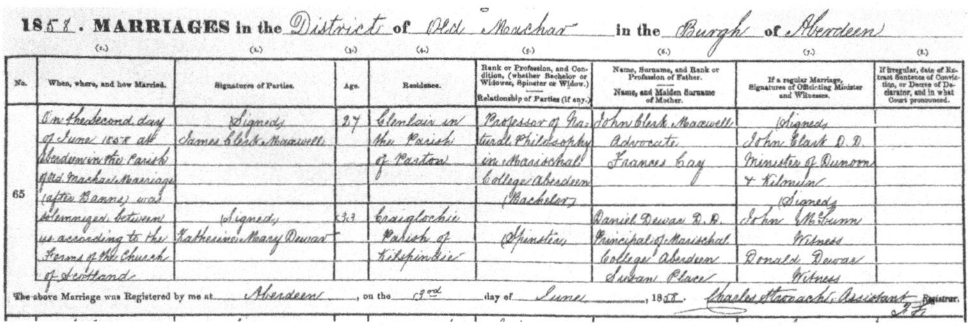

18*58*. **MARRIAGES** in the *District* of *Old Machar* in the *Burgh* of *Aberdeen*

No.	(1.) When, where, and how Married.	(2.) Signatures of Parties.	(3.) Age.	(4.) Residence.	(5.) Rank or Profession, and Condition, (whether Bachelor or Widower, Spinster or Widow.) — Relationship of Parties (if any.)	(6.) Name, Surname, and Rank or Profession of Father. — Name, and Maiden Surname of Mother.	(7.) If a regular Marriage, Signatures of Officiating Minister and Witnesses.	(8.) If irregular, date of Extract Sentence of Conviction, or Decree of Declarator, and in what Court pronounced.
65	On the second day of June 1858 at Aberdeen in the Parish of Old Machar Marriage (after Banns) was solemnized between us according to the Forms of the Church of Scotland	(Signed) James Clerk Maxwell	27	Glenlair in the Parish of Parton	Professor of Natural Philosophy in Marischal College Aberdeen (Bachelor)	John Clerk Maxwell Advocate Frances Cay	(Signed) John Clark D.D. Minister of Dunoon & Kilmun	
		(Signed) Katherine Mary Dewar	33	Craiglochie Parish of Kilspindie	(Spinster)	Daniel Dewar D.D. Principal of Marischal College Aberdeen Susan Place	(Signed) John McGunn Witness Donald Dewar Witness	

The above Marriage was Registered by me at *Aberdeen*, on the *19rd* day of *June*, 18*58*. *Charles Stronach Assistant* Registrar.

Fig. 2.8 Maxwell's marriage certificate, reproduced with the kind permission of the Registrar General for Scotland.

a property that had been gifted to Principal Dewar some years earlier and to which he and his wife retired in 1861. Brief announcements of the marriage appeared subsequently in the *Aberdeen Journal* and the *Aberdeen Free Press*. Maxwell was 27 years old and his bride 33. At least Lewis Campbell, as a representative of his Edinburgh and Cambridge friends, was able to attend the ceremony.

Katherine nourished Maxwell's spiritual and artistic side. Although Maxwell had a hugely productive academic life, he also had a life outside academia. His pastimes included serious literature, plays, theology, walking and riding. In all of these Katherine and James shared. Campbell and Garnett published a substantial selection of Maxwell's poetry and verse as an appendix to their biography. Some poems had already given Maxwell a public following through their appearance in the pages of *Blackwood's Magazine*. Most poems are either humorous or philosophical in overtone, excepting his romantic poems of the late 1850s addressed to Katherine. 'Will you come along with me,/ In the fresh spring tide,/ My comforter to be,/ Through the world so wide?/ . . .'.

It is well documented that Katherine assisted in Maxwell's colour-matching experiments in Aberdeen and in his experiments in London to measure the viscosity of air over a range of temperatures. Both of these activities were carried on in their home. There is no evidence, though, that she involved herself with any of the technicalities of Maxwell's science or took an interest in the science of Maxwell's circle. It is hard to avoid the conclusion that she never escaped from the shadow of her father and her father's preaching. She was brought up in a strongly religious household by comparatively elderly parents. At least two of her brothers became Ministers of the Church (Donald was ordained in 1860 in the Church of Ellon, 20 miles north of Aberdeen). She became a woman of exceptional piety. Nothing announces this more strongly than the much reproduced studio portrait of Maxwell, Katherine, and their dog Toby in which over a dark dress hangs her bright cross that would not be out of place in a convent.

When Daniel Dewar died in 1867, his estate that included 637 acres of farmland near Perth[81] and land in Rubislaw, Aberdeen, was valued at £1180 (about three times his former annual salary and not exactly the fortune one might expect of a man of 'avarice'). It passed into a trust with his wife effectively sole trustee while she lived, and upon her death (which happened in 1876) the Maxwells and the McCunns shared as trustees. No sons are mentioned in either the trust deed or in his will, for it is likely the Principal had outlived all four. The daughters were close to their father and vice versa. Maxwell felt he had a mutual understanding with Katherine that there was no need to explain. A letter in London's *Daily News* following his own death commented that although he was by then recognized as a scientist without equal 'to those who knew him his social virtues outshone all'.[82] Katherine would have agreed, but to the outside world, Maxwell had chosen a wife who was deeply religious, conventional, dutiful, private, tolerant and supportive, a wife to share his private life. They had no children.

In later years Katherine was accused of being possessive about Maxwell. It was all gossip, though there may have been some substance to it. Katherine clearly married Maxwell for his personal qualities, not because she wanted to display herself in society on the arm of Professor Maxwell. The more famous he became, even if it was mainly in the circles of the scientific elite in Britain and beyond, the harder it would be to find the privacy that they both enjoyed. To make matters worse, Katherine is said to have suffered prolonged spells of ill health, so her home life became her only life. In short, Maxwell's marriage was strong but in the scientific circle in which he lived, almost invisible.

Maxwell and The British Association at Aberdeen

The British Association for the Advancement of Science (frequently abbreviated BAAS in the nineteenth century and often, simply, BA or, as Maxwell and his circle jocularly put it, British Ass) was one of the great successes of nineteenth-century science in Britain. It was re-branded in

2009 as the British Science Association. The BAAS was founded in 1831 in response to a perception that had grown during the first quarter of the century that science in the round was largely unappreciated by government, the public and the state, and that such scientific organizations as existed were disconnected and had little influence on the political process.[83] The association was modelled on the Deutscher Naturforscher Versammlung, founded in 1822 and still thriving as the GDNÄ. David Brewster was particularly influential in the association's founding, and his vision for it was that it would possess no funds, property or artefacts but would organize annual meetings that would bring together scientists from all disciplines to overview the state of their subjects, bring up-to-date science to public notice, and propose measures for advancing science.[84] These broad aims, differently worded, were more or less made the objects of the association.

Following its initial meeting in York it quickly began to achieve all that Brewster and others had hoped for – not least because almost all the leading scientists in every discipline made a point of participating regularly. It became the place to go to hear the latest developments in subjects familiar and unfamiliar, to meet those responsible, to listen to speculation, to debate in a national forum issues and trends in specific areas of science and, for those engaged in science, to present one's work. The BAAS attracted all who had an interest in science: academics, manufacturers, the gentry, medics and the clergy, civic and private individuals. For the aspiring scientist there was no better place to network.

Maxwell must surely have attended his first BA meeting in Edinburgh in 1850, where the president was Brewster himself. Maxwell is recorded in Campbell and Garnett's biography as speaking impromptu from the floor on the subject of Haidinger's brushes (a visual phenomenon within the eye) and producing a small polarizing analogue that he had made himself. He spoke formally on this topic in the 1856 BA meeting in Cheltenham. He certainly attended the Glasgow meeting of 1855 (the year he was elected to life membership of the BA) but there is no indication he went to Dublin in 1857, and with his marriage in the summer of 1858 there was no question of going to the meeting in Leeds that year; 1859 was different. The BA meeting was to be in Aberdeen.

The final decision to meet in a town was not made until the meeting of the previous year. Aberdeen had two universities but it had no large meeting hall. Prior to the Leeds BA meeting, a proposal had been promoted to build a new hall at the back of the Assembly Rooms in Union Street under the aegis of the Music Hall Company, established for that purpose. Maxwell was a shareholder. The proposal would go ahead quickly if Aberdeen were granted the venue. It was, and within the year the 'Music Hall' was built of granite, 150 feet long, 50 feet high internally and able to accommodate over 2400 people. The proposers said the building would be a 'permanent benefit' of locating the BA in Aberdeen. Over a century and a half later, the Music Hall is still one of the city's principal entertainment venues, restored to close to its original 1859 appearance, excepting that the gas lights have given way to computer-controlled electrical illumination.

The Music Hall and Marischal College were the main meeting places for the 1859 BA. Had Maxwell wished to advertise his work and his credentials he could not have done better. The world would come to him, or at least the British scientific community would, and there was scarcely a branch of science at the time that did not have leading representatives in Britain. Would they actually come to remote Aberdeen? Certainly. H.R.H. Albert Prince Consort had accepted the BA Presidency. He could combine his genuine interest in science with his equally strong interest in the Upper Deeside royal estate of Balmoral where he had invested a substantial amount of time and energy in building a new castle that had been finished only three years earlier. As it would turn out, the week-long Aberdeen BA meeting would be the largest and possibly the highest profile meeting in the history of the BA until that date. Over 2500 tickets were sold and

Fig. 2.9 The Music Hall in 1859, with His Royal Highness the Prince Consort delivering the opening address of the BA meeting where Maxwell announced the *Maxwell distribution* of velocities of molecules in a gas. From a contemporary print in possession of the author.

the Music Hall was packed for the opening ceremony, and on the two subsequent evenings it hosted public lectures (Fig. 2.9). Among the participants were over 60 Fellows of the Royal Society of London, some 30 FRGSs, and FGSs and representatives from every university in Great Britain and Ireland.[85] Such a presence might have been expected in Oxford; it was truly exceptional in Aberdeen.

Modern attendees at a large scientific conference would find the organization of the Aberdeen BA meeting very familiar. The opening address by Prince Albert lasted an hour and was subsequently published. There were two keynote evening public lectures, one by Sir Roderick Impey Murchison[86] on the geology of the north of Scotland and one by Dr Thomas Romney Robinson of the Armagh Observatory, past president of the BAAS, etc. on electrical discharges in highly rarefied media. This talk was illustrated by 'magnificent experiments' conducted by Mr Gassiot (who was a vice president of the Royal Society and had given the Society's Bakerian Lecture on this subject the previous year) and the optician and instrument maker William Ladd. At parallel daytime sessions, 361 papers were read,[87] including the valuable reports on progress in each area of science; there were evening *conversaziones* in the Assembly Hall rooms, visits to local sites and manufactories of interest, commercial displays of the latest scientific instruments, a room with experiments displayed, and the spacious Marischal College Museum collection was augmented by donations and a special geological display open to all participants, including public ticket holders. H.R.H. Prince Albert was given a four-hour tour of Marischal College and attended some of the Section A papers.

Maxwell was not only a local host and a secretary to Section A (Mathematical and Physical Science) at this cornucopia of science, and one of the few local life members of the BA, he was also the principal local presenter in the mathematical and physical section who really had something to say. He presented three of his papers to the meeting: 'On the Dynamical Theory of Gases', 'On the Mixture of the Colours of the Spectrum' and 'On an Instrument for Exhibiting the Motions of Saturn's Rings'. Harman has published the summaries that appeared in the 1859 annual report of the British Association for the Advancement of Science.[88] The first of these papers contained the announcement of the *Maxwell distribution law* of molecular velocities in a gas, probably the only fairly fundamental law of physics that can be associated with Aberdeen. It was a nice Maxwellian touch that he chose to announce it at a public meeting of the BA.

The session proceedings were reported at some length in the local press and even more extensively in three editions of the weekly *Athenæum*.[89] Physical topics of the moment were notably optics, meteorology and telegraphy. One curiosity of the meeting was a talk by James Bowman Lindsay of Dundee on 'Telegraphing Without Wires', listened to with interest by Faraday, Lord Rosse, Airy, William Thomson and others. It was followed up by a successful demonstration of Lindsay's proposals in which two stations communicated across the river Dee without wires.[90] This was not a demonstration of 'wireless' some 30 years before Marconi, but it was a demonstration that if some other medium could conduct a signal (water in this case) then wires were not a necessary part of electrical signalling.

The attendance at the Aberdeen meeting was some 750 greater than the previous year's meeting at Leeds. When Sir Roderick Murchison summed up in the concluding meeting he remarked 'little could I have anticipated that we should have had such a meeting as we have had, and which, I will say, has been second to no one held by the Association'.[91] Maxwell had been in the spotlight on the stage of British science in September 1859. There is no indication he had any inkling that he would be considering a new job before the year was out, but it would be so. In the following year of change, testimonials would be collected, academic papers listed; it was a fortunate coincidence but the hugely successful BAAS meeting in 1859 would have done as much as the more formal requirements to raise Maxwell's profile, not among those already familiar with his work but among those who were not.

Maxwell's Departure from Aberdeen

During the summer break of 1859, Maxwell had heard that the Chair of Natural Philosophy at St. Andrews was becoming vacant but, even though the Commissioners' report on the merger of the two Aberdeen universities could not be long in appearing, he did not apply. News broke in early November 1859 of the imminent resignation of his old mentor, J. D. Forbes, in Edinburgh to take up the principalship at St. Andrews. This spurred Maxwell to collect testimonials from the Aberdeen professoriate and influential parents of some of his pupils, one of whom was John Webster, Lord Provost of Aberdeen and Vice President of the BA in 1859. All were of course strongly approving. It was likely a 'once-in-a-lifetime' opportunity for Maxwell to move to arguably the most prestigious of the four Scottish universities, one that was, into the bargain, significantly closer to Glenlair. In the event, he was up against two contemporaries who had been Senior Wranglers and Smith's Prizemen: P. G. Tait and E. J. Routh. Tait was appointed over him this time and stayed in the chair until 1901. (Tait had also been a candidate for the Marischal College post.)

The myth that Maxwell was sacked by the University of Aberdeen deserves to be buried. Nonetheless, it lives on because we like to hear stories that seem to bear witness to the stupidity of authority. Maxwell did lose his job in 1860, but it was thanks to the Royal Commission that oversaw the fusion of Marischal College and King's College into a single university, with one professor in each discipline. Fuller details of the circumstances have been related by Reid.[92] The single Chair of Natural Philosophy in the new University of Aberdeen was awarded by the Commission to David Thomson, Maxwell's counterpart at King's College. It was a remarkable piece of politicking by Thomson. Both chairs were royal appointments, technically not in the hands of the two universities concerned, who were therefore in a weak position to dispute the assignation had they wanted to. Curiously, though, the regius status of the chair was relinquished by the Crown, thereby losing their right to make subsequent appointments. What fuelled the myth about Maxwell was that he was the only younger incumbent to lose his job among the pairs of professors in each subject in King's and Marischal Colleges.

One aspect of the loss of post for Maxwell that has not been clear to historians was whether Maxwell received any compensation for the decision. It seems that he did, for his case was considered separately by the Commissioners under the Universities (Scotland) Act 1858 who reported in 1863. Their decision on Maxwell's pension, made in 1861 after requesting details of Maxwell's new appointment, was that 'the amount should be an annuity during his life, equal to the average income derived by him, both from endowment and from fees, at the time when he was deprived of his office in Aberdeen.'[93] This would have been close to £400 per annum.[94]

Maxwell could have retired to Glenlair in 1860, but he chose to apply for the post of Professor of Natural Philosophy at King's College, London, that had been vacated by T. M. Goodeve. He secured the position from a field of five considered applicants (according to the College Council minutes), and before the newly named 'University of Aberdeen' had opened for business in its inaugural session under that title, Maxwell found himself in London with new colleagues.

CHAPTER 3

Maxwell at King's College, London

JOHN S. REID

London and Maxwell's Appointment

London was a place of completely different character to Aberdeen, where Maxwell had spent the past four academic years. London was the chief town of the British Empire, 'the emporium of England, the centre of its great monetary transactions, the home of its science, literature, and art, and the yearly resort of its aristocratic and landed proprietor classes' according to one roughly contemporary description.[1] It is debatable whether it was the home of British science. It was the home of the Royal Society of London (founded 1660), a meeting place for a wide variety of scientists of distinction, and a publishing house, but not an institution with significant scientific facilities; it was the home of the Royal Institution (founded 1799) whose successive resident professors had a laboratory and made very good use of it, though the main mission of the Institution was promoting science. Gresham College had existed since the mid sixteenth century with a lecturing professoriate. The Royal Observatory at Greenwich had maintained its reputation as one of the world's premier observatories since its foundation in 1675; but what almost beggars belief is that, as late as the beginning of the nineteenth century, London, the wealthiest and most populous city in the world, had no universities at all. By any comparison with the great academic cities of Europe whose universities went back more than 500 years, London was not an academic city.[2] Did this make any difference? It's a curious fact that in spite of a population of some 2.8 million in Maxwell's time (14% of the total population of England and over half a million more than the city of Paris in 2010), very few nineteenth-century natural philosophers of note were born in London. The inference from these statistics is that, in the spectrum of London's achievements, science was not near the top. In complete contrast to Aberdeen, however, it *was* a place one happened to pass through on many a journey. It was a place that embodied change in general and exchange of ideas in particular. Maxwell could hardly have moved to a better city to maintain himself on the stage where science was exhibited.

By the time Maxwell arrived in 1860, London had two university colleges, founded within a few years of each other. In brief, the University of London was founded on paper in 1826 and University College opened for business in new buildings in 1828. The significant motivation for its founding was not the perceived want of a university in the great capital city, but the fact that the other English universities of Oxford and Cambridge were so closely tied to the United Church of England and Ireland (as it was then) that non-conformists and others of different faiths were at the time prohibited from graduating or becoming fellows.[3] University College was to be explicitly non-denominational. Indeed, theology was excluded from a portfolio that was otherwise as wide as any long-established university.

King's College, London, was founded just a year later (1829) with a similar plan of instruction but with a remit to include 'the doctrines and duties of Christianity, as the same are inculcated by the United Church of England and Ireland'. The religious link is echoed in the college motto '*Sancte et Sapienter*' and theology was included in the subjects on offer. King's took in its first students in 1831, initially in premises in a wing of Somerset House before building their college on an adjacent site. Notwithstanding the establishment slant that had motivated its founders, dissenters were allowed to attend classes from the beginning. Both institutions, as oases in the desert, had no difficulty attracting students in the early decades. In 1836, the University of London was redefined as simply an examining body that would oversee the academic output of the two separate colleges; a merger that at the same time maintained much of their autonomy. It was a model that Aberdeen could have followed with profit, but the Aberdeen Royal Commissioners chose not to.

For historians, the Aberdeen Commissioners did leave one important legacy. They opened up the working of the two universities there to the scrutiny of the public and posterity. No such transparency exists for King's College, London. It was a new university, scarcely 30 years old when Maxwell arrived, still finding its academic and financial feet, still making major changes to its portfolio of professors and courses, still building its reputation and place in the capital.[4, 5] Moreover, it was a 'new university' in its outlook too, emphasizing explicitly its rôle in providing training for the rising professional classes. The want of and need for science-based education in England had been highlighted by Lyon Playfair in an impassioned discourse[6] to the Society of Arts following the success of the Great Exhibition in 1851. This exhibition had made it clear that to sustain a competitive place in the world of invention and manufacture required a scientific and technical base that the traditional education in England was providing inadequately. King's College was at least moving in the right direction in the view of the modernizers, though Lyon Playfair would have had it providing post-graduate education too.

Just as Maxwell brought with him diverse interests from Cambridge to Aberdeen, so he brought his Aberdeen interests to London for continuation. His King's College years are notably characterized by the gelling of his thoughts on the mathematical representation of Faraday's electric and magnetic fields. This led to what is now commonly seen as his greatest achievement, the full unification of electric and magnetic phenomena, with his prediction of electromagnetic waves. His contemporaries, though, regarded his work on molecular science as more fundamental, and this, too, developed as he was leaving King's College (Fig. 3.1).

Before arriving in London, Maxwell had followed up his 1859 BA papers with further presentations on colour vision and the dynamics of gases at the Oxford BA meeting in June 1860 (the still famous meeting where Thomas Huxley met Bishop Wilberforce in an influential exchange on Darwin's theory of evolution). He was appointed to King's College as Professor of Natural

Fig. 3.1 The Institute of Physics blue plaque at King's College.

Philosophy in July 1860, some two months after losing out in Edinburgh. His duties were 'To instruct **all** the Students of the Department of Applied Sciences in the principal Mechanical Sciences, including Statics, Dynamics, Hydrostatics, Pneumatics, Hydraulics, together with Optics and Astronomy'.[7] He was required to be in college three mornings a week from 10 am to 1 pm and to teach an evening class for an hour a week. The salary was 5 guineas (£5.5.0) per annum for each matriculated student, £0.18.0 for each evening class student, and £2.7.3 for each occasional student (mentioned later) per term. Though we don't know exact numbers of students in the last two categories, this works out at around £450 per annum, a moderate step up from his Aberdeen income. Out of the five named candidates[8] for the post, Maxwell was appointed unanimously.

After a troubled September, when Maxwell caught smallpox at Glenlair and was nursed through his illness by Katherine, the Maxwells took up residence in their leased home at 8 Palace Gardens Terrace, Kensington (renamed as number 16 after the Maxwells left) (Fig. 3.2). The Kensington

Fig. 3.2a Palace Gardens Terrace, London, in spring 2011. Maxwell's residence at No. 16 (formerly No. 8 in his day) is at the centre of the picture. A London County Council blue plaque shown in Fig. 3.2b is above the balcony but hidden by the foliage in this picture. Photograph by Anne Ross Muir.

Fig. 3.2b London County Council blue plaque on Maxwell's house at No. 16 Palace Gardens Terrace.

district alone had a population in 1861 twice that of Aberdeen. The address was within easy access of Kensington Gardens and Hyde Park but a good hour's walk from King's College, when Maxwell chose to walk. As recorded in the 1861 census, they had a cook and a Scottish housemaid. It would have made sense to employ Glenlair staff.

On a personal level, a London residence made it more likely he would see friends passing through and gave him ready access to activities at the Royal Institution and, soon, the Royal Society of London. He was awarded the Rumford Medal of the Royal Society in November 1860 for his work on colour vision, and he was elected Fellow of the Royal Society in May 1861. His King's College years are frequently described as his most productive, but the fact is he received all his major honours (Smith's Prize, Adams' Prize, Rumford Medal and election to the Royal Societies of Edinburgh and London) largely for the work he carried out before he came to London.

Colour Vision and Colour Photography

One very important application came out of Maxwell's work on colour perception, one that he had foreseen as early as 1855. He realized that to achieve almost full colour reproduction one only needed to combine together views taken in three primary colours. This could be done in principle by photographing the scene on a black and white plate through a red filter, to record the red component, and likewise making two separate photographs of the same scene through green and blue (or violet, as Maxwell mentioned) filters. Positives are made from the plates (or they are made directly by suitable processing). Projecting the three positives simultaneously into a composite image on a screen using three separate projectors, each including the same filter the positive was taken through, would reconstruct the original coloured scene.

The implementation doesn't sound too difficult, but the fact is that it required a photographer of skill, perseverance and, as will be seen, with some luck to pull it off. No one had reported attempting it since Maxwell first mentioned the idea in a Royal Society of Edinburgh paper of 1855. Maxwell is correctly credited with being the father of colour photography, but his efforts did not herald the onset of colour photography. For example, Charles Fabre's eight-volume encyclopaedic treatise of photography mainly issued over the 1890s[9] has no references to Maxwell's work and virtually nothing on colour photography. Maxwell's 'proof of concept' that he was about to demonstrate in 1861 remained just that until almost into the twentieth century, an academic demonstration.

The exceptional photographer Maxwell needed was Thomas Sutton. He had been a Cambridge Wrangler and established a successful photographic business based in Jersey that claimed the patronage of Prince Albert.[10] In 1859, Sutton had come to the British Association meeting in Aberdeen to present a paper on his newly invented, distortion-free triplet photographic lens (a pre-cursor to the highly successful Cooke triplet of some 34 years later). Maxwell may have spoken with him but, in one of life's coincidences, King's College, London, had run photographic instruction since 1857 and Thomas Sutton was appointed the college photographer, in October 1860, upon the resignation of the incumbent. Maxwell and Sutton must have spent many hours in the college sunshine and in the college basement getting the technique to work. Maxwell had a commission to give a Friday evening discourse to the Royal Institution on the theory of three primary colours in May of 1861. The subject they chose to photograph was a suitable multi-coloured ribbon, probably one that belonged to Katherine. It is often described as a 'tartan' ribbon, though tartans usually have only two dominant colours. Since their results were projected onto a screen at the Royal Institution, only the audience knew exactly what the image looked like. We do know, though, that the colour reproduction wasn't of the quality we now expect, and Maxwell himself in the published note of the talk says 'by finding photographic materials more sensitive to the less refrangible rays [i.e. green and red in particular], the representation of the colours of objects might be greatly improved'.[11]

Coloured versions of the projected image of the ribbon can be found on the web and elsewhere. They come from reproductions made using a copy of the original positive plates by Kodak to mark the centenary of the event.[12] Kodak also tried to reproduce Sutton's technique. They realized that the original plates were not sensitive to green and red and hence Maxwell's demonstration should not have worked at all. Mid-nineteenth century photographers were pioneering the whole technique of photography and did not seem to be aware of the very limited spectral sensitivity of the wet collodion plates then in use, sensitivity that was restricted to the blue and violet end of the spectrum and continued into the ultraviolet. The plates better represented the spectral sensitivity of a bee's eye than a human's, though that was not known at the time. What Maxwell and Sutton actually achieved was the world's first false-colour image, the recorded ultraviolet image being shown in red. False-colour imaging is now the most common imaging technique used in remote sensing from space, where X-ray, ultraviolet and infrared images can be recorded by instruments sensitive well beyond the limits of human vision. Maxwell was certainly correct in his reasoning of how a full colour image could be made and in suggesting what was wrong with the demonstration.

Sutton left the employ of King's College after only a year. At the Manchester BA meeting, in 1861, he described his unusual panoramic lens (an extreme development of his triplet) and at the BA meeting in Cambridge, in 1862, he described a rapid-drying collodion process. The panoramic lens attracted significant attention in the 1862 exhibition mentioned below, and is illustrated and described at length in the official record of the exhibition cited later. His name is remembered by some photographers as the patentee, in 1861, of the first single-lens reflex camera though, like the colour photograph, the idea took decades to catch on. Sutton deserves more credit than he has been given for enabling Maxwell's ideas to be realized, albeit imperfectly.

Photographic emulsions that were sensitive across the range of colours that we see did not appear until the second half of the 1870s, which was the earliest that a genuine demonstration of three-colour photography could have been made. The first partially successful technique was developed by John Joly in Dublin, in the mid 1890s, using a three-colour additive process with colour filters that mimicked on a sub-millimetre scale Maxwell's method. The colours were approximate and muted; the practical details tricky. In the event, Maxwell's three-colour additive technique was finally incorporated into commercial film, in products such as Autochrome and Dufaycolor, early in the twentieth century, and is of course the basic colour reproduction mechanism used in colour television, mobile phones and today's three-colour CCD-based imaging systems.

Maxwell's King's College Teaching

Maxwell arrived at King's College with experience and courses in hand for teaching students new to natural philosophy, for teaching advanced students and for evening classes. King's College could hardly have expected any better experience in their appointee, for he would have to undertake similar courses there. The workload was lighter than at Aberdeen but still significant, and only someone who valued the whole teaching experience would take it on. One story from his Aberdeen years that was related by Campbell and Garnett is this: 'The professors had unlimited access to the library, and were in the habit of sometimes taking out a volume for the use of a friend. The students were only allowed two volumes at a time. Maxwell took out books for his students, and when checked for this by his colleagues explained that the students were his friends'.[13] When you see students as friends, you value the opportunity to teach. Maxwell had done the hard work in Aberdeen of thinking about his subjects, but the context at King's was different and his courses needed to be tailored to the new circumstances, circumstances that were in detail different from those in Aberdeen.

King's College was divided into four 'Departments', better considered as 'Faculties' in twentieth century-university language. They were 'Theology', 'General Literature and Science' (which included botany, zoology, Wheatstone's Chair of Experimental Philosophy and also a wide range of foreign languages), 'Applied Sciences' and 'Medicine', the largest department in terms of student numbers. Maxwell was in the Applied Sciences Department that hosted the 'Engineering Section' and the 'Military Section' as far as student programmes were concerned. Most students in this department enrolled in the Engineering Section. Its objectives were 'to provide a system of general education, practical in its nature, for the large class of young men who, in after life, are likely to be engage in commercial and agricultural pursuits, or in professional employments, such as Civil and Military Engineering, Surveying, Architecture, and the higher branches of Manufacturing Art'.[14] Thirteen subjects were taught in the Engineering Section, of which Maxwell's was 'Natural Philosophy and Astronomy'. The minimum entrance age in this section was 16. For students in the Military Section, it was just 15.

It is interesting to compare Maxwell's professorial colleagues in Applied Sciences when he arrived with the Marischal College staff discussed in the previous chapter. They were Prof. Rev. Thomas Grainger Hall in Mathematics, in post since 1830, Dean of the Engineering section and author of several textbooks on calculus, descriptive geometry and trigonometry, with Rev. T. A. Cock as Lecturer in Mathematics and Rev. Walter Howse as Assistant Lecturer in Mathematics; Prof. William Hosking, Civil Engineering and Architecture, railway engineer, bridge designer and architect, appointed in 1840 but died in 1861,[15] with A. Moseley as Lecturer in Civil Engineering; Prof. Charles Percy Bysshe Shelley, Manufacturing Art and Machinery, later author of textbooks on applications of heat, workshop appliances and co-author with T. M. Goodeve of a treatise on Joseph Whitworth's measuring machines; Henry James Castle, Lecturer in Land Surveying and Levelling, present since 1839, whose book with F. W. Simms on mathematical drawing instruments, the theodolite and levelling, first published in 1837, is, amazingly enough, still in print; Professor Thomas Bradley, Geometrical Drawing (author of a detailed work on Practical Geometry, Linear Perspective and Projection) with Fredrick Alfred Bradley as Lecturer; Professor William Allen Miller FRS, Chemistry (medical doctor, author of a popular textbook on the elements of chemistry, notable for his wide-ranging scientific interests, particularly in spectroscopy) with K. A. Haddow as demonstrator; Professor Charles Loudon Bloxam, Practical Chemistry (whose own long-running textbook on inorganic and organic chemistry, first published in 1867, would complement Miller's elements of chemistry);[16] Professor James Tennant FGS, Mineralogy and Geology, a highly respected mineralogist and geological collector who had been with King's since the inauguration of its engineering course in 1838. Tennant's collection of over 5000 minerals is now with Royal Holloway College, University of London.

As at Aberdeen they were competent university men. They covered a wider range of technical subjects than Maxwell's Aberdeen counterparts and were more willing to publish textbooks in their areas of expertise. Indeed Shelley, Miller, Bloxham and Maxwell would all be authors in Longmans' 'Text-books of Science' series in the early 1870s under the editorship of T. M. Goodeve. Given there were on average only 80 matriculated students[17] in the department during Maxwell's tenure, the staff list is impressive and suggests why the college always seemed short of money. The average number of matriculated students in the three non-medical departments of King's College during Maxwell's years, excluding evening class students, was 215, very similar to the size of the Marischal College MA cohort.

As an aside, London was not a substantial centre of manufacturing and construction in 1860s Britain. Trade and finance were its strengths, brewing its biggest manufacturing industry. Nonetheless, trade supported a large dock industry, and in providing effectively a degree in engineering from their early years, both London colleges were ahead of most of the other British universities. Outside London, only Glasgow had a Chair of Engineering by 1860. Its occupant then was the notable W. J. Macquorn Rankine, a strong advocate of the importance of scientific principles in underpinning engineering practice and, indeed, the reverse synergy of engineering solutions guiding the development of science.

All matriculated students at King's College were obliged to attend daily chapel at 10 am. Maxwell would have had no issue with this. Indeed his willingness to attend Church of England services was a pre-requisite for the job and he almost certainly attended on the days he was in college. All matriculated students also had to attend the chaplain's divinity class. King's College was not authorized to issue MA or BA degrees for their students, but instead offered the AKC, the Associate of King's College. There were of course various recognitions and exemptions in special cases, but the standard requirement was attendance and satisfactory performance for nine terms, namely three years minimum, for the academic year comprised three terms. The Michaelmas term began in early October and ended just before Christmas; the Lent term began a month later after the middle of January and ended in early April; Easter term began after a break of 10 days and ended towards the end of June. Maxwell and the students then had a break of three months. Fees for engineering students were £12.17.0 per term, excluding books, the compulsory gown and cap, and excluding boarding fees for those who wished to make use of the limited living-in facilities. Compare this with Marischal College fees of about £5 per year. Students who did not wish to matriculate could enrol in individual classes, the fee for natural philosophy being 3 guineas (£3.3.0) per term. Such students were called 'occasional students'. The average number of occasional students spread over all subjects in the three non-medical departments was 50.

This was the broad context of Maxwell's King's College teaching, according to the college calendars. However, the implication that students began in October, attended for nine terms, and then departed with their associateship is completely misleading. The student registers[18] reveal a different story. Taking a sample of 150 students who were in attendance during the first years Maxwell taught, the average length of time a student stayed was just 4.3 terms. Approximately 40 stayed for just one year, 30 for two years and others for every other possibility. Only a very small minority emerged with the AKC. (Compare this with Marischal College, where almost 50% of students completed their four years of study.) Not all started in the Michaelmas term either. For example in the 1861/62 session, 30 began in the Michaelmas term, 19 in the Lent term and 7 in the Easter term. It is hard to avoid the conclusion that King's College was seen from the student perspective as providing not so much a degree as pre-professional further education that opened doors to professions where formal knowledge had more significance than manipulative skill. When a student felt he had acquired sufficient knowledge, or when a door of opportunity opened or, perhaps, when he could no longer afford the fees, he left.

Applied science students had a fuller timetable of coursework than most modern university students. The register recorded an engineering student's performance in 12 areas each term, usually just with the letters G (good) or VG (very good), though other entries dealt with special circumstances. The areas were: divinity, chapel, maths, mechanics, arts of construction, chemistry, geology, mineralogy, manufacturing art, geometrical drawing, surveying and workshop. Mechanics covered Maxwell's natural philosophy courses; arts of construction was not available to first-year students; photography was a compulsory part of the AKC for final-year students but not recorded explicitly in the register. Workshop practice for two three-hour sessions per week was part of each term's course. The associated timetable included such complications as some activities every fortnight or activities restricted to certain terms, but overall a student's programme involved about 13 lectures a week and 12 hours of practical work a week in four 3-hour afternoon sessions.

Maxwell gave two separate courses, one for first-year students and one for students in the second or third years. Each course was taught for three hours a week. With the addition of his evening class, this made for seven hours a week in front of a class (about half his Aberdeen commitment). The pattern of student enrolment on applied sciences discussed above suggests that the three single-term courses had to be fairly self-contained. The first-year class would have been much the larger class. It was given from 11.30 to 12.30 on Mondays, Wednesdays and Thursdays; the senior class preceding it from 10.15 to 11.15 on the same days. Cyril Domb, a successor to Maxwell at King's College, has written particularly[19] about Maxwell's courses, quoting syllabus details from the annual university calendars and outlining how his course content evolved. No student notes have been found of his King's College lectures, but some examples of the detail Maxwell included in his course have been given in the preceding chapter on his Aberdeen years. Maxwell continued his teaching practice of regularly posing the class a set of 10 questions. Almost 40 sets of these are contained in a notebook[20] now in the possession of the college. These questions were given once a fortnight, Maxwell interleaving the questions for his 'Junior' class and his 'Senior' class in successive weeks. Broadly speaking, it would appear from these questions that his first-year class received a simpler introduction to natural philosophy than did his Aberdeen classes. This is not surprising, for his Aberdeen course was given to third-year students who had already received a year of mathematics and, in addition, his Aberdeen course was longer.

The first-year class texts at King's College were Galbraith and Haughton's manuals on mechanics, hydrostatics and optics, the same texts Maxwell had adopted at Aberdeen and the same ones his predecessor at King's, the able T. M. Goodeve, had used to support a course that differed from Maxwell's in many details. Maxwell's introductory lectures at Aberdeen and King's College have both been published and examined.[21] They are quite similar. The general coverage of Maxwell's first-year course at King's College was also similar to his Aberdeen course, namely mechanics and dynamics, followed by some hydrostatics and then heat. The course finished in the Easter term with optics, at King's College illustrated with demonstrations using an arc lamp or oxyhydrogen lamp. These are the only demonstrations explicitly mentioned in the calendar, but Maxwell did receive an expenses allowance of £15 per annum.

The college had a museum founded upon the George III collection of scientific instruments and apparatus that had been donated by Queen Victoria, in 1841, to support the teaching of engineering. Much of this collection can now be seen in the Science Museum[22] and, although it included models of machinery, by 1860 it was somewhat dated. However, additions had been made by the college council and private donations and the museum was overseen by a curator. Nineteenth-century museum practice was to exhibit items in packed glass cases. The idea of a 'hands-on' museum was well in the future and it is likely that the museum artefacts were there mainly to be observed. This was still a significant privilege for the College, since London had no Science Museum as such in 1860, though it did have the educationally oriented South Kensington Museum. This was a conglomerate institution, mainly a museum of art and artefacts that had evolved from the Museum of Manufactures spun off from the Great Exhibition of 1851, but it also including other components such as a Museum of Education that had spun off from the 1854 Exhibition of Education.

We have to rely on the university calendars for the contents of Maxwell's senior course, and on the sets of questions still extant in Maxwell's notebook. The calendar entries look the more impressive. The course began with theoretical lectures on applied mechanics and continued with the motion of an incompressible fluid that included 'solution of questions in heat, electricity, and magnetism', the analogies he was pursuing in his papers on electromagnetism discussed later on. After a section on waves relevant to sound and light, the course ended with the first principles of astronomy.

When Maxwell had been an undergraduate himself, in Edinburgh and Cambridge, he had belonged to student societies that supported discussion on physical topics. The King's College Engineering Society was founded in 1847, the year of inauguration of the Institution of Mechanical Engineers (I.Mech.E.), the second oldest of the British professional engineering institutions. The King's College Society can now rightly claim to be one of the oldest engineering societies in the world. It had been vibrant in the late 1850s with a varied programme of lectures,[23] before hitting a rocky patch in the early 1860s, but it came back to life and, in 1864, Maxwell was listed as an honorary President along with the Professors of Maths, Civil Engineering, Manufacturing, and Surveying. This suggests that Maxwell had carried his student-friendly persona to London from Aberdeen.

The college's biggest teaching success story during Maxwell's years was its evening classes. In 1855 these had begun by offering 15 subjects. The choice rose almost year on year, so that in Maxwell's final year 30 subjects were offered. Each class lasted an hour in the week and a course ran over 20 weeks. Students could take a portfolio of courses that included divinity plus four others, and matriculate in the college, for an extra fee, counting the passes accumulated over three successive years towards an AKC. 'Occasional students' could enrol in one or more classes. During Maxwell's five years at college, an average of about 80 evening class students matriculated annually and almost 500 additional evening class students annually were classed as 'occasional students'. Evening class students were given access to the college library. Maxwell's annual contribution was 20 lectures on experimental physics, given from 8 to 9 pm on Wednesday evenings, running from mid October to the end of March, with the one month's winter holiday breaking the course in two. In the advert for his successor, the number of students attending experimental physics was put at 'from 20 to 30'.

For the first two years, Maxwell presented a similar, wide-ranging course covering 'the effects of heat, the measurement of temperature, the gas laws and the efficiency of engines, conservation of energy, basic electrostatics, magnetism, current electricity and batteries, the telegraph, radiation, optical instruments, the eye and eyesight, properties of light including its wave nature and interference, waves in general, optical polarization phenomena, physiological optics, the measurement of light and colour'.[24] The final subjects fit closely with Maxwell's own colour researches. This was clearly 'experimental physics' well supported by the concepts needed for understanding. The syllabi for the following three years were all different, a reasonable strategy if one is targeting a fixed pool for whom the classes were accessible. In 1862/63 the course concentrated on the effects of heat on bodies and electricity and magnetism, the 1863/64 course on properties of matter as affected by pressure and heat – effectively a course on thermodynamics – and the 1864/65 course covered magnetism and electricity. Cyril Domb has published this last syllabus in full.[25] Campbell and Garnett state that Maxwell 'continued lecturing to the working men' in the winter of 1865/66, more than half a year after he had retired, but the college calendar explicitly states that the class will be taken by his successor. Perhaps Maxwell made a contribution without being formally in charge.

One difference in the college from the Aberdeen situation was that larger subjects at King's College employed assistants. During Maxwell's second year, the first 'Lecturer in Natural Philosophy' was appointed, George Robarts Smalley. Smalley had been assistant to Her Majesty's Astronomer at the Cape of Good Hope (Thomas McLear) for five years, then Professor of Mathematics at the South African College, and from 1854 in London a respected mathematics master for seven years at King's College School (that had an annual intake of about 400 pupils). He had given the King's College evening class in mechanics on at least one occasion and been an applicant for Maxwell's post. Smalley assisted Maxwell's day classes for two years before being appointed Government Astronomer at the Sydney Observatory, beginning in 1864. He died in this post aged 48 in 1870. The post of Lecturer in Natural Philosophy was refilled in October 1863, after Smalley's departure, by William Grylls Adams, brother of Cambridge astronomer John C. 'Neptune' Adams, whose work the Adams Prize commemorates. Grylls Adams was five years younger than Maxwell and succeeded to his chair when Maxwell retired in 1865. He kept the appointment for 40 years.

Maxwell and the 1862 Exhibition

Following the great Crystal Palace 'Exhibition of the World's Industry' of 1851, it had been in British minds to hold a sequel a decade later. The success of the London exhibition spurred others, such as the Irish Industrial Exhibition of 1853 and the New York Exhibition of the Industry of all Nations, also in 1853, designed to let America measure its industrial strength against that of the rest of the world. Since 1798, France had been holding national exhibitions of industry more frequently than every decade and these led to its own International Exhibition in 1855. However, the 1850s had not been a particularly happy decade in Britain. The Crimean War drained national resources from 1853 to 1856; the first Indian War of Independence (the Indian Mutiny) made a large psychological impact even though in the end Britain retained control of India. The planned

Fig. 3.3 View of the exterior of the 1862 International Exhibition from Exhibition Road giving a flavour of London in Maxwell's day. He contributed the explanatory account in the exhibition covering the optical instruments exhibited.

repeat of the Great Exhibition was finally staged in London as The International Exhibition of 1862 in a specially constructed exhibition building in South Kensington (Fig 3.3). It ran from 1 May to 15 November, attracting some 6 million visitors. In a number of ways it was considered less successful than the Great Exhibition of 1851: it was in a larger hall that was open longer but attracted fewer visitors per day and barely paid for itself, whereas the Great Exhibition had made a substantial profit. In addition the Queen and Court had not attended its formalities, for in 1862 Victoria was still in mourning for Prince Albert, who had died in December 1861. It was more of an industrial exhibition than its 1851 precursor, with a large Western Annexe occupied by machinery.

One section of the exhibition was devoted to 'philosophical instruments'. Maxwell was involved in the exhibition at least as the author of the explanatory document that accompanied the subsection 'instruments connected with light'. This summary was published separately but also appeared in the volume that acted as a record of the International Exhibition of 1862.[26] The account is not included in his collected papers, but it is a superb example of didactic writing by a gifted expositor. The temptation for many given the task would simply be to describe what the visitors would see. Maxwell goes much deeper. He separates the instruments into their different kinds, clearly explains principles behind their operation, includes key aspects of their historical development, technical difficulties encountered in improving instruments of a given type and describes applications of interest, all in the context of the examples on display.

For example, in his opening section on glass and, in particular, exhibits of parallel-sided glass plates for sextants, he describes how he has tested for the flatness of glass using an arrangement we now recognize as Fizeau's fringes. The remarkable aspect of this is that Fizeau only published his method of obtaining fringes in the same year, possibly after Maxwell wrote his guide. Maxwell's experimental knowledge and practice was clearly totally up to date. In the next section on an exhibit of lenses, he highlights how cylindrical lenses (on display) exhibit the technology of correcting for astigmatism of eyesight, an aspect realised by G. B. Airy. Following on, his discussion on microscopes explains why the design of microscopes is a much harder problem than the design of a good telescope, and how the relatively newly invented binocular microscopes on display give a three-dimensional view with exaggerated relief. This naturally leads on to stereoscopes, although in historical terms it was the stereoscope that gave rise to the binocular microscope. Visitors were encouraged to look at Warren de la Rue's stereoscopic pictures of the Moon, taken using the Moon's libration to provide the necessary parallactic views. The three-dimensional views enabled the viewer to see the Moon as no astronomer could directly, with or without a telescope – an early example of imaging technology extending our ability to explore the natural world beyond the provision of just magnification.

This is not the place to describe all the fields of optics Maxwell covers in his exhibition account. In the next decade Maxwell would become Professor of Experimental Physics at Cambridge and people have wondered why he was chosen for that post, since many of his most outstanding contributions to physics had been conceptual and mathematical. It is clear, though, that Maxwell's boyhood love of making things grew into a real expertise with laboratory equipment and experimental demonstration. His teaching experiences gave him the opportunity to pursue both aspects and the opportunity to encourage the interests of others and pass on his knowledge. Fourteen years after the 1862 exhibition, under the auspices of the South Kensington Museum of Science, an international 'Special Loan Collection of Scientific Apparatus' was exhibited on almost the same site as the 1862 exhibition. Maxwell wrote the introductory article on the principles and function of scientific instruments in the accompanying handbook.[27] He also wrote the article on 'Instruments Connected with Fluids' in the section on Molecular Physics.

Maxwell's exhibition articles provide clear evidence that Maxwell was an accomplished experimentalist.

The British Association Electrical Resistance Standard

Maxwell did not present any papers at the BA meeting of 1861 in Manchester, but at that meeting the BA set up its subsequently famous 'Committee on Electrical Standards' to which Maxwell would be invited in the following year. For reasons that are about to be given, the first major report of the committee concentrated on electrical resistance. Maxwell was one of the principal contributors to their first report[28] at the BA meeting of 1863 in Newcastle-upon-Tyne. Their 65-page offering was an important step in the future establishment of the international definitions and standards for the volt, amp, and ohm, later defined by the International Electrical Congresses, the first of which was in 1881. The problem was this. Historically, measurement of widely used quantities like length, volume, weight and so on had been encumbered by almost arbitrary standards that varied from nation to nation. Electricity was increasingly pervading industrialized society, and it was desirable to avoid a repetition of this state of affairs. Telegraphy, in particular, was in the process of encircling the world with little regard to national boundaries, and an international consensus on electrical units would clearly benefit everyone. At least, that was the altruistic motivation. Commercially, Britain was the leading nation pushing the trans-global telegraphic revolution, and her electricians perceived that it was advantageous in terms of quality control, fault-finding and other practical operations to develop a clear and unified system of electrical units. In practice, the ultimate results of this work would affect every man and women in the country and, in due course, everyone in the world. The desirability of an 'absolute' system of units, one where electrical units are as far as possible related to units of other quantities, as opposed to units 'relative' to arbitrary standards, was outlined at the beginning of the report.

Absolute systems were based on firmly established relationships between different quantities. For example, through Coulomb's Law (mid 1780s) relating the force of attraction or repulsion experienced by two charges at a given separation, charge can be defined in terms of force and distance. Current is defined as charge passing in unit time, and hence currents defined in terms of force, distance, and time, all basic, pre-defined quantities. The power dissipated in a resistance, which is a rate of doing work, is a product of its resistance and the square of the current passing through it. This brings electrical resistance into view. Finally, Ohm's Law relates the electromotive force (now called the voltage) across a resistor to the value of its resistance and the current through it. So a system of electrical units can be established based upon the electrostatic attraction or repulsion of charges and the underlying system of mechanical units. Using the centimetre, gramme, second (cgs) mechanical system produces the electrostatic system of electrical units, later called the *statcoulomb*, *statamp*, *statvolt* and *statohm*.

There are two issues. First, all the units except the statvolt so defined are unsuitable in their sizes for day-to-day electrical work. Second, there is an alternative way of going about things that produces a different set of electrical units. Force is also related to the force of repulsion between two coils carrying a given current (as discovered by Ampère early in the 1820s). Starting here, current can be defined in terms of force and distance (the distances involved being the size of the coils and their separation). Charge is now defined through current and time, and so on. This system produces a second set of units, 'electromagnetic units', later called the *abamp*, *abcoulomb*, *abvolt* and *abohm*, only the first two of which are reasonably practical for everyday use. As a short digression, it is worth saying that it was Maxwell himself who showed that the ratio of charge measured in electrostatic units and electromagnetic units must equal the velocity of electromagnetic waves in space (e.g. light). In 1868 he published his own completely electrical measurement of this velocity. That, though, takes us to a time in another chapter.

Now aspects of reality enter. Some quantities are more easily measured than others. Current is one of the most easily measured through the interaction of a coil carrying the current and an external magnetic field. For example, the moving coil meter that was the everyday electrical measuring instrument for over a century basically measures current. To measure voltage or resistance, their effects are converted into equivalent effects on current. However, to measure current absolutely by this means one needs to know the value of the magnetic field involved, either the Earth's field or one specially created by a magnet inside the instrument. William Thomson had hit upon an alternative doorway into absolute electrical units. The BA report spells out the basis of the method. The theoretical result is that by rotating a thin closed-circuit coil over a suspended small magnet, the magnet is observed to deflect by an amount that depends on the speed of rotation and the dimensions of the coil (all mechanical quantities) and the resistance of the coil. Hence, when the mechanical factors have been determined, including the deflection of the magnet, the resistance of the coil can be measured in absolute units. Once this is done for a standard coil, then using a Wheatstone bridge (promoted by Wheatstone in the 1840s) any other resistance can be measured as a ratio to the standard resistor. With resistance known, then the other quantities can be deduced from the relationships mentioned above. The measuring rod, the balance and the clock, the fundamental measuring tools that underpin all mechanical quantities, can therefore be made to underpin all electrical units.

Maxwell was central to the BA committee's effort. He, Fleeming Jenkin and Balfour Stewart undertook the experiments needed, in King's College. These continued for over a year beyond those reported at the 1863 BA meeting. Maxwell calculated the corrections necessary because of the departure of the actual equipment from the theoretical ideal that yielded the relatively simple formulae involved. They all considered at great length minor factors that had to be given attention in order to achieve an accurate result. For example, August Matthiessen, who had been appointed Professor of Chemistry at St. Mary's Hospital, was especially concerned with the observed variation of resistance with temperature, and chose an alloy ('German silver') that would be most suitable as a working standard of resistance. Appendices A and B by Matthiessen single out this aspect of the work. Appendices C and D by Maxwell and Fleeming Jenkin are a 30-page paper, almost half of the report, including a valuable and innovative section 'On the Elementary Relations between Electrical Measurements'. In this they introduced in modern notation the method of dimensions for analysing the relationship between units of different quantities, a method that quickly became one of the fundamental tools of basic physics. For example, in the electromagnetic system electrical resistance has the dimensions of velocity, a fact that some telegraph engineers at the time found hard to appreciate, for it was not the velocity of transmitted signals. The second part of their appendices described in detail their King's College measurements. Surprisingly, these appendices, too, do not appear in Maxwell's collected works.

Fleeming Jenkin was also responsible for the governor that kept the coil turning at a constant speed and, spurred by efforts to improve this device for later measurements, Maxwell was led to reflections on the stability induced by governors, from which he produced a paper in 1868 that is usually quoted as the founding paper of the subject of cybernetics.[29] Another spin-off from the apparatus was Maxwell's own experimental determination of the viscosity of air at varying pressures and temperatures, as indicated in a passing remark to Stokes in 1863. The experiment described above was conducted by rotating the coil at around 6 revolutions per second for some time in an electrically open-circuit configuration, and noting the position of the central suspended magnet, and then repeating this with the coil in a closed-circuit configuration, during which time a current induced in the coil deflects the magnet. Its deflection was measured by an optical lever attached to the suspension wire. When the coil rotation was stopped, the magnet oscillated back to equilibrium, being damped by the viscosity of the air and internal friction within the suspension. In a fine example of lateral thinking, this gave Maxwell the idea for measuring the viscosity of air by observing the damping of the torsional oscillations of a disc swinging between fixed plates. This method was to form the basis of his 1865 paper to the Royal Society of London, describing experimental results that would confirm his 1860 predictions on the variation of viscosity with pressure and temperature. In the BA resistance determinations, one would be hard pressed to find a better example of how academic work on an apparently obscure problem could provide the spark for a diverse range of knowledge and also provide results that had immense practical importance (Fig. 3.4).

The work continued in 1864, with the 1864 Bath meeting report[30] of the committee outlining their philosophy that 'every experiment should be made afresh and every element of the experiment should be varied'. The 1865 committee report[31] at Birmingham was a mere six pages, confirming that the committee's work had been accomplished (as far as resistance was concerned), but importantly describing the construction of the standard resistors that would form the practical outcome of the whole exercise. Figure 3.5 shows the design of the 'BA standard

Fig. 3.4 James Clerk Maxwell pictured with the spinning coil apparatus that he developed with his collaborators at King's College to establish the British Association standard resistance. This portrait is a copy of the posthumous original owned by the IET. It has been gifted to the James Clerk Maxwell Foundation by Professor David Ritchie and hangs in the Royal Society of Edinburgh's Maxwell Room. Image courtesy of David Ritchie, the James Clerk Maxwell Foundation and the RSE.

resistor'. Within the casing, silk covered wire is wrapped around a hollow brass former and the whole embedded in paraffin wax. Thick copper wires take the connection to the outside circuit which is made by dipping them into mercury cups. The resistors are used in a constant temperature water bath and are defined accurate at a specific temperature. A set of 10 resistors using different materials was made to investigate aging effects, but the preferred wire material for copies was a platinum/silver alloy and the accuracy expected was better than 0.01%. The purchase price of a copy was £2 10s (approximately £250 in 2010 currency). All these fine details and more illustrate what had to be got right if a credible international standard was to be established. Figure 3.6 shows a late ninteenth-century water bath with two resistance standards developed from the BA archetype.

The experiments may have been carried out in a King's College University laboratory, but the business was a hugely practical matter. Wheatstone, Varley, Thomson, Siemens, Jenkin and Bright on the committee all had strong telegraphic interests. Telegraphy was the foundation of the electrical engineering profession. Electrical resistance was hugely important, both in connection with the telegraph lines themselves and with the ancillary equipment. The BA unit of resistance defined by the committee was a velocity. The particular German silver coil that was at the centre of the 1863 experiments had a resistance of 1.0762×10^8 m s^{-1}. The BA unit of resistance was defined as 10^7 m s^{-1} and came to be called the ohm, after Georg Ohm, an electrical pioneer who had died in 1854. (Before the 1860s were out, today's symbol Ω for ohms had been introduced by William H. Preece, a former pupil at King's College by coincidence, and later to rise to Engineer in Chief for the Post Office in Britain's nationalized telegraph and telephone service.) Through Maxwell's contact with the instrument making firm of Elliott Brothers, particularly with Carl Becker, secondary standard coils were carefully made and calibrated with reference to the BA master coil. Copies were distributed in the first instance to Directors of Public Telegraphs in 11 countries and to other bodies. One result was that electrical resistors in ohms were spread rapidly among the telegraphic community and from there outwards. Once the unit became widespread its

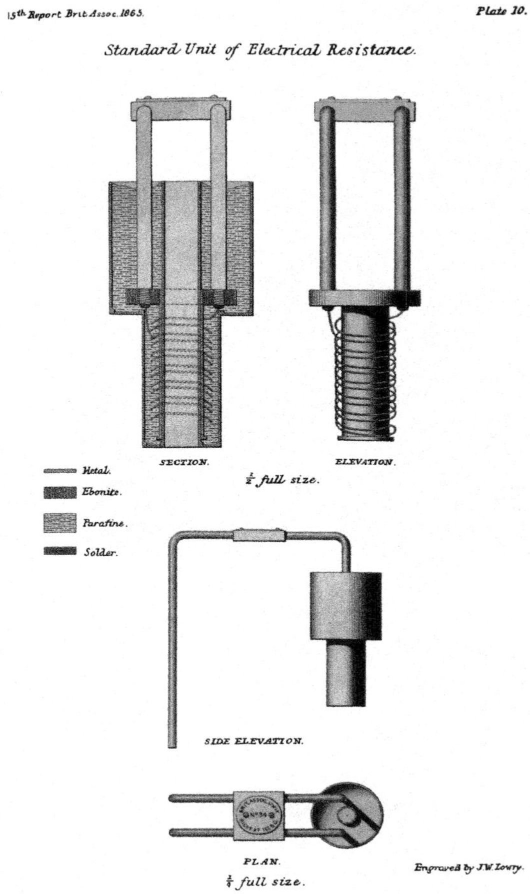

Fig. 3.5 Plan of the British Association (BA) standard resistor of 1865, from the 35th report of the British Association, 1865.

Fig. 3.6 Late nineteenth-century standard resistors in a constant temperature bath, from the Natural Philosophy Collection of Historical Scientific Instruments, University of Aberdeen. The central hole in each resistor is for the insertion of a thermometer that could read to 0.1 °C. Courtesy Natural Philosophy Collection of Historical Scientific Instruments, University of Aberdeen. Permission of the Hon. Curator (the author).

incorporation into the international electrical standards was not surprising, though its definition has changed more than once.

One other legacy of this work is that, thanks to the involvement of scientific men in the enterprise, the system of electrical units was tied from the very beginning to the metric system of mechanical units. It has taken the whole of the twentieth century to persuade most of the world to use metric units for commerce and everyday life, and the task is not yet quite complete in the twenty-first century, but electrical units have needed no change, for they have been universal and tied to the metric system from the beginning. For example, the unit of electrical power, the Volt-Amp, is identical to the metric mechanical unit of power, the Watt.

The corollary to this episode is that it involved particularly careful work by Maxwell and colleagues, extending over two years, to determine the absolute value of the resistance of the master coil in absolute units to the required precision. No one who wanted to reproduce as exactly as possible one ohm of resistance could contemplate such effort. One of the secondary standards made by the BA was a column of mercury representing one ohm, and for practical purposes it was decided by the International Electrical Congress in Chicago in 1893 to redefine the ohm as the resistance of a specified column of mercury at 0 °C. In the twentieth century, the emphasis moved to considerations of how a standard resistance could be defined so that it could be reproduced and used most accurately in comparisons. This led to the abandonment of the mercury column and the adoption of a more erudite definition involving the quantum Hall effect that was made the international standard in 1990. The redefinitions were intended to be backward compatible, so that within the physical uncertainties of the old standards they maintained the value of the ohm as that specified by the original BA Committee. In practice they didn't quite succeed in this aim.

Maxwell on Electromagnetism

Maxwell's work on electromagnetism is now seen to outshine all his other efforts, but it was certainly not so in his day. It was while at King's College that he finally worked his way through the mist of uncertainty that had obscured his vision until then, revealing a new landscape in which were situated the complete laws of electricity and magnetism. Maxwell's laws were formulated in terms of differential equations (an important difference from some earlier work); the quantities involved were electric and magnetic fields; the laws contained some addition to previous knowledge; the laws predicted the possibility of transverse electromagnetic waves that travelled at the speed of light. By implication, Maxwell took it that light and radiant heat were therefore electromagnetic waves. It would take another quarter of a century before experiments were directed at the now obvious question of whether other electromagnetic waves existed or could be generated. Maxwell would no longer be alive to hear the answer.

Maxwell's work was a visionary effort. After starting with William Thomson at least alongside him, there was no one leading him through, no one at his heels threatening to discover a few months later what he himself found. Even Maxwell's own involvement in the issue is a long story, beginning at Cambridge, running through his Aberdeen years, when the issue was clearly gestating in the background of his other work,[32] to his major papers produced at King's College. Electricity was not, and is not, an easy science. In Maxwell's own words on the state of the subject as he found it in 1855: 'In order therefore to appreciate the requirements of the science, the student must make himself familiar with a considerable body of the most intricate mathematics, the mere retention of which in the memory materially interferes with further progress.'[33] Many a student since would share this sentiment. To appreciate the King's College work, a brief introduction to the mist Maxwell was immersed in is necessary. It must be said that the next few paragraphs are but a very brief outline.

Ask the question 'what is electricity?' or 'what is magnetism?' and even today no one will give a straight or simple answer. There isn't one to give, for the subjects are intrinsically complex and interlinked. Some connection between them had been suspected for years, but it was Oersted's demonstration in 1820 that a steady electric current (namely, a steady movement of charge) produced a steady magnetic field that initiated the concept of electromagnetism. This discovery was soon followed, in 1821, by Faraday's that a magnetic field acting on a current-carrying conductor exerted a force on the conductor that could, in a suitable arrangement, make the conductor rotate. This was 'Faraday's motor', though it was not a device that could provide any useful power. A decade on, Faraday discovered electromagnetic induction, the property that a changing current in one circuit can induce a changing current in a neighbouring circuit. Faraday's experimental researches on electricity and magnetism were extensive, driven by his wish to explore new phenomena and to understand what these phenomena were about. For understanding, one needs a grasp of underlying concepts and Faraday's visualization slowly crystallized in terms of his own concepts: the magnetic field and the electric field.[34] In this picture, the causes of electromagnetic phenomena are electric and magnetic fields. Bodies in a field react with the field that is present where they are in a way that depends on the strength and direction of the field. The source of the fields that influence surrounding space are electric charge and currents. (Magnetic poles, it turns out, are generated by currents.) Because this is the picture drawn nowadays in all textbooks on the subject, it seems to us a natural description, but the reason it seems so natural now is down to the work of Maxwell.

Faraday was only one of many experimenters and synthesizers of the phenomena of electricity and magnetism in the first half of the nineteenth century. Their story is a long one, covered in varying detail in textbooks, in histories such as that of Darrigol,[35] in articles and in Maxwell's cited biographies. Well-known laws of electricity and magnetism, such as those of Ampère, Biot-Savart, Lenz and others, are a recording of this work but Faraday's researches were the explicit source of Maxwell's inspiration. To continental investigators, Faraday's ideas were mere hand-waving metaphor, devoid of depth and, ultimately, reality. Charge was the source of electrical phenomena and charges influence each other at a distance, via Coulomb's Law and, in more complicated circumstances, with a dependence on their motion. Interacting points made for a mathematical basis of electricity and magnetism (such as laid out by Wilhelm Weber) and set the paradigm within which quantitative formulae were developed. Maxwell had digested Weber's treatise on electrodynamics and was highly respectful of his work, summarizing its foundations in his own 1855 paper. Notwithstanding, it was Faraday all the way for Maxwell, particularly in terms of the *processes of his reasoning*. It is hard to visualize the effects of moving interacting points. 'Lines of force' are easier to visualize and the electric field, for example, is simply defined as the force on a unit charge at a given location. An analogous definition works for magnetism. Place a magnet under a sheet of paper and sprinkle iron filings over the paper and it is not hard to visualize magnetic field lines. Faraday and Maxwell might have argued that the filings make the field lines visible, in the way that a brass rubbing makes visible the corrugations in a sheet of plane paper smoothed over a brass plaque.

For Faraday and Maxwell, field lines weren't just visualizations; field lines had properties. For example, they never crossed, they repelled each other and, to invoke a more modern attribute, it took energy to squeeze them together. A charge interacted with the field at the charge's location, not with other charges through some 'spooky action-at-a-distance'. In placing a charge somewhere, one altered the field everywhere. When magnetic field lines within a conducting loop were changed, then a current was induced in that loop. Maxwell's self-appointed task was to express the properties of these fields in mathematical terms that embodied the known behaviour of interacting electric and magnetic systems.

It is easy in principle to follow the line of reasoning that, if fields exist, then space must be filled with a medium whose properties reflect those of the fields we observe everywhere. It is not so easy in practice to follow the implied mathematics. Maxwell's first major paper on the subject, written in 1855 and using mathematical techniques developed by William Thomson to discuss the flow of heat, made the fully quantitative analogy between electrostatic and magnetostatic field lines and the hydrodynamic flow of an imagined incompressible medium that carried their influence. It's just an analogy, Maxwell emphasized, the basic idea of which came from William Thomson. The analogy between heat and electrostatic attraction is not obvious but is at the heart of Maxwell's entry into the field. Heat is conducted away from a small source so that the energy per unit area falls off as the inverse square of the distance from that source. Surfaces of constant temperature are spherical surfaces around the source. The analogy is that the electrostatic force at a distance from a small charged object changes as the inverse square of the distance away; equi-potential surfaces are concentric spherical surfaces. Maxwell argued that, at least as far as the mathematics is concerned, we can consider electrostatic attraction and heat flow as analogous. Heat conduction is imagined as the flow of energy through the interaction between neighbouring parts of a medium, so this is suggestive that electrostatic interactions can be modelled by flow. From heat flow, Maxwell moved quickly to *an imaginary fluid* and so began what would turn out to be a long-running analogy between electrical phenomena and incompressible fluid flow. Maxwell was at pains to point out that his approach was not offered as an explanation: 'I do not think that it contains even the shadow of a true physical theory; in fact, its chief merit as a temporary instrument of research is that it does not, even in appearance, account for anything'.[36] Perhaps not, but as a fruitful idea it had more to offer. Many people have since implied that Maxwell was trying to explain electromagnetism in mechanical terms. As he himself said repeatedly, he was not.

At King's College he returned to the analogy in 1861 and 1862, now making it sufficiently involved to represent interlinked electrical and magnetic phenomena. His model of the mechanics of the medium imagined vortices, representing the magnetic field, between which were 'idlers' (a mechanical concept referring to free wheels in a gear train) representing the electric field, that allowed neighbouring vortices to rotate in the same direction. The idlers could slip between the vortices with some resistance.[37] It all sounds almost bizarrely clunky to a modern ear, yet our description of a current passing through a magnetized material as involving the passage of electrons past spin up and spin down atoms involves philosophically very similar concepts, worded in a different language. Unlike Maxwell, our language claims to describe what is really there in a magnetic material. Maxwell was trying to describe the constituents of the fields themselves, but it is not as unlikely as it may seem at first that Maxwell did come up with the correct mathematical description of electric and magnetic fields. He was led down this path by trying to put into mathematical form already established experimental results. This is not the place to go into detail, but it is the place to describe one of his great flashes of insight embedded within these papers, the revelation of a missing piece in everyone's understanding of electricity.

Electric current is traditionally described as the rate of flow of charge. In modern units, a current of one ampere is a flow of one coulomb of charge per second. Through a suggestion thrown up by his analogy, Maxwell realized that the concept of current had to be extended. Currents, he said, were also caused by a rate of change of the electric displacement field (**D** in today's standard notation). This term had to be present in the equations alongside the traditional 'conduction current'. Maxwell called it 'displacement current'. Modern textbooks introduce the concept as if it were almost 'obvious', but this was not so in Maxwell's time. He was led to include it by a consideration of what happens to charges in a dielectric material when an applied electric field is changing. Since the displacement field reduces to the electric field (within a constant) in a vacuum, displacement current must also exist in a vacuum or, as Maxwell would have put it, in the medium that supports fields in a vacuum. With the inclusion of displacement current, then some manipulation of the differentials in his mathematics that can be done nowadays in only a few lines using modern vector notation, he was able to show that electric and magnetic fields could support waves obeying a simple wave equation, even in a vacuum. Moreover, their speed of propagation is given by the ratio of the measures of charge in electrostatic and electromagnetic units, a constant known then to be close to the velocity of light, but which needed further, more precise, measurement. Also, the waves are transverse waves that can exhibit the full range of polarization phenomena observed in optics. In Maxwell's own much quoted words 'we can scarcely avoid the inference that light consists in the transverse undulations of the same medium which is the cause of the electric and magnetic phenomena'.

Without the concept of displacement current, no waves are predicted by the remaining equations. The displacement current depends on the property of the medium in which the fields occur and hence can't be represented by any property of sources alone in the action-at-a-distance formalism. Maxwell's insight reminds one of the insight of Einstein, who recognized that mass wasn't the only source of gravity, as it is in Newtonian physics, but that gravity has other causes in space. Without that extension of an apparently fundamental understanding, there would be no gravitational waves. Without Maxwell's extension of the idea of current, there would be no electromagnetic waves in electromagnetic theory.

Maxwell's final stroke of genius (to use a justifiable cliché) in relation to this work was to sweep away all use of the mechanical analogy, presenting electric and magnetic fields as existing in their own right, obeying the relationships that he made clear in his 72-page, multi-part paper presented to the Royal Society of London in 1864 and published in 1865.[38] Maxwell does think of space as filled with a medium 'capable of receiving and storing up two kinds of energy', for example, but there is no mention of specific mechanism, no vortices, no lines of idlers. 'In the present paper I avoid any hypothesis of this kind' he said, echoing Newton's famous Latin phrase '*hypotheses non fingo*'. This is essentially the modern view that the electromagnetic field is a fundamental entity of nature, not explicable in other terms. His statement is essentially: sources generate fields, fields obey the following laws 'the general equations of the electromagnetic field' as Maxwell called them, now known as *Maxwell's equations*. The mist had cleared, the landscape revealed. Physics had been introduced to a new world, or at least a new paradigm, but it was one that very few of his contemporaries could see clearly in. There were some extenuating circumstances. For example, the *general equations* were not in a form or a notation that today's students of electromagnetism would recognize easily. In addition, Maxwell had emphasized the field descriptions in terms of scalar and vector potentials. This was mathematically elegant, but his successors found that representing the field equations directly in terms of the electric and magnetic fields themselves (as usually presented in modern textbooks) was more fruitful for their application.[39] This development had to wait until the work of Oliver Heaviside was published in 1885.

Maxwell presented new evidence for the idea that light and radiant heat are electromagnetic waves, linking the previously separate studies of electricity, magnetism, and optics. This wasn't too big an idea to swallow, since others, Faraday in particular, had felt there might be a connection. Over his years at King's College, Maxwell continued his correspondence with Thomson, but in so far as can be judged from surviving letters, it was all in relation to experiments and apparatus, not to his 'Dynamical Theory of the Electromagnetic Field'. In fact it is hard to pin down any immediate influence of his 1865 paper, because eight years' later he published *A Treatise on Electricity and Magnetism*[40] that included a major exposition of the whole field and some reformulation and expansion of his own ideas. The 'Maxwellians' that promoted his ideas later in the century drew their inspiration from the Treatise rather than the King's papers. Thomson and Tait were significantly involved in corresponding with Maxwell over this production – it was in many ways the sequel treatise to their own *Treatise on Natural Philosophy* that dealt with kinematics, statics, dynamics and related topics – but their attitudes to it differed markedly. At least for a couple of decades after Maxwell's 1865 paper, and arguably for the rest of his life, Thomson considered that Maxwell had wandered into the wilderness; Tait considered the *Treatise on Electricity and Magnetism* revolutionary in the very best sense. In the *Treatise*, Maxwell wanted to lay out more clearly and more fully his locally acting field theory, in contrast to the action-at-a-distance theory still promoted, particularly on the continent. Maxwell had seen further than others, but he himself would have acknowledged it was because he was sitting on the shoulders of Faraday. The influence of Maxwell's ideas gathered pace slowly[41] after the publication of the *Treatise*, but that takes us far from his King's College years.

Maxwell on Gases and Other Topics

There is a breed of scientist who spots an opening to a little explored field, enters it and picks the low-lying fruit before moving on to new pasture. This was not Maxwell's way. He had wide interests, it is true, but his interests were deep. He brought with him from Cambridge and Aberdeen research interests in colour, in optics, in mechanical systems, in electricity and magnetism, in the kinetic theory of gases. He developed all of these while at King's College without finishing with any of them. His final two years' research at King's College was almost entirely occupied with electrical matters. As he was contemplating resignation from his chair, he began to take up again his work on the statistical treatment of gases, aspects of which he continued until the end of his life. It was the area he was best known for in his last decade.

As a closure on his 1860 predictions that the viscosity of gases would be independent of density and would depend on temperature in a way that was determined by the inter-molecular force law, Maxwell began his own experiments, carried on in the attic of 8 Palace Gardens Terrace. In a letter to Tait, in early March 1865, he mentioned that he had been set back a month by an implosion of his glass vacuum vessel. By the end of the month, he had begun measurements in earnest using his method of observing the damping of a stack of torsional discs interleaved between a stack of fixed discs. He was able to present the analysis of the experiment and his results to the Royal Society of London in November 1865, and read his much postponed Bakerian Lecture to the Society in February 1866, now entitled 'On the Viscosity or Internal Friction of Air and Other Gases'. This was Maxwell's *experimentum crucis* in the field. The results did more than any other single paper to establish him as the leading 'molecular scientist' of his age. In 1866, he followed this up with another long and significant paper on the dynamical theory of gases, reworking some of the controversial aspects of his 1860 paper, but this again takes us beyond his King's College years. Francis Everitt,[42] among the Maxwell biographies, provides the best summary account of this aspect of his work. Fuller accounts with supporting documentation have been published by Garber, Everitt and Brush[43] in their two volumes that cover all of Maxwell's related science.

One theme of Maxwell's interests that is usually underplayed is his continuing interest in elasticity and mechanics. His first extensive paper on the subject ('On the Equilibrium of Elastic Solids') was read to the Royal Society of Edinburgh while he was still a teenager, before he went to Cambridge, and was published by the RSE in 1853. This largely mathematical paper includes an explanation of the stress patterns seen in polarized light shone though slabs of glass, a subject that Maxwell had experimented with at length. Maxwell followed this by another mathematical paper written at the end of his Cambridge undergraduate career 'Transformations of Surfaces by Bending'. His strong background in the mathematics of mechanical systems played an important part in seeing through his mechanical analogy of electromagnetic phenomena. At King's College, he turned his attention to the graphical solution of mechanical problems and wrote the first of several papers on graphical statics. Maxwell developed the generality of a method that Rankine had introduced in 1854.[44] He laid out clearly, for the first time, the mathematics underlying the reciprocal relationship between forces in a framework (when these are not over-determined) and the polygons of forces at each intersection point of the members of the framework. At an intersection the lines of force (tensions or compressions of the frame members) all come to a point and sum to zero. The forces when drawn out on paper as vectorial lines form a closed polygon. The piecewise polygonal framework and the composite of the polygons of forces form reciprocal figures. The geometry of the reciprocal figures enables the stress in each member to be determined graphically when the framework is suitably supported.

He followed this up after he left King's College with two further papers extending the concepts. The final paper on this subject 'On Reciprocal Figures, Frames, and Diagrams of Forces' published in the *Transactions of the Royal Society of Edinburgh* in 1870 earned him the Keith Prize of the RSE. The prize was a gold medal and sum of around £50 awarded biennially for the best paper in science published by the RSE. Maxwell did not win many prizes or secure many awards during his lifetime, but the Keith Prize rightfully added him to a list of recipients that already included Brewster, Forbes, Graham, Rankine, Boole, Thomson, Piazzi Smyth and Tait, a list strongly biased towards Scots but nonetheless a list of outstanding scientists.[45] E. J. Routh's widely used two-volume *Treatise on Analytical Statics*[46] devotes a whole chapter to reciprocal figures, crediting Maxwell with supplying the underlying theory. The algebraic solution of such problems using computer packages has now completely replaced graphical methods.

Maxwell's Departure from King's College

London was undoubtedly a hub for scientific intercourse, and in various ways Maxwell made good use of his time at King's College. He particularly involved himself with the Royal Society of London, including refereeing papers in much more detail than necessary simply to sanction them for publication in the *Transactions*.[47] In April 1864, John Tyndall asked if he would join the Philosophical Club, a group of leading fellows pledged to promote the work of the Royal Society. However, looking over the whole of Maxwell's scientific output, he published little joint research. His work on the BA standard of resistance carried out at King's College was his principal collaborative undertaking.[48] Only the few British Association committee reports included in his collected works have a joint author, out of a hundred or so publications, though Maxwell certainly corresponded with his contemporaries on issues he was gestating. His four books published in his lifetime, on heat, on electricity and magnetism, on matter and motion and on the collected works of Henry Cavendish, were his alone. His work on the stability of Saturn's rings was, arguably, a fifth book and not a paper. It is hard to resist the conclusion that Maxwell's inspiration came from deep within himself.

London was also a magnificent opportunity to widen one's social life, but this aspect of the capital Maxwell does not seem to have drawn on much. It's true that absence of evidence is not convincing evidence for absence, but it is certainly the case that neither by his biographers Campbell and Garnett, who cited a large number of Maxwell's letters, nor in other published correspondence, are the Maxwell's recorded as attending a ball, a concert, a play, a literary reading, or any social relaxation in the city. Nor is it mentioned that they had any soirées at 8 Palace Gardens Terrace. Of course they were not monastic, but the fact was that Maxwell spent the formative years of his youth in the country and was at heart the country laird that he was now in name. Glenlair was his real and his spiritual home. City life was business life, not home. This was reflected in the leased nature of 8 Palace Gardens Terrace. Katherine was not a socialite and neither was Maxwell, though given the invitation undoubtedly he would have been charming and erudite.

Maxwell resigned his King's professorship in February 1865, mid session. The Easter term classes in 1865 were taken by his successor. He made a passing comment in a letter of 1866 that he could undertake more experiments and theoriszng now that he had no public duties. This was obvious in a way, but suggests that the motivation for his resignation was at least in part the feeling that his research was building to such a level that it needed his full-time commitment. Given all he had already achieved while at King's College, and that he had in mind to produce a treatise on electricity and magnetism, then that was almost certainly a major reason for his decision. Another less obvious contribution may have been the way things were going at the college. Maxwell took the King's College job because he valued the experience of teaching. Yet he had now a much narrower range of students than he had encountered at Aberdeen and even at Cambridge. He was, in effect, teaching mainly engineering students. Moreover, most did not last the course, for their priorities lay elsewhere. His students' decided ambitions were echoed in the utilitarian ethos of the college as a whole. If Maxwell's satisfaction in his teaching was declining, then it made good sense to re-group his resources. Already he was doing less than earlier and there was a competent second in command in the shape of Grylls Adams, fully capable of stepping into his shoes in front of his classes. Resigning his post was the obvious move, given all the circumstances.

There may also have been unspoken reasons. For example, the centenary historian of the college asserts that there was a marked loss in quality of the student intake to the Department of Literature and Science from the mid 1850s, resulting in cohorts that could not follow the lecture courses. Possibly this situation affected applied sciences too. Maxwell's successor was offered the post at a significantly reduced salary,[49] perhaps a sign that the college could scarcely afford the Lecturer in Natural Philosophy who was appointed during Maxwell's tenure. Maxwell himself may have been faced with a similar salary cut. We do know that one consequence of the fusion of the Aberdeen colleges was that Maxwell would continue to receive his full Aberdeen income for life, adding to his financial independence. Any continuing concerns over Katherine's poor health would have made Glenlair much more attractive than central London.[50] Into the bargain, Maxwell and Katherine had been planning an enlargement of Glenlair house for some years, and hence the move from London may have been encouraged by Katherine and foreseen for much longer than Maxwell ever confessed.[51] We're not likely to weigh correctly all the factors that might have influenced their decision. Clearly, though, Maxwell had not lost his intellectual interest in teaching, for it would be well after leaving King's College that he produced his long-running teaching textbooks.

As far as Maxwell's research was concerned, this time was not a natural break for any of his interests, even those on colour vision. These issues are taken up in other chapters. In fact, Maxwell spent the winters of 1865 and 1866 at 8 Palace Gardens Terrace and the early months of 1868. However, these were not his King's College years. It is tempting to speculate on a few counterfactual scenarios for his life. If Maxwell had followed up less diffidently Stokes' 1857 comment suggesting Fellowship of the Royal Society of London, it's hard to see how the Aberdeen Royal Commission could have preferred David Thomson had Maxwell been a fellow by 1859; it would have made it less likely that P. G. Tait would have been preferred at Edinburgh in 1860. (Tait never became an FRS, though that was the Royal Society's loss.) Had Maxwell settled in Edinburgh, would he have resigned in 1865 to retire to Glenlair? Possibly not, but then we may not have had the *Treatise on Electricity and Magnetism* as it emerged and Maxwell's Cavendish contributions. How Maxwell's life actually turned out is related in other chapters.

CHAPTER 4

Cambridge and Building the Cavendish Laboratory

ISOBEL FALCONER

It will, I think, be a result worthy of our University . . . if, by the free and full discussion of the relative value of different scientific procedures, we succeed in forming a school of scientific criticism, and in assisting the development of the doctrine of method.[1]

In his inaugural lecture in 1871, Maxwell set out the task of the new laboratory for experimental physics at Cambridge, then known as the 'Devonshire', but christened the 'Cavendish' two years later.[2] He carefully positioned it within the Cambridge tradition of liberal education, downplaying the industrial connotations that had driven experimental physics to prominence. The lecture acts as a guide to what he was trying to achieve during what turned out to be the last eight years of his life.

Following his resignation from King's College, London, in 1865, Maxwell maintained a presence on the stage of science through a voluminous correspondence and through extended trips to London and Cambridge in the winter months. Pressure for reform of Cambridge's dominant Mathematical Tripos (honours degree course) was mounting, and in 1865 he was invited to be an examiner, to ensure the proper examination of new physical subjects (spherical harmonics, and the shape of the Earth) that were to be included from 1867. His first experience in the exams of January 1866 brought home to Maxwell the need both for adequate textbooks in physical subjects, and for a more complete update of the physical content of the syllabus. He urged his friends William Thomson and Peter Guthrie Tait to rapid completion of their *Natural Philosophy* (1867) so that he could set questions based upon it, campaigned for the re-introduction of electricity and magnetism to the syllabus, and decided to write a major text book to support the new topics. The book was to become his *Treatise on Electricity and Magnetism*.[3]

The efforts of the reformers were rewarded in 1868 when new regulations saw the re-introduction of electricity, magnetism, and heat to the Mathematical Tripos, to be examined for the first time in January 1873. The University Senate also approved recommendations to establish a professorship of experimental physics with a physical laboratory, to teach the new subjects. They failed, however, to find a way of financing this initiative; the university's constituent (and wealthy) colleges were unwilling to contribute the money, and a subsequent proposal to increase the capitation fee on all members of the university was turned down by the Senate on 7 May 1870.[4] This prompted the university's Chancellor into action. William Cavendish, Seventh Duke of Devonshire, was not only a wealthy landowner and industrialist, but also a former Second Wrangler (i.e. came second in the competitive Mathematical Tripos) and related to the famous eighteenth-century scientist, Henry Cavendish. In the summer of 1870 he was chair of the Royal Commission on Scientific Instruction and had just interviewed, among others, Sir William Thomson about his facilities in Glasgow, and Robert Clifton about the new physical laboratory in Oxford.[5] On 10 October 1870, he wrote to the Vice Chancellor offering to finance the building of the laboratory.

I am desirous to assist the University in carrying this recommendation into effect and shall accordingly be prepared to provide the funds required for the building and apparatus, so soon as the University shall have in other respects completed its arrangements for teaching Experimental Physics, and shall have approved the plan of the building.[6]

Faced with an offer it could not refuse, the university had little choice but to establish a professorship, though it did so cautiously, stipulating that the chair was to terminate with the tenure of the first incumbent unless decided otherwise.[7]

After an initial approach to Thomson, the most eminent British physicist of the day, failed to persuade him to leave Glasgow, Maxwell was put under increasing pressure to stand, most notably by John Strutt (later Lord Rayleigh and Maxwell's successor at Cambridge) and by George Gabriel Stokes the Lucasian Professor of Mathematics.[8] Initially disinclined to agree, Maxwell was soon convinced that he 'could do some good by it', and put his name forward on 23 February 1871. In their biography of Maxwell, Campbell and Garnett claimed that he was persuaded to stand, 'on the understanding that he might retire at the end of a year, if he wished to do so.'[9] This may have been an informal understanding, but there is no indication of it in his letter of application, reproduced in Fig. 4.1.

He was elected unopposed on 8 March. Kim has suggested that the informal search for a candidate, which started before the post was advertised on the 14 February, the short time between advertisement and election, and the unusual lack of any alternative candidates, indicate that the election was a put up job. All resident members of the Senate were entitled to vote, but only 13 of around 300 did so.[11]

Maxwell set to work immediately. He and his wife, Katherine, took up residence in Cambridge at 11 Scroope Terrace, which was to be their winter home for the remainder of his life. They continued to spend summers at Glenlair (Fig. 4.2).

So, what did the terrain look like to Maxwell in 1871? The laboratory would be disruptive, and he summed up the issues in a letter to Strutt:

. . . it will need a good deal of effort to make Exp. Physics bite into our University system which is so continuous and complete without it . . . If we succeed too well, and corrupt the minds of youth till they observe vibrations and deflexions, and become Senior Ops instead of Wranglers, we may bring the whole University and all the parents about our ears.[12]

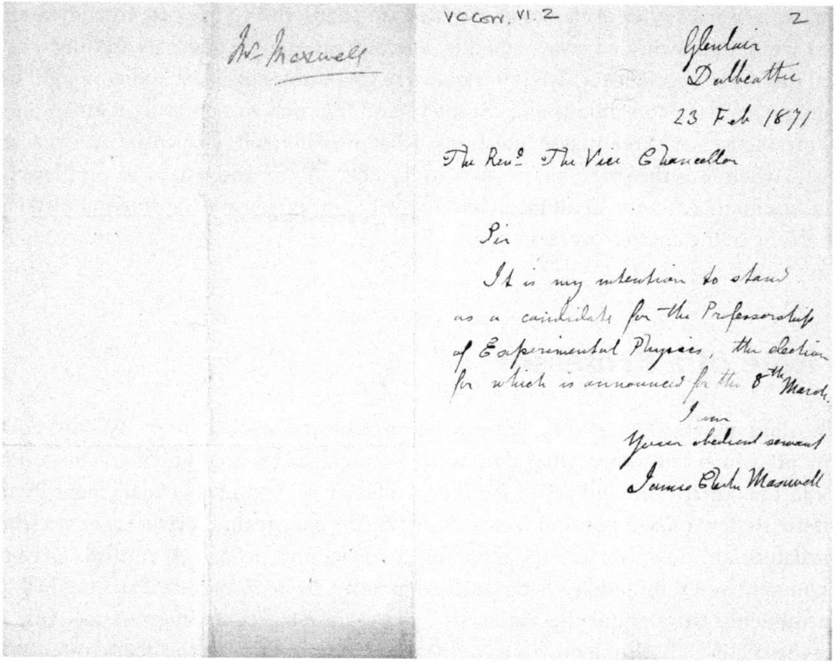

VC Corr. VI. 2

2

Mr Maxwell

Glenlair
Dalbeattie
23 Feb 1871

The Revd The Vice Chancellor

Sir

It is my intention to stand as a candidate for the Professorship of Experimental Physics, the election for which is announced for the 8th March.

I am
Your obedient servant
James Clerk Maxwell

Fig. 4.1 Maxwell to John Power, Vice Chancellor of Cambridge, 23 February 1871. Reproduced by kind permission of the Syndics of Cambridge University Library.[10]

Fig. 4.2 Scroope Terrace, where Maxwell and his wife lived when in Cambridge. Number 11 is at the far end.

The vibrant voluntary science classes of the 1820s to 1840s, the vestiges of which he himself had benefited from, had dwindled away, killed by a combination of economics and the increasingly technical discourse of science which put it out of reach of non-specialist audiences.[13] The formal university system was 'continuous and complete' and Maxwell was entering it at a point defined by the intersection of three major questions: what was the role of a professor in Cambridge University; where was the space for 'physics' in the curriculum; and what was the place of experiment in teaching? Answers to all three had to evolve in unison if experimental physics was to flourish. None of the answers were obvious.

The Role of a Professor

Until the mid nineteenth century, Cambridge professorships had been established through endowments which laid down what they were to teach; increasingly they got out of line with changes in the curriculum and relied for their audience on voluntary engagement by the most enthusiastic students. Useful tuition was provided by the colleges and private coaches who taught to the syllabus laid down for exams, retaining financial and pedagogic control. Driven by the highly competitive examination system in mathematics, by 1850 private coaching had 'attained such an influence throughout the University, as to cause a serious interference with all other forms of instruction'.[14] Although strongly criticized for undermining the liberal intentions of the Tripos, the coaches filled a gap that arose from mixed ability college teaching. Only the richest colleges could afford enough tutors to provide differentiated teaching and, with college positions awarded for achievement in mathematics or classics, they lacked men who could teach science subjects. The solution, according to the 1850 Royal Commission on the 'State, Discipline, Studies and Revenues of the University and Colleges of Cambridge' was to appoint a body of university professors and lecturers on terms and conditions that aligned their teaching with the curriculum, and who were overseen by Boards of Studies. But, as noted above, the colleges were unwilling to contribute to university positions; 20 years later, Maxwell's was the first of this new type of professorship in science. His was followed by appointments of James Stuart in engineering (1875) and Michael Foster in physiology (1883). University lectureships were even slower in coming. Instead, in the prevailing system where students generally paid fees to attend any lectures outside their own college, including those by university professors such as Maxwell, the richer colleges went their own way. Trinity appointed Thomas Danby as a college lecturer in physical sciences in 1867, where he was succeeded by Coutts Trotter in 1869. St. John's and Gonville and Caius colleges both established their own chemical laboratories and lecturers. So, while in 1875–76 Maxwell was lecturing, at various times, on heat, elasticity, electricity and magnetism, Trotter was also lecturing on electricity and magnetism, heat, sound and light at Trinity, Richard Pendlebury on theory of chances including errors of observation at St. John's, Norman Ferrers on hydrodynamics at Gonville and Caius, James Stuart on electrostatics at Trinity and George Pirie on thermodynamics at Queens'. In 1876, William Niven began teaching Maxwell's field theory in his electrostatics lectures at Trinity.[15]

In 1875 the university attempted to gain some control and appointed a 'Syndicate [committee] to consider the Requirements of the University in different Departments of Study'. The ensuing report strongly recommended regular meetings between professors and college lecturers to coordinate their activities, and officially sanctioned expansion of the inter-collegiate lecture scheme, initiated by Stuart in 1868, which mitigated lecture fees within small groups of collaborating colleges. Maxwell's input came through two different channels, as a member of the Natural Sciences Board, and a member of the Board of Mathematical Studies. The needs of physics dominated the responses of the latter. Through both boards he pushed for differentiation in teaching with the establishment of separate elementary and advanced lectures – the elementary lectures might be given by a demonstrator.[16] This position was to leave Maxwell, as professor, ultimately responsible for university physics teaching and with an influence on college teaching, but giving only the advanced lectures himself.

The 1868 report to the Senate recommending establishment of a Professorship of Experimental Physics had made quite clear that the professor was not expected to do all the practical teaching himself. He had other important duties as well. They quoted the 1850 Commissioners,

> . . . the Professor, who ought to be at least partly occupied in original research, and whose attention must at all events be sufficiently occupied in preparing himself to impart and in actually imparting in the most luminous manner the scientific principles of his subject, is not likely to have much time at his disposal for the instruction of tyros in the use of their tools.[17]

They therefore recommended appointing a demonstrator as well as a professor.

Maxwell tried early on to make the appointment, writing to the Vice Chancellor in September 1871 to suggest a candidate. John Hunter had published on 'the absorption of vapours by charcoal', had worked with Thomas Andrews in Belfast, Thomson in Glasgow, and Tait in Edinburgh, and had recently retired from a Chair at Windsor College, Nova Scotia, suffering from the cold. Maxwell would, he said, 'be well satisfied if we got Mr Hunter, especially at the beginning of the work . . . His devotion to science is manifest, and I believe he is a great teacher'.[18] At the same time, he wrote to Tait for a reference in a typically whimsical letter:

> I only know him as the man who charges charcoal with bad smells. Would he be a good demonstrator at Cambridge? I have no doubt that a man who could occlude a fishy fume in a burnt stick could also floor a demon which I suppose to be the essential part of the office. But I doubt if our laboratory would be sufficiently absorbent to occlude so volatile a spirit, for he seems to have improved the shining hour in every laboratory under heaven.

Tellingly, he considered, 'that we would be the better of a man who has seen many professors and known their manners rather than one who represents simply the continuity of University life'.[19]

It is not clear why nothing came of this proposal, whether because Hunter represented too great a discontinuity for Cambridge to accept, because until the laboratory was built and there were 'tyros' to instruct the university considered the expense unnecessary, or because Hunter's health deteriorated further – he died the following year.[20] But it seems likely that Maxwell sensed a negative reaction for he did not suggest another candidate until the laboratory was ready, and this time it was one who *did* represent the continuity of university life. William Garnett was Fourth Wrangler in 1873 and impressed Maxwell with his knowledge of physics. A year later, having reminded the university of its obligations, Maxwell appointed him as demonstrator.[21]

Garnett conceived his role as a supportive one. He assisted Maxwell's lectures, gave the elementary lectures, developed and built lecture demonstrations, helped the other researchers and contributed to the laboratory's social mission by marking endless exams for the Science and Art Department which promoted education across Britain. William Napier Shaw, a student of Maxwell's and subsequently Director of the Meteorological Office, dubbed him 'A typical

"Administrator"'.[22] Maxwell, though, held him in high regard for his character and teaching ability.

Whatever he undertakes, he does it with his might. He has out of his own energies developed a course of Nat. Phil. for elementary men, in which the experiments are not only exhibited but are reasoned from, and the intellectual status of those who attended them was raised pretty considerably, so that the results contrasted favourably with those obtained by mathematical coaching ... If [things] have to be done he always does them, and that within the specified time. In short he is one of the few men I know who has a strong will and knows how to govern it.[23]

With the laboratory built and Garnett to help out, Maxwell could turn more attention to experimental work. He had occupied the years since leaving King's College in developing and writing up his ideas on the kinetic theory of gases, thermodynamics, electricity and magnetism. He continued to make significant contributions to thermodynamics and statistical mechanics, and to electromagnetism, throughout the 1870s, as described elsewhere in this volume. He worked also on an eclectic variety of subjects, publishing on soap bubbles, vision, the diffusion of gases and various types of electrical measurement. In 1873–74, he returned to his earlier work on the theory of optical instruments, generalizing his mathematical method through use of projective geometry and developing William Rowan Hamilton's concept of the characteristic function of a lens system.[24]

This was an individual research programme such as other professors at Cambridge pursued. But Maxwell's position was unprecedented. Unlike the other professors he headed an expensive university institution. What were the implications? To implement an institutional research programme entailed assembling, in a single place, trained scientific and technical personnel, instrumental resources and a shared view of the role of the laboratory.

Maxwell's inaugural lecture set his vision for an egalitarian institution embedded in a wider physics community that contributed to society through its scientific results, its teaching, and its public engagement. He was inspired by the romance of Humboldt, Gauss and Weber's Magnetic Union, through which investigators at 50 stations across the world simultaneously recorded the declination of magnetic needles, the results of which were published while he was a boy. Not only were the results of great significance, for they 'opened up fields of research which were hardly suspected to exist by those whose observations of the magnetic needle had been conducted in a more primitive manner', but the process furthered his democratic and moral beliefs, 'the scattered forces of science were converted into a regular army, and emulation and jealousy became out of place, for the results obtained by any one observer were of no value till they were combined with those of the others'.[25] Such experiments, to which he applied the Baconian term, 'Experiments in concert', should be the ultimate aspiration of the new laboratory. It would be part of a worldwide network of shared endeavour.

For this to work, though, experimental methods had also to be developed, and shared through training, and standards produced that embedded the methods and enabled valid comparison between investigators. The British Association work on electrical standards, which he transferred to the laboratory, was a necessary part of this programme, but it addressed only one area of physics, and was not an end in itself. 'We have no right to think thus of the unsearchable riches of creation, or of the untried fertility of those fresh minds into which these riches will be poured.' The purpose of measurement was to open up fresh areas of science. 'The history of science shows that even during that phase of her progress in which she devotes herself to improving the accuracy of the numerical measurement of quantities with which she has long been familiar, she is preparing the materials for the subjugation of new regions, which would have remained unknown if she had been contented with the rough methods of her early pioneers.'[26] Before experiments in concert could be widely implemented, more than measurement was needed, both in scope and intellectually, and it was this that made the endeavour worthy of a university rather than a workshop. He aimed to establish a 'school of scientific criticism' and develop 'the doctrine of method' 'worthy of our University'. This was a social as well as a scientific responsibility, for 'the most absurd opinions may become current, provided they are expressed in language, the sound of which recalls some well-known scientific phrase. If society is thus prepared to receive all kinds of scientific doctrines, it is our part to provide for the diffusion and cultivation, not only of true scientific principles, but of a spirit of sound criticism . . .'.[27] Thus, in the immediate future, 'Our principal work ... in the laboratory must be to acquaint ourselves with all kinds of scientific methods, to compare them, and to estimate their value'.[28]

Understanding this throws a great deal of light on the way Maxwell ran the laboratory. There were fundamental tensions in his vision of experiments in concert between the egalitarianism of the endeavour and the managerial control of the training and process required to enact it; between the broad scope of physics and the exclusive focus required to advance any particular area. The latter Maxwell attempted to address by classifying physics through experimental method (weighing, spinning, etc.) rather than by topic, embedding this classification into the very building of the laboratory, as discussed below.[29] The former, though, was exacerbated by the Cambridge system that enshrined individual rather than group achievement.

Faced with either tension, if it came to a choice, Maxwell, unlike his successor Rayleigh, always erred on the side of freedom and egalitarianism rather than imposed discipline and control. He might not have gone quite as far as Isaac Todhunter in believing that, 'the experimenter is born and not manufactured', but he did not believe in manufacturing physicists.[30] Experimental work was arduous and could not be imposed – but workers might be inspired to undertake it. He relied on individuals to take responsibility for their own self-discipline, approving those that did so as 'real' men; it was their effort and dedication that counted, rather than any external judgement of the significance of their contribution, as evident in his remark to Schuster that 'The question whether a piece of work is worth publishing or not depends on the ratio of the ingenuity displayed in the work to the total ingenuity of the author'.[31]

The upshot was that Maxwell lacked a flow of skilled collaborators, and the Cambridge system mitigated against retaining or collaborating with the few he had. The men who came to work with him comprised natural sciences undergraduates, for whom there were no positions in Cambridge once they graduated – in 1870 the *University Reporter* commented that some colleges would consider themselves 'guilty of extravagance' in appointing scientific fellows – and Mathematical Tripos graduates who hoped to develop a dissertation that would gain them a college fellowship.[32] These dissertations had to be presented as individual work. Richard Glazebrook, for example, abandoned his collaboration with George Chrystal and Samuel Saunder on testing Ohm's Law because, 'it did not seem likely to afford material for a Fellowship Dissertation'.[33] Even when obtained, college fellowships were designed primarily as a stepping stone into careers outside Cambridge; many were short term and they had to be resigned upon marriage. In 1877, Chrystal moved to a professorship at St. Andrews because his fellowship came to an end.[34]

Enthusiasm was what counted and Maxwell operated an open door policy. Although the laboratory had few workers, it was notable for the breadth of its research, although with a core of work on electrical standards. Maxwell had good reason to pursue standards further. In 1868, he compared the ratio between electrostatic and electromagnetic units, finding a velocity for magnetic waves that differed significantly from that of light. Exhaustive investigation of experimental errors and procedures, and repetition in other laboratories, failed to bring the values in line and he blamed the value of the British Association resistance unit. He needed to take the standards work forward to resolve the difficulty, and Schaffer has suggested that need for resource to do so may be one reason he applied for the Cambridge chair.[35] Maxwell arranged for the British Association apparatus to be transferred to the Cavendish, providing the instrumental foundation for ongoing work. When Arthur Schuster reported experiments that suggested a deviation from Ohm's Law, Maxwell suggested to Chrystal and Saunder extensive tests which confirmed the accuracy of the law, and encouraged Schuster to work in the laboratory. At different times, Alexander Freeman, John Lord and Richard Glazebrook contributed by investigating the characteristics of the voltaic cells used to supply power. The tests extended into a comparison of the British Association units of resistance, but the programme faltered when Saunder became a master at Wellington College, in 1876, and Chrystal moved to St. Andrews. In 1878, Ambrose Fleming took up the work and designed a resistance balance for comparing the British Association standards.

But electrical standards were only one element of the wider school of criticism that Maxwell was trying to build; no one was compelled to follow this line. Other work was on related topics in electricity and magnetism. William Hicks, the first graduate student in the laboratory, attempted unsuccessfully to measure the velocity of electromagnetic waves. James Gordon was one of the most prolific publishers from the laboratory. He worked first on the newly discovered Kerr effect of light on the conductivity of selenium, and then on the specific inductive capacities of transparent dielectrics which, it was hoped, would provide support for Maxwell's electromagnetic theory. He was unable to obtain satisfactory agreement, ascribing this to dispersion in the dielectrics, and moved on to another experiment with a significant bearing on electromagnetic theory – measuring the rotation of the plane of polarised light by a magnetic field, initially through water and subsequently through carbon bisulphide which had a greater effect. Arthur Sunderland worked on the anomalous dispersion of fuchsin, trying to measure the refractive indices as close to the absorption band as possible. Having abandoned the electrical standards programme, Glazebrook established his own investigation, on the form of wave surfaces in biaxial crystals, providing confirmation and support for the undulatory theory of light.

Meanwhile, Shuster's interests veered towards gas spectroscopy and solar physics, while Charles Heycock and Arthur Clayden investigated the spectrum of indium. John Marshall measured the viscosity of liquid films. Sedley Taylor, who was writing a book on sound constructed a 'kaleidophone' which made visible the characteristics of a sound wave by the beautiful coloured effects arising from variation in the thickness of a soap film acting as an elastic membrane. In 1879 Maxwell welcomed to the laboratory an experimental programme that was instigated by the Meteorological Council but was right in tune with his aims. This was a comparison of the chemical method of determining the absolute amount of water in the air with those using Dine's, Regnault's and Saussure's hair hygrometers, and also with wet and dry bulb thermometers, which Shaw undertook after Maxwell's death, laying the foundations of his career in the Meteorological Office. Another experiment that excited Maxwell, but that he did not live to see completed, was John Poynting's on the mass of the Earth using a common balance. The two men shared an enthusiasm for the sheer instrumental challenge of this work. Maxwell wrote of Poynting that, 'it is rare, especially in this country, to find the requisite mathematical ability combined with mechanical ingenuity made available for so delicate and tedious a research by the still higher endowment of patience and constant pertinacity', and to Joule, 'You see that the age of heroic experiments is not yet past'.[36]

This range of investigation required a corresponding range of apparatus. Lack of access to skilled instrument makers had been the main reason Thomson gave for not wanting to take the Cambridge chair. This was cutting edge research, and although the Duke of Devonshire had paid to equip the laboratory initially, new apparatus had to be both designed and built. The attendant appointed to the laboratory in 1873 had little technical skill, and although Garnett had installed a few tools in the laboratory workshop, he was no instrument maker. Maxwell relied initially on ordering instruments from London, Glasgow, and Germany. But standards instruments required, 'a considerable amount of supervision during their construction, for their whole value depends on their fulfilling conditions which can as yet only be determined by trial'.[37] In 1877, Maxwell and Stuart (who was setting up the engineering department) solved this problem by installing Robert Fulcher as a technician in the laboratory, employing him on a freelance basis, which 'greatly improved the efficiency of several pieces of apparatus'.[38] Thus, in 1878, when further verification of the resistance standard was contemplated, Chrystal proposed, 'having the coils wound by Fulcher in my immediate presence so that I can give account of every layer if necessary'.[39] But Maxwell soon found himself competing with other departments for Fulcher's services, and lacked financial resources to employ a dedicated technician. When Rayleigh succeeded, he found the laboratory significantly under-equipped.[40]

Understanding Maxwell's vision for the laboratory also explains why he devoted so much time in the 1870s to two more activities of his own. He furthered the doctrine of method in editing the works of Henry Cavendish, and fulfilled the social obligation to cultivate scientific principles and a spirit of sound criticism among the public as a scientific author of books, reviews for *Nature*, articles for the *Encyclopaedia Britannica*, and referee reports for the Royal Society. In this activity also can be counted his Rede Lecture on the telephone in 1878.

Maxwell's devotion to editing the works of Henry Cavendish has often been dismissed as a work of duty to the Cavendish family, and an unfortunate waste of his time. But Maxwell regarded things differently – he had sought the task out himself and it was a valuable contribution to science and to critical evaluation of method. Back in 1849 William Thomson had drawn attention to the papers as a 'valuable mine of results' whose publication was desirable. By 1869, the results were assuming potential significance for Maxwell's electromagnetic programme and he had tracked the papers down in private hands. In 1873, he enlisted the Duke of Devonshire's help in recovering the papers.

Once recovered, Maxwell did indeed see editing the papers as a duty – but to physics, writing to Thomson, 'I am just going to walk the plank with them in the interest of physical science'.[41] He soon became fascinated, not only by Cavendish's results, but by the methods used to obtain them. He sought out eighteenth-century instruments to test the experiments – his acquisition of William Hyde Wollaston's collection for the laboratory may be part of this endeavour – and went much further than just reproducing or editing the text. His attempts to understand Cavendish's ideas threw light on his own and led to changes in the second edition of the *Treatise on Electricity and Magnetism*. They inspired an improved method of verifying the inverse square law of electrostatic attraction, conducted and published by Donald MacAlister. Cavendish's use of his own body to measure the 'power' of electricity by the strength of shock, before the invention of galvanometers, Maxwell deemed, 'the most wonderful of them all and well worth verification'.[42] As with his earlier work on colour vision, his interest was in understanding the role of the body of the observer – the extent to which through standardization and discipline their observations would be the same, and the variation necessarily engendered by differences in individual physiology. He was still experimenting on this, with himself as subject, in the summer of 1879, published as an extended note on the 'Intensity of the sensation produced by an electric discharge' in *The Electrical Researches of Cavendish* shortly before his death.[43] He conscripted students and visitors alike to try the method. Schuster recalled, 'a young American astronomer expressing in severe terms his disappointment that, after travelling on purpose to Cambridge to make Maxwell's acquaintance and to get some hints on astronomical subjects, the latter would only talk about Cavendish, and almost compelled him to take his coat off, plunge his hands into basins of water and submit himself to the sensation of a series of electrical shocks'.[44]

Maxwell's concern to promulgate true scientific principles is evident in his literary work. He was one of the scientific editors of the ninth edition of the *Encyclopaedia Britannica* (1875–89), commissioning articles from associates, including Garnett and Crystal, and contributing a number himself, most famously those on 'atom' and 'ether'. 'Atom' reviewed the problems of explaining mass and gravitation, repeating his view that while science was about the form of matter, its initial creation lay outside the realm of scientific explanation – an attack on the materialism promoted by John Tyndall. 'Ether' contained a suggestion for detecting the Earth's motion through the ether that inspired the Michelson–Morley experiment.

During this period, Maxwell rose to prominence as a scientific author, beginning with the popular reception of his *Theory of Heat* in 1871. Even *The Ironmonger* noted in its review that, 'the language throughout is simple and the conclusions striking'.[45] The *Treatise on Electricity and Magnetism* was published in 1873, and Horace Lamb recalls, 'there was a great rush to the booksellers for copies'.[46] By 1874, both books went to new editions, and were translated into other languages. Meanwhile, Maxwell was already at work on another introductory text, *Matter in Motion*, published in 1876, which emphasized energy as a unifying concept. After two year's experience as an examiner in the Natural Sciences Tripos, he was convinced that physics could

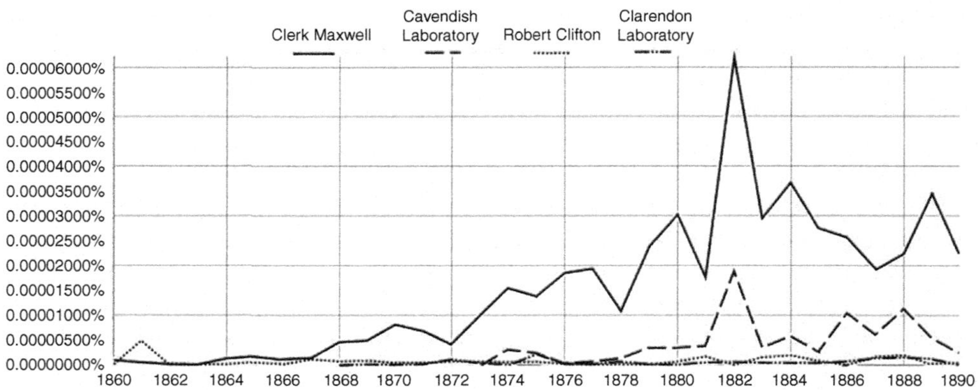

Fig. 4.3 Google Ngram comparing the phrases 'Clerk Maxwell', 'Cavendish Laboratory', 'Robert Clifton' and 'Clarendon Laboratory'. The Ngram measures the yearly count of a phrase, normalized by the number of phrases published in the year, in a corpus of over 5.2M books digitized by Google up to 2009.[47] The Ngram shows how Maxwell's public profile rose through the 1860s and 1870s and was followed by the profile of the Cavendish Laboratory. To evaluate the significance of the yearly counts, they are compared with the phrases 'Robert Clifton' and 'Clarendon Laboratory'. These were chosen, not because of any traditional Oxford–Cambridge rivalry, but because the laboratory and its professor have distinctive names that make them equally suitable for this type of analysis.

be profitably studied experimentally without calling on advanced mathematics. In 1875 he began planning an *Elementary Treatise on Electricity and Magnetism* that drew on Faraday's lines of force to provide a non-mathematical model of the electric field. The completed portion of the book was edited and published in 1880 by Garnett, following Maxwell's death.

Maxwell's work on electrical standards at King's College, London, had embedded him in the British scientific network. His activities were reported in the scientific media of the day, and his literary output in the 1870s raised his profile still further and garnered publicity for the Laboratory. The Ngram in Fig. 4.3 demonstrates clearly how the Cavendish's public profile followed that of Maxwell in becoming established in the public consciousness.

On his death there seems to have been no question but that the professorship be continued. Maxwell had set a precedent as an institutional leader that subsequent Cavendish professors were to follow. Regardless of how they went about achieving them, university-wide oversight of teaching, leadership of research and public engagement were part of their remit.

Space in the Curriculum for Physics

Where was the space for physics in the curriculum? It was not that physics was not taught – it appeared already in various places, most influentially in the Mathematical Tripos – but that as a discipline in its own right it was not recognized. This reflected the wider landscape; physics was an emergent discipline at the time, differentiating itself from an all-encompassing natural science, and its boundaries had yet to be established. In 1848, the Natural Sciences Tripos was founded at Cambridge with no explicit mention of physics. When the 1850 Royal Commission recommended a Board of Natural Science Studies, they 'considered the more exact Natural Sciences as included in the application of Mathematics', while noting that, 'in foreign Universities they are considered, under the name of Physics, as a branch of experimental rather than of mathematical science'.[48]

Although topics that we might recognize as physics, in non-mathematical form, began to appear in the Natural Sciences Tripos in the 1850s, they were not distinct from other subjects. As late as 1866, the entry in the *Student's Guide to the University of Cambridge* stated that, 'The Examination for the Natural Sciences Tripos embraces the greater part of ... natural science; that is to say, the following subjects: 1. Chemistry, including the laws of Heat and Electricity. 2. Mineralogy, including Crystallography. 3. Geology, including Palaeontology. 4. Botany, including Vegetable Physiology. 5. Zoology, including Comparative Anatomy and Animal Physiology'. Chemistry assumed primacy, 'For Chemistry teaches laws of matter which are universal, and which find their applications whenever the structure of natural objects is under consideration ...'.[49]

Thus, when in 1869, the need for a professor of experimental physics was summarized to Senate, it was as a service, '(1) To certain new subjects introduced as optional branches of study into the Mathematical Tripos; (2) To the Natural Sciences Tripos; (3) To the optional branch of Mechanism and Applied Sciences, and in that of Chemistry for the Ordinary Degree; to which may be added the introduction of Heat and Electricity into the final Examination for the M.B. [medical] degree'.[50]

The primacy accorded to the Mathematical Tripos epitomizes the place it held in Cambridge at the time. Maxwell, in his letter to Strutt of 1871 about the issues facing the new professor, placed experimental physics firmly in the context of the Mathematical Tripos. Indeed, he was bound to do so; one of the conditions of the professorship was that the subjects of his lectures be approved by the Board of Mathematical Studies. During the eighteenth century, under the influence of Newton's followers, mathematics had become the dominant subject at Cambridge. It was seen as core to a liberal Anglican education, providing both training for the mind and a means of deciphering the mysteries of God's creation. Although rising commercialism and industrialization in the latter half of the eighteenth century challenged Cambridge's role as a training ground for Anglican clergy, they went hand in hand with a move to competitive written examinations as a means of ranking students, consolidating the place of mathematics as the ultimate discriminator between candidates. As Shaw remarked caustically:

> No doubt it requires a little time to appreciate the idea of classifying the competence and capacity of a hundred young men by what they could put on paper in three hours without any opportunity for discussion or reference to any authority . . . [However,] as a formula for providing a class list in order of merit the Mathematical Tripos was practically unrivalled, and so the three-hour examination paper satisfied the University ideal.[51]

A high place in the Mathematical Tripos was greatly prized and had entered the national consciousness; the results were published in national newspapers and the Senior Wrangler's home town or school often celebrated with a procession or holiday.[52]

The Mathematical Tripos was the only course that qualified students for an honours degree until 1850, when the Classical Tripos was added as an option. Less mathematically inclined students (referred to as the 'oi polloi' or 'poll men') might get an ordinary degree through an exam that, nevertheless, included arithmetic, algebra, parts of Euclid and elementary principles of mechanics and hydrostatics, followed by a special exam in another subject.[53] Even after honours degrees in other subjects were opened up in 1861, the Mathematical Tripos remained the most prestigious and the most clearly rewarded with scholarships and college fellowships, forming a strong disincentive to study natural sciences. As George Bettany, a recent natural sciences graduate, judged in a frustrated article in *Nature* in 1874,

> The great hindrance to the success of the Cavendish Laboratory at present is the system fostered by the Mathematical Tripos. The men who would most naturally be the practical workers in the laboratory are compelled to refrain from practical work if they would gain the best possible place in the Tripos list. Very few have courage so far to peril their place or to resign their hopes as to spend any valuable portion of their time on practical work . . . For a man to do practical work in physics at Cambridge implies considerable exercise of courage and self-sacrifice.[54]

In the face of the dominance of the Mathematical Tripos within Cambridge, and its prominence in accounts of the nineteenth-century university, it is easy to overlook the fact that it represented less than a sixth of undergraduates during the 1870s. Natural sciences accounted for even fewer; the vast, and increasing, majority of students in Maxwell's time did not take an honours degree, demonstrated in Fig. 4.4.

The exclusion of advanced mathematics (i.e. using calculus) from the Natural Sciences Tripos, and experimental work from the Mathematical Tripos, was to bedevil physics teaching at Cambridge for at least the next 50 years.[56] It was a system that Maxwell appears to have accepted and worked within, while raising the prominence of physics in the curriculum as an examiner, by his membership of various committees, and through teaching.

Maxwell was an examiner for natural sciences in 1873, when physics was listed for the first time as a separate subject, and in 1874, when a laboratory exam in physics was introduced. By this

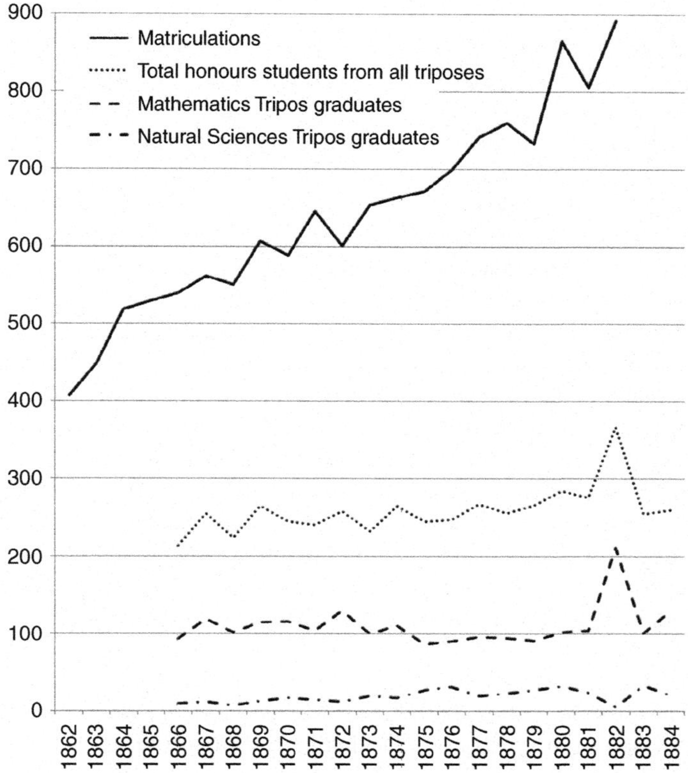

Fig. 4.4 Numbers of students enrolling at Cambridge (matriculations), and those graduating with honours (all triposes), graduating in the Mathematical Tripos, and in the Natural Sciences Tripos, 1862–84. Typically, students graduated in the fourth year after matriculation. The peak in Mathematical Tripos numbers in 1882 represents a change from holding the exams in January of the fourth year to June at the end of the third year. Thus two cohorts graduated this year. The dip in Natural Sciences at the same time represents a brief hiatus when parts I and II separated by a year were introduced. Data is taken from the *Historical Register of the University of Cambridge*.[55]

time the proportion of the Natural Sciences Tripos devoted to physics, and the range of topics taught, had increased significantly, enabled by a range of new text books resulting from the nationwide concern for science education.[57] However, Maxwell and his fellow examiners found the syllabus so broad that many students attempted a nodding acquaintance with the full range, rather than mastery of a few, topics. 'The knowledge of [physics] as shown by their answers was in a good many cases of a very unsatisfactory kind, arising partly from an unintelligent use of popular textbooks and partly from a familiarity with the appearance of instruments without any knowledge of the principles on which their action depends.'[58] Maxwell was encouraged, though, that, 'several of the candidates . . . sent up answers which show that Experimental Physics, treated without the higher mathematics, may be learned in a sound and scientific manner'. The examiners recommended that the two stages of the exam be separated by six months, with candidates being restricted in the number of subjects they could take at second stage. These changes were introduced in 1876.

After 1870, Maxwell appears only once as an examiner of the Mathematical Tripos, in 1873, the year that electricity and magnetism were introduced. Far from securing the position of physics, its introduction led to problems of breadth here also, with candidates spreading their efforts too thinly. In 1877, Maxwell was a member of the committee that addressed this problem. But whereas in natural sciences breadth was retained at lower levels and choice of specialism enforced at higher, the Mathematical Tripos saw a controversial narrowing of the two-stage honours degree, and introduction of an optional 'Part III'. The new physical subjects were moved into this third stage and so were taken only by the most able students as post-graduates.[59] These differing solutions were to have a considerable effect on the makeup of the student body studying physics.

Maxwell was not a methodical record keeper, but in 1871–72 he kept lists of the students attending his lectures. In his first term, 16 of the 19 students (a mix of post-graduates and undergraduates) were mathematicians while one was a natural scientist and one a lawyer (see Appendix). A similar pattern held the following term. By Michaelmas 1873, numbers in his list had dwindled to seven – five mathematicians and a classicist – viewed against a backdrop of slightly reducing numbers in the Mathematical Tripos generally. At the same time numbers graduating with honours in natural sciences fluctuated but increased – from 14 in 1871 to 27 in 1879 (see Fig. 4.5). However, after 1876, an influx of students taking natural sciences as a route to an ordinary degree nearly doubled the honours numbers. Almost all these students took physics in the first part of the new two-stage exam. For a steadily increasing number, physics was either their major or a significant option, even if they were not ultimately successful in achieving honours.[60]

Maxwell also drew in students formerly taught elsewhere. Following his election, he had taken over the teaching of heat and electricity from George Liveing, Professor of Chemistry, and in 1876 similarly agreed to provide for teaching heat, light, electricity, and magnetism to the

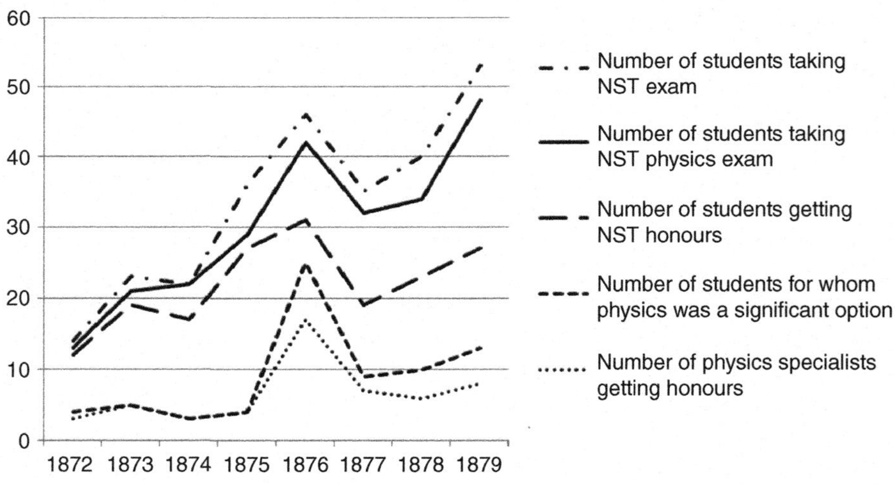

Fig. 4.5 Numbers of students taking the Natural Sciences Tripos (NST), and taking physics as a significant option. The data comes from the Natural Sciences Tripos mark book.[61] Those whose mark in physics was greater than 2/3 of their mark in their highest subject were deemed to be taking it as a significant option. For over half of these, physics was their highest mark. From 1876 onwards, in order to include all those studying physics, calculations were for the first stage of the exam in June; those successful went on to the second stage six months later and generally achieved honours.

engineering students.[62] Additionally first M.B. students were taught and examined on heat and electricity. By 1879, 42 students in total were attending the lectures on heat recommended for natural science and first M.B. students given by Maxwell's demonstrator, William Garnett; large numbers also attended Trotter's lectures in Trinity. By contrast, Maxwell's advanced lectures attracted only between three and nine.[63]

Inserted into Maxwell's notebooks is a list of 'gentlemen at practical work' in the Lent term 1878. The picture was very different to those attending lectures in 1871–72. Of the fourteen 'gentlemen' only four were mathematicians, one of whom studied natural sciences at the same time; there were five more natural scientists, three medics and three visiting scientists.[64] It seems that physics was establishing a place for itself, but not necessarily among the elite mathematicians as initially envisaged.

The Place of Experiment

Conceptually, Maxwell believed that physics occupied all that space between 'the abstract sciences of arithmetic, algebra and geometry', and chemistry, and for which 'dynamics, or the doctrine of the motion of bodies as affected by force', provided a fundamental explanation. Although chemistry was expanding rapidly in the nineteenth century, dynamical explanations for some chemical phenomena were 'reclaiming large tracts of good ground' for physics.[65]

The place of *experimental* physics was more concrete. Maxwell explicitly utilized the creation of a physical space to define the discipline, 'The main feature, therefore, of Experimental Physics at Cambridge is the Devonshire Physical Laboratory'.[66] The laboratory was the locus of the interplay between the question of the disciplinary position of physics and the role of experiment in teaching as Maxwell manoeuvred between the university ideal of a liberal education and the workshop requirements of precision experiment. Resistance in Cambridge to practical instruction in physics was strong, epitomized by the mathematician, Isaac Todhunter's, contention that it was unnecessary for students to see experiments performed: 'If he does not believe the statements of his tutor—probably a clergyman of mature knowledge, recognised ability, and blameless character—his suspicion is irrational, and manifests a want of the power of appreciating evidence.'[67]

The move to establish practical science teaching in Cambridge extended back over 40 years. Plans for a university complex to include science workshops, lecture rooms, and museums had been given the go-ahead, but in 1831 fell victim first to a bitter internal dispute, and then the economic slump of the late 1830s.[68] By 1850, the Royal Commissioners were informed by the gradual development of laboratory teaching in chemistry, beginning in the 1820s with Liebig at Giessen and Thomas Thomson at Glasgow. Their report compared Cambridge unfavourably with King's College, London, Birkbeck College, the Royal College of Chemistry, and unnamed 'foreign institutions'. They recommended trenchantly that,

> Besides museums containing collections of natural objects . . . and in addition to . . . experiments performed before a class, something further is requisite to enable the Student in his hours of private study to digest and utilize the knowledge he has gained at lectures and in reading . . . He should have access . . . to a collection of instruments . . . which are in daily use in the hands of working men of science; which instruments also, under appropriate regulations, might be intrusted to him for private use, either in his own rooms, or in buildings expressly constructed for the purpose by the University.[69]

Evidently, in 1850, the idea of formal instruction in practical science was not firmly established; indeed the very value of practical science lay in its voluntary and private nature. There is a suggestion that experiment might take place in institutional rather than personal space, but it is not yet called a laboratory. Over the next 30 years the idea became better accepted, and suitable lecture theatres and museums of instruments and specimens were gradually provided in Cambridge for Zoology, Anatomy, Chemistry, Mineralogy and Botany, conveniently close to each other on the site of the old Botanic Gardens (now known as the New Museums Site).

Both the term 'laboratory', and practical teaching within them, had originated in chemistry and transferred only slowly to other sciences.[70] William Thomson's pioneering practical physics teaching at Glasgow, in the mid 1850s, had originated in his need for help with his own research – he recruited student volunteers and trained them informally through working together. By 1870, Edinburgh, Manchester, Oxford and London had also established practical teaching laboratories and Cambridge was lagging behind. Led by Thomson, the function of these laboratories became one of precision measurement, by which the practice of physics was increasingly defined. The spaces they occupied, however, were makeshift conversions from other rooms, and as such were prone to interference and vibration and not well suited to precision measurement. Only Thomson in Glasgow and Clifton in Oxford had purpose-built laboratories, both opened in 1870.

A week after his election, in March 1871, Maxwell was appointed to the committee overseeing the building of the laboratory and plunged into work. He had only his own experience of extending Glenlair, advice from Tait at Edinburgh and Thomson's and Clifton's laboratory precedents to draw on. As Coutts Trotter, another member of the committee, remarked regarding architects,

> . . . I hope it will not be a great swell from London, there is I take it no one who is likely to have the faintest idea of what is wanted from a physical laboratory and the only chance of a convenient building seems to be the getting of some one who will not be above taking hints as to the arrangements . . .[71]

Following this advice, and Maxwell's own practice at Glenlair, the committee engaged a local architect, William Fawcett, known mainly for his church restorations. Fawcett was evidently not above taking hints, for, despite his predilection for mouldings, the outcome was 'plain and substantial' and, 'in no case was convenience sacrificed to architectural effect'.[72] His medievalism was evident mainly in the carving on and around the main door which famously bore the inscription '*Magna opera Domini exquisite in omnes voluntates ejus*', the Vulgate version of Psalm 111 verse 2, surmounted by the arms of the university and the Devonshire family, and a statue of the Duke holding a model of the laboratory.

Despite the backing of the Chancellor and his money, Maxwell and the committee were careful not to push the Senate too far in their demands; the laboratory represented a threat to the power of the colleges, financially and pedagogically, and to the role of the long-established science professors. Apart from ensuring that the site was easy to access and free from vibration, Maxwell was content with cramped surroundings sidelined by the main science complex. In contrast to the parkland setting of Clifton's Clarendon Laboratory, the Cavendish faced the medieval wall of Corpus Christi College across the narrow confines of Free School Lane – so narrow that Corpus attempted to sue the university for right of ancient lights (Fig. 4.6).[73]

At one level, Maxwell clearly envisaged a largely do-it-yourself laboratory, writing to Tait, 'Present notion is – Avoid smooth plastered walls & ceilings where no wood can be found for the thread of your screw, but introduce freely wooden pilasters and beams in ceiling not plastered over.'[74] His attitude originated in his experience of experimenting at home without access to

Fig. 4.6 The front of the Cavendish Laboratory, showing the gate with statue of the Duke of Devonshire above, the wide windowsills for heliostats, and the wall of Corpus immediately opposite.

craftsmen or instrument makers, but it was also characteristic of some of the growing number of country gentlemen such as Rayleigh and Lord Salisbury, who were building private physical laboratories on their estates.[75]

But Maxwell was also part of a tradition that was driving physics towards precision measurement and the 1860s had seen a change in his approach to experiment. His time in London, and involvement with the British Association Committee on Electrical Standards, had shifted his attention to science that, if not big, was complex enough to require two pairs of hands for precision, and increasingly took place in institutional or at least semi-public spaces. After his departure from King's College in 1865, there is very little evidence of him performing experiments at home, and he does not appear to have taken the opportunity of extending Glenlair to improve his workshop there.[76] Instead, his experimental work took place in London on his winter visits. He used John Peter Gassiot's well-equipped private laboratory on Clapham Common, 'always open to his fellow workers' for his comparison of electrostatic to electromagnetic units in 1868, which established a value for the velocity of electromagnetic waves.[77]

Now Maxwell had to work out how to reconcile inspiration for all with the requirements of precision. He wrote to Katherine that,

> There are two parties about the professorship. One wants popular lectures, and the other cares more for experimental work. I think there should be a gradation – popular lectures and rough experiments for the masses; real experiments for real students; and laborious experiments for first-rate men like Trotter and Stuart and Strutt.[78]

He embedded gradation in the design of the laboratory. As well as a lecture theatre capable of seating 180, he copied Thomson's practice and included a large general laboratory, 60 feet by 20 feet, which contained ten workbenches. But, following Clifton, he added small rooms 'in which advanced Students can conduct experiments requiring instruments of precision'.[79] Even before his election, Maxwell had been concerned that students should be allowed to leave their apparatus set up undisturbed.[80] Now he responded to Trotter's insight that, 'There is no doubt much to be said for natural selection but will not the struggle for existence between the men who want their rooms darkened and the men who want their rooms light, the men who want to move about magnets and the men who want to observe galvanometers, be unduly severe', by organizing the small rooms by type of measurement rather than by topic as Clifton had done.[81] Thus there was a room for magnetic measurements, one for experiments involving rotation, a balance room, and rooms devoted to electricity, heat, optics and acoustics. The ideal of precision measurement was designed into the fabric of the laboratory: benches were isolated from floors to avoid vibration, pipes were exposed (and were banned completely from the area of magnetic measurements), shutters could black the rooms out while heliostats could stand on the unusually broad window ledges on the south and east sides to bring light in.[82] This last trick was one Maxwell had learnt from Stokes, and that was implemented also at Oxford. A tower provided for a Bunsen's water pump with 'a vertical fall of considerably more than 50 ft . . . used to exhaust a large receiver, from which pipes will communicate with the different rooms, so that if it be desired to exhaust the air from any vessel it will only be necessary to connect it with one of these pipes and *turn on a vacuum*' (Figs. 4.7 and 4.8).[83]

Maxwell also took care to equip the laboratory with high-quality precision instruments; his correspondence becomes full of discussions of the merits of magneto-electric engines compared to Grove's cells, and of enquiries about where best to source the latest instruments. The Duke of Devonshire had undertaken to equip the Laboratory and, in April 1875, Maxwell reported that, 'Instruments of the best construction . . . have accordingly been made and placed in the Laboratory' and, in May 1877, that 'The Chancellor has now completed his munificent gift to the University by furnishing the Cavendish Laboratory with apparatus suited to the present state of science'.[85] But Maxwell also drew apparatus in from elsewhere. Workers at the Cavendish became part of a resource network that extended beyond the laboratory walls, to other parts of the university and the British physics community. When Glazebrook wanted to test Fresnel's theory of the form of the wave surface in a crystal, Stokes lent him his own goniometer, Schuster, an academic visitor, sourced a suitable crystal, and Maxwell provided a dark room in the laboratory and access to a workshop.[86] Maxwell's transfer of the British Association electrical standards instruments has already been mentioned, as has his acquisition of the Wollaston collection. He took over the late Professor Willis' collection from the engineering department, and he donated some himself. An increasing proportion was made on the premises by students and Fulcher. After 1877, he relied totally on what could be made in the laboratory, purchased from fees or his own pocket, and the amount acquired dropped significantly. The apparatus covered electricity, magnetism, heat, and optics, but there are very few multiple purchases – evidence of an absence of mass teaching.[87]

But while his ideas about the role of precision measurement in physics, and the space in which the measurements took place, might have been up to date, Maxwell's ideas of the social practices within the laboratory remained rooted in the ideals of democracy and voluntary science fostered by his Scottish upbringing and experience at Cambridge as a young man. His remark to his friend Cecil Monro that, 'The desideratum is to set a Don and a Freshman to observe & register (say) the

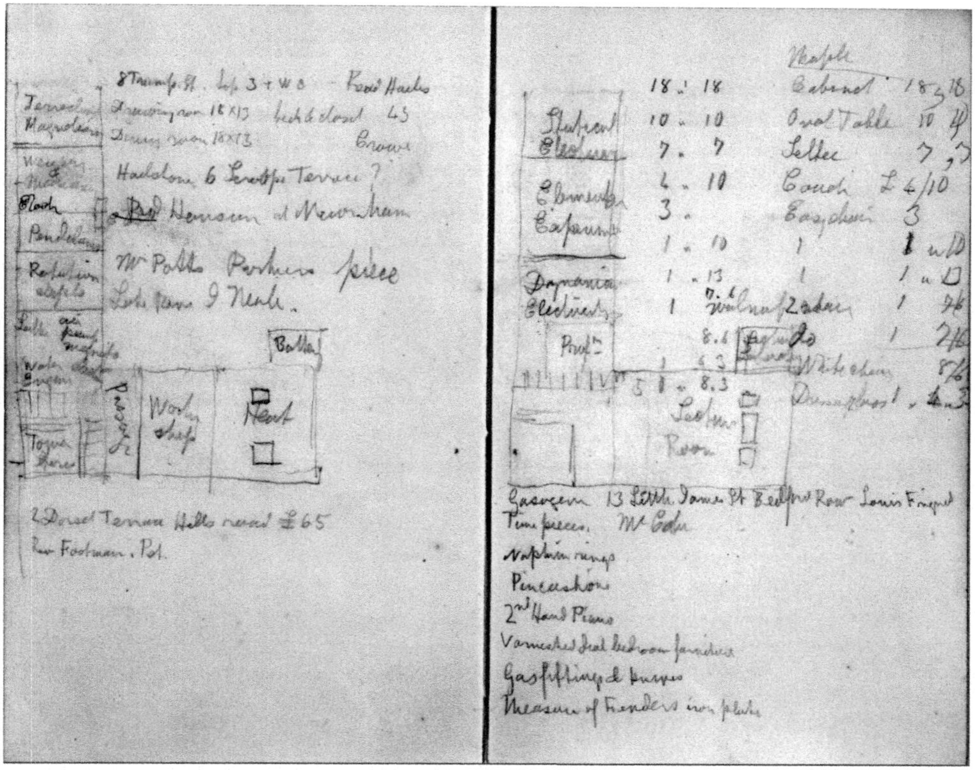

Fig. 4.7 Photograph of one of Maxwell's notebooks, showing his sketch plans for the laboratory alongside notes on his search for a house to rent and prices of furniture.[84] The L-shaped building, the tower in the angle of the building, the ground floor workshop and first floor lecture theatre are evident. At this stage it seems that Maxwell was planning only a series of rooms for experiments categorized by the type of apparatus used. In the event the three rooms shown on the first floor (right hand plan) were replaced by a large general lecture theatre on Thomson's model. Photo courtesy of the Cavendish Laboratory.

vibrations of a magnet together, or the Don to turn a winch & the Freshman to observe and govern him' was more than a quip; his laboratory teaching model presumed close and informal relations between interested students, graduates, and professors.[88] His enjoyment of these relations is evident in the relish with which he reported to his friend Lewis Campbell, 'I have had a pupil quite innocent of mathematics who has learned to measure focal lengths of lenses, and has found the electro-motive force from the water-pipes to the gas-pipes, and from either set of pipes to the lightning-conductor'.[89] His approach was perfectly in tune with the recommendations of the 1850 Commissioners 25 years previously – even to the lending of equipment for private study (to James Gordon for his experiments in Caius College on Kerr's effect).[90]

The laboratory came into use in October 1873, and was formally opened in June 1874. In March 1874, Maxwell advertised that, 'The Cavendish Laboratory will be open daily from 10 am to 6 pm under the superintendence of the Professor of Experimental Physics, for the use of any Member of the University who may desire to acquire a knowledge of experimental methods, or to take

Fig. 4.8 Photograph of the back of the Cavendish Laboratory, showing the tower which provided a head of water for the vacuum pump. Note also the wide windowsills for heliostats, and that the rear wing of the laboratory, on the right of the photo, was built from brick as an economy measure.

part in physical researches'.[91] Entry to the laboratory and its facilities was as informal as this notice suggests and Maxwell made no distinction between undergraduates and graduates – or even non-members of the university. Schuster and Henry Middleton both visited for several years, eventually regularizing their positions by becoming 'fellow-commoners' of a college. Heycock was seemingly still at school when he collaborated with Clayden on the spectrum of indium and did not enrol as an undergraduate for another year.[92]

Maxwell made no distinction between members and non-members of the university – as long as they were male. The first women's colleges were opening, but women were not yet members of the university. Garnett recalled that, 'For several years Maxwell would not open the Laboratory to women. At last he gave permission to admit women during the Long Vacation, when he was in Scotland, and I had a class who were determined to go through a complete course of electrical measurements during the few weeks for which the Laboratory was open to them.'[93] However, in October 1878, at the end of a long list of lectures across the university for women, which included lectures and practical classes in chemistry, there was a brief announcement that 'A class is also being formed in Experimental Physics'.[94] Numbers were, no doubt, small. Maria Anelay, the first woman to get honours in natural sciences took her Part 1 exams in 1881; the following year there were six.[95]

The Appendix details 30 men whose work in the laboratory between 1875 and 1879 is easily documented. There were undoubtedly more, and Maxwell reported that there were up to 20 on any one day engaged in practical work in 1878–79.[96] However, the 30 are enough to suggest some general patterns. Half of them were undergraduates; the Cavendish under Maxwell was not a graduate school, but it was one that largely assumed no distinction between practical teaching and independent – or semi-independent – research. In the first few years, undergraduates were a scattered handful. When, in 1875–76 Maxwell coerced five people into having their resistance measured (between iron rods held in each hand), two were undergraduates, Edmund Sargant and William Sell; both subsequently gained first class degrees in physical sciences suggesting that this may have been more than a casual visit to Maxwell and the laboratory. The other three were graduates who were laboratory regulars.[97] Neither Clayden nor Heycock in the summer of 1876 had graduated – they preceded by a few months Hugh Newall who was not, as he believed, the first undergraduate in the laboratory.[98] Newall was scathing about the lack of direction he experienced as an undergraduate in the Cavendish in 1876–77. However, despite his later eminence as a Fellow of the Royal Society and Professor of Astrophysics, Newall had difficulty adjusting to the Cambridge system in general if his undistinguished undergraduate record is anything to go by. Working as a pair, Clayden and Heycock had fared better. They also came with a specific question in mind; they did not need direction. While attending Liveing's chemistry demonstrations they (presumably Clayden) noticed that the spectrum of indium appeared more complex than the three lines given in the textbook. They came to use the Cavendish's powerful, four-prism, spectroscope and found and measured 16 lines.

Given that the large laboratory with its 10 benches appears specifically designed for practical classes, the delay in establishing formal classes may seem surprising – but perhaps not. Thomson's general laboratory, which Maxwell probably copied, was not used for regulated practical classes either. Instead, Thomson's 'corps' of volunteers were set to making measurements that would contribute to his current investigations – the more experienced and able students helping the novices.[99] The early pairings of the recent graduate John Lord with the much older Alexander Freeman, the third year Clayden with Heycock, and Garnett and Hicks with the younger Marshall, suggest that Maxwell adopted a similar strategy, running against the grain of the individual fellowship system. This was definitely so in the case of Glazebrook, who entered the laboratory in 1876 just after his Mathematical Tripos exams, along with Poynting and Shaw. Glazebrook recalled, 'The first morning Maxwell explained to us, standing round a table in the big Laboratory, Wheatstone's Bridge method of measuring resistance ... Soon we were measuring resistances and learning to set up and adjust various pieces of electrical apparatus ... After a short time I was set a task which I think any student with only a few weeks' Laboratory experience would consider alarming.' This was to measure periodically the emf of the Daniell's cells that Chrystal and Saunder were using in their test of Ohm's Law.[100] Here, again, was the novice working alongside the more experienced and contributing to their research.

Although Maxwell had reported, in May 1874, that, 'an experimental class is now at work on magnetic measurements and on the principles of optical instruments', it was not until the Lent term, 1878, that he formally advertised a practical class and the list of 'Gentlemen at Practical Work' was inserted into his notebook.[101] The 'gentlemen' varied widely in experience; eight were undergraduates, though one of these, Ambrose Fleming, was a mature student who had already taught physics himself. If this is a list of the advertised class, Maxwell seems, again, to be relying on informal supervision by the experienced. However, it seems more likely that it represents increased voluntary research activity among motivated undergraduates, perhaps stimulated by the introduction of a formal class. Either way, the undergraduate presence in the laboratory was on the rise.

It is far from clear how many of the more elementary physics students were doing practical physics in the late 1870s. When the Natural Sciences Tripos was separated into two stages in 1876, Maxwell was emphatic that, 'During the last three days [for honours students] the object of the examination should be to give credit to those who can show a practical as well as a theoretical knowledge of the nature and results of physical research. The questions should ... therefore ... give full occupation to the real student of physics during the hours of examination while they afford no chance of marks to the men of hearsay information.'[102] The eight to ten a year taking honours correlate well with the numbers reported by Maxwell in 1878–79 and recorded as at practical work in 1878. Maxwell said nothing, though, about whether practical exams were to continue at elementary stage. By the late 1870s there were 30 to 40 such elementary students, probably prompting the need for the more organized practical classes that Newall claims were instituted in 1879 and may have started the previous year.[103] Prior to this, it is possible that they learnt purely from Garnett's increasingly popular demonstration lectures.

Maxwell had, all along, argued for gradation in teaching. With the separation of the Natural Sciences Tripos into two stages in 1876, he finally had an opportunity to distinguish between elementary and advanced lectures. He gave the elementary ones to Garnett, encouraging him to develop illuminating experimental demonstrations for them. These lectures were directly aimed at those taking stage 1 of the Natural Sciences Tripos, the M.B., and the General Examination, and grew rapidly in popularity, becoming among the largest science lectures in the university.[104] In comparison, Maxwell's audiences were tiny, yet both Fleming and Newall recall the care with which he prepared and gave the advanced lectures.[105]

The overall numbers are very comparable to those reported in physical laboratories in other universities. Witnesses to the Royal Commission in 1870 agreed that six to eight students was the maximum that one member of staff could handle in a practical class at once, and this, and the amount of apparatus that would be required for larger classes, imposed a limit on numbers. Even Thomson, whose laboratory was far better established, had only 15 to 20 at practical work in any one year.[106]

Under Maxwell's leadership, experimental physics in Cambridge became centralized in the Cavendish, and it is significant that none of the colleges set up their own physical laboratories, as they had done in chemistry, or did in Oxford.[107]

Summary

The Cambridge system had an impact, not only on the makeup of the body of workers at the Cavendish, but also on the way its early history was written. Although lists of workers, and lecture numbers, show a strong presence of undergraduates in the laboratory, even doing research, these were mostly natural sciences students. There was no subsequent place for them in Cambridge and they left upon graduation. Those that entered the laboratory as post-graduates, obtained fellowships and subsequently university positions and, in 1910 wrote the history of the laboratory from which most subsequent accounts are derived, were mathematicians who were successful in the Rayleigh era that succeeded Maxwell's.[108] For these authors, Maxwell's contribution was judged against a benchmark of a hierarchical distinction between research and teaching, and systematized production of men trained in precision measurement for the benefit of industry and empire.

While they praise the inspiration Maxwell provided, they regard his teaching as an unsystematic failure and cite the low numbers attending his own lectures.

Following their accounts, the Cavendish under Maxwell has often been characterized as a graduate research school. It was, indeed, a research school, but it was not an exclusively graduate one; it was also a teaching school where practical research and practical teaching were integrated in Maxwell's eyes. He himself had contributed to science as a schoolboy, before going through the rigours of a Cambridge education, and he seems to have seen no reason why others should not do likewise. He believed that the best way of learning experimental physics was through independent investigation, and his teaching model assumed this, relying on a network of relations between interested students, graduates, and professors that extended beyond the walls of the Cavendish. Even the evidently shy Newall was on visiting terms with the more experienced researchers.[109] Maxwell's approach was typical of many British scientists of his generation, as reported by the numerous witnesses to the 1870 Royal Commission on Scientific Instruction. Indeed, when Rayleigh later introduced more directed teaching, he looked to the United States for a precedent.[110] In the case of Thomson the voluntary approach worked well; with the subjects of measurement directed by Thomson, the evolutionary competition in the general laboratory predicted by Trotter may not have been felt. Direction, however, was not Maxwell's way. Resistant, himself, to the imposition of discipline, he was reluctant to impose it on others, and famously remarked that, 'I never try to dissuade a man from trying an experiment. If he does not find what he wants, he may find out something else'.[111] His system generated enthusiasm amongst those who succeeded in it, produced a hugely diverse range of research, a high public profile and undergraduate numbers comparable to other universities. It may, though, have begun to falter as student numbers began to rise. Certainly it did not take Rayleigh long to decide there was a need to control the space by introducing more directed and organized teaching. He put a stop to the elementary lectures for medical students and instituted regimented practical classes.[112] The difference between his announcement of the laboratory's facilities and Maxwell's is striking. Under Rayleigh the laboratory was open only, 'for more advanced practical work . . . for those who have had the necessary training'.[113] The immediate need for such control was a measure of Maxwell's success.

Appendix: Maxwell's students and co-workers 1871–79

Source codes:

1 = 'students michaelmas 1871'[114]
2 = 'students lent 1872'[115]
3 = 'students michaelmas 1873'[116]
4 = 'gentlemen at practical work, lent 1878'[117]
5 = 'resistances'[118]
6 = Museums and Lecture Rooms Report 1875[119]
7 = *A History of the Cavendish Laboratory*[120]
8 = Charles Whitmell to John Couch Adams 27 Jan 1875[121]
9 = *James Clerk Maxwell: A Commemoration Volume, 1831–1931*[122]

Abbreviations:

MT = Mathematical Tripos; NST = Natural Sciences Tripos; ug = undergraduate; pg = postgraduate

Name (Source)	Activity associated with Maxwell or the Cavendish	Matriculation year	Tripos, class & year	Status[123]	Career
Benson, George Vere (1)	Attended lectures Michaelmas 1871	1870	MT, 35th Wrangler, 1873	ug	Called to the Bar, 1878. Coroner for East Sussex 1899–1927
Carpenter, Philip Herbert (1)	Attended lectures Michaelmas 1871	1871	NST, 1st class, 1875	ug	FRS. Assistant Master at Eton, 1877–91. Palaeontologist and zoologist.
Drury, Henry D'Olier (1)	Attended lectures Michaelmas 1871	1869	MT, 22nd Wrangler, 1873	ug	Mathematical Master at Marlborough College, 1873–1909
Henderson, John (1)	Attended lectures Michaelmas 1871	1870	1874	ug	High Sheriff of Surrey, 1884
Laing, John Gerard (1)	Attended lectures Michaelmas 1871	1869	MT, 26th Wrangler, 1873	ug	Called to the Bar, 17 Nov. 1874. Judge's Associate on the South Eastern Circuit.
Pendlebury, Richard (1)	Attended lectures Michaelmas 1871	1866	MT, Senior Wrangler, 1870	pg	Cambridge University Lecturer in Mathematics, 1888–1901.
Peter, John Franklen (1)	Attended lectures Michaelmas 1871	1870	MT, 15th senior optimes, 1874	ug	Called to the Bar, 13 June 1877
Ritchie, William Irvine (1)	Attended lectures Michaelmas 1871	1869	MT, 8th Wrangler, 1873	ug	Senior Examiner in the Education Office, 1893; Assistant Secretary.
Salmond, David Murray Lyon (1)	Attended lectures Michaelmas 1871	1870	1875	ug	Mathematical Master in Montreal. Died 1875
Swanwick, Frederick Tertius (1)	Attended lectures Michaelmas 1871	1870	MT, 18th Wrangler, 1874	ug	Fielden Lecturer in Mathematics, Manchester University, from 1907

Tennant, John (1)	Attended lectures Michaelmas 1871	1866	MT, 12th Wrangler, 1870	pg	Called to the Bar, 26 Jan. 1875
Ball, Walter William Rouse (1,2)	Attended lectures Michaelmas 1871 & Lent 1872	1870	MT, 2nd Wrangler, 1874	ug	Trinity College Lecturer in Mathematics, 1878; Director of Mathematical Studies, 1891
Bovey, Henry Taylor (1,2)	Attended lectures Michaelmas 1871 & Lent 1872	1869	MT, 12th Wrangler, 1873	ug	Professor of Civil Engineering and Applied Mechanics, McGill University, 1887–1909
Lamb, Horace (1,2)	Attended lectures Michaelmas 1871 & Lent 1872	1868	MT, 2nd Wrangler, 1872	ug-pg	FRS. Professor of Pure Mathematics, Owens College, Manchester, 1885–1920
Nanson, Edward John (1,2)[124]	Attended lectures Michaelmas 1871 & Lent 1872	1869	MT, 2nd Wrangler, 1873	ug	Professor of Mathematics at Melbourne University, Australia, 1875–1922
Radford, George (1,2)	Attended lectures Michaelmas 1871 & Lent 1872	1869	Law, 2nd class, 1874	ug	In business
Vinter, Arthur (1,2)	Attended lectures Michaelmas 1871 & Lent 1872	1869	MT, 14th Wrangler & NST, 2nd class, 1873; LLD 1894	ug	Headmaster of Woodhouse Grove School, Leeds, 1883–1913
Webb, Robert Rumsey (1,2)	Attended lectures Michaelmas 1871 & Lent 1872	1868	MT, Senior Wrangler, 1872	ug-pg	St. John's College Lecturer, 1877–1911; well known as a 'Coach' for the Mathematical Tripos

Name (Source)	Activity associated with Maxwell or the Cavendish	Matriculation year	Tripos, class & year	Status[123]	Career
Hicks, William Mitchinson (1,2,6,7)	Attended lectures Michaelmas 1871 & Lent 1872, working in laboratory 1875 on the velocity of electromagnetic waves, and on Lippmann's capillary electrometer (with Garnett and Marshall)	1869	MT, 7th Wrangler, 1873	ug-pg	Professor of Mathematics and Physics, Firth College, Sheffield. First Vice-Chancellor Sheffield University 1905
Arnett, Braithwaite (2)	Attended lectures Lent 1872	1866	1873	ug	Headmaster Truro Grammar School
Finlay, William Henry (2)	Attended lectures Lent 1872	1869	MT, 34th Wrangler, 1873	ug	No further details
Gaskell, Walter Holbrook (2)	Attended lectures Lent 1872	1865	MT, 26th Wrangler, 1869; MD, 1879	ug	FRS. Cambridge University Lecturer in Physiology, 1883–1914
Gregory, Philip Spencer (2)	Attended lectures Lent 1872	1869	MT, 10th Wrangler, 1873	ug	Called to the Bar 1875. Conveyancing Counsel to the Court of Chancery 1902, Bencher of Lincoln's Inn 1907
Harding, Thomas Olver(2)	Attended lectures Lent 1872	1869	MT, Senior Wrangler, 1873	ug	Called to the Bar, 9 June 1880. Equity Draftsman and Conveyancer
Hart, Harry (2)	Attended lectures Lent 1872	1867	MT, 4th Wrangler, 1871	pg	Professor of Mathematics, Royal Military Academy, Woolwich

Hawks, Robert Mitchell (2)	Attended lectures Lent 1872	1869	1873	ug	No further details
Hayden, Christopher John (2)	Attended lectures Lent 1872	1865	MT, 10th Wrangler, 1869	pg	No further details
Johnson, James Magnus(2)	Attended lectures Lent 1872	1868	MT, 11th Wrangler, 1873	ug	No further details
Marshall, Leonard (2)	Attended lectures Lent 1872	1870	MT, 16th Wrangler, 1874	ug	Master at Charterhouse, 1874–1904
McMahon, Henry (or Harry) Eugene (2)	Attended lectures Lent 1872	1868		ug	No further details
Shephard, Stuart (2)	Attended lectures Lent 1872	1871	1876; LLM, 1879	ug	Called to the Bar, 25 June 1879. On the South Eastern Circuit
Taylor, Henry Martyn (2)	Attended lectures Lent 1872	1861	MT, 3rd Wrangler, 1865	pg	FRS. Trinity College tutor, 1874–84. Amplified the Braille alphabet and extended its use to textbooks on geometry, trigonometry, etc.
Winslow, James Stephen (2)	Attended lectures Lent 1872	1869	1872	ug	V. of Great Burstead, Essex, 1880–1904
Brown, Francis Faulkner (2,3)	Attended lectures Lent 1872 & Michaelmas 1873	1870	MT, 46th Wrangler, 1874	ug	In his father's business (furniture warehouse) at Chester 1898
Garnett, William (2,5,6,7)	Attended lectures Lent 1872, appointed demonstrator 1874, Maxwell measured resistance May 1876	1869	MT, 5th Wrangler, 1873	ug-pg	Professor of Mathematics and Principal Durham College of Science, Newcastle-on-Tyne, 1884–93; Educational adviser to the L.C.C., 1904–15

Name (Source)	Activity associated with Maxwell or the Cavendish	Matriculation year	Tripos, class & year	Status[123]	Career
Gordon, James Edward Henry (2,6,7)	Attended lectures Lent 1872, working in laboratory 1875, published on a repetition of Kerr's experiment using apparatus lent by Maxwell 1876, on Verdet's constant 1877, and on electrical constants 1879	1872	MT, 12th junior optimes, 1875	ug-pg	Electrical inventor (Gordon dynamo; Gordon divided main system) and engineer
Darwin, George Howard (3)	Attended lectures Michaelmas 1873	1864	MT, 2nd Wrangler, 1868	pg	FRS. Plumian Professor of Astronomy and Experimental Philosophy at Cambridge 1883
Savielle, George Abreo (3)	Attended lectures Michaelmas 1873	1872		ug	Engineer, Bombay Baroda and Central India Railway
Tottie, Oscar Joseph (3)	Attended lectures Michaelmas 1873	1866	Classics, 3rd class, 1871	pg	Died 1879
Marshall, John William (3,6)	Attended lectures Michaelmas 1873, working in laboratory 1875 on Lippmann's capillary electrometer (with Garnett and Hicks) and on viscosity of liquid films	1871	MT, 8th Wrangler, 1875	ug-pg	Assistant Master at Charterhouse, 1875–1909

Saunder, Samuel Arthur (3,5,7)	Attended lectures Michaelmas 1873, Maxwell measured resistance May 1876, published on emf of Leclanche cell 1875, on BA resistance 1876 (with Chrystal)	1871	MT, 13th Wrangler, 1875	ug-pg	Assistant Master at Wellington College, 1876–1912. FRAS, 1894. Gresham Professor of Astronomy, 1909–12
Chrystal, George (3,7,9)	Attended lectures Michaelmas 1873, in laboratory 1875–77, published on galvanometers 1876, and on BA resistance 1876 (with Saunder)	1871	MT, 3rd Wrangler, 1875	ug-pg	Professor of Mathematics Edinburgh University, 1879–1911. Royal Medal of the Royal Society 1911 for his work on oscillations in lakes
Freeman, Alexander (6)	Working in laboratory 1875on properties of voltaic cells (with Lord)	1857	MT, 5th Wrangler, 1861	pg	Fellow of St. John's College, 1862–82; Deputy for Plumian Professor of Astronomy, 1880–2
Lord, John William (6)	Working in laboratory 1875 on properties of voltaic cells (with Freeman)	1871	MT, Senior Wrangler, 1875	pg	Fellow of Trinity College. Died 1883
Whitmell, Charles Thomas (8)	Experiments on highly refractive liquids 1875	1868	NST, 1st class, 1872	pg	HM Inspector of Schools 1879–1910. FRAS 1898
Niven, William Davidson (4,7)	In laboratory 1875, at practical work Lent 1878, published on trajectory of shot 1876	1862	MT, 3rd Wrangler, 1866	pg	FRS. Director of Studies at the Royal Naval College, Greenwich 1882–1903

Name (Source)	Activity associated with Maxwell or the Cavendish	Matriculation year	Tripos, class & year	Status[123]	Career
Pirie, George (5,9)	Maxwell measured resistance May 1876	1862	MT, 5th Wrangler, 1866	pg	Queens' College Tutor, 1870–78. Professor of Mathematics Aberdeen, 1878–1904
Sargant, Edmund Beale (5)	Maxwell measured resistance May 1876	1874	MT, 7th Wrangler & NST, 1st class, 1878	ug	Adm. at the Inner Temple, 25 Jan 1878. A Civil Service Examiner
Sell, William James (5)	Maxwell measured resistance May 1876	1873	NST, 1st class, 1877	ug	FRS. Senior Demonstrator and Cambridge University Lecturer in Chemistry, 1904–15
Heycock, Charles Thomas (7,9)	Published on spectrum of indium 1876 (with Clayden)	1877	NST, 1st class, 1881	ug	FRS. Goldsmiths' Reader in Metallurgy at Cambridge 1908–29
Newall, Hugh Frank (7)	In laboratory 1876–77	1876	MT, 5th junior optimes & NST, 2nd class, 1880	ug	FRS. Professor of Astrophysics, 1909–28
Clayden, Arthur (7)	In laboratory 1876–78, published on spectrum of indium 1876 (with Heycock)	1873	NST, 2nd class, 1877	ug	Science Master at Bath College, 1878. Principal of University College, Exeter, 1894–1920
Trotter, Alexander Pelham (4,7)	At practical work Lent 1878, in laboratory 1876–79	1876	NST, 2nd class, 1880	ug	Dynamo manufacturer. Electrical Adviser to the Board of Trade, 1899–1917
Schuster, A (4,7,9)	At practical work Lent 1878, in laboratory 1876–81, published on spectra, electric discharge and solar physics 1877, 1879, 1880	Fellow commoner at St. John's College, Dec. 1876–			FRS. Professor of Physics at Manchester University, 1888–1907

Sunderland, Arthur Wellesley (4,7)	At practical work Lent 1878, in laboratory 1877, investigated anomalous dispersion of fuchsin	1872	MT, 8th Wrangler, 1876	pg	Manager and Actuary of the National Life Assurance
Macalister, Donald (7)	In laboratory 1877–78, published on the inverse square law 1879	1873	MT, Senior Wrangler, 1877; MB 1881; MD 1884	pg	Cambridge University Lecturer in Medicine, 1884. President of the General Medical Council, 1904–31. Vice-Chancellor of Glasgow University, 1907–29
Fleming, John Ambrose (7)	In laboratory 1877–82, published on a resistance balance 1880	1877	NST, 1st class, 1880	ug-pg	FRS. Professor of Electrical Engineering at University College, London, 1885–1926
Bennett, James Alfred Beresford (4)	At practical work Lent 1878	1876	1881	ug	No further details
Day (4)	At practical work Lent 1878	Cannot identify with certainty; possibly Richard Evan Day BA (17th Wrangler) 1868 and Physics Lecturer at King's College, London, in 1876			
Hewitt, Frederick William (4)	At practical work Lent 1878	1877	NST, 3rd class, 1881; MB, 1883 MD 1886	ug	Physician and Anaesthetist at St. George's Hospital. Anaesthetist to Edward VII and to George V
Pigeon, Henry Walter (4)	At practical work Lent 1878	1877	NST, 1st class, 1881; MB & Bchir, 1884	ug	Surgeon, Hull Royal Infirmary
Taylor, Edward Francis (4)	At practical work Lent 1878	1874	NST, 2nd class, 1878	pg	Diocesan Inspector, Truro, 1887–1932
Vicars, George Rayleigh (4)	At practical work Lent 1878	1877	1884	ug	Physician at the Brompton Hospital, subsequently Vicar of Wood Dalling, Norfolk, 1897–1936.

Name (Source)	Activity associated with Maxwell or the Cavendish	Matriculation year	Tripos, class & year	Status[123]	Career
Whittam, William Goldsborough (4)	At practical work Lent 1878	1876	NST, 1st class, 1880	ug	Headmaster of the Isle of Wight College, Ryde, 1892–1905. Vicar of Messing, Essex, 1921–35
Poynting, John Henry (7,9)	In laboratory 1876 & 1879–80, published on the sonometer, and on a saccharimeter 1880	1872	MT, 3rd Wrangler, 1876	pg	FRS. Professor of Physics at Birmingham University, 1880–1914
Glazebrook, Richard Tetley (4,7,9)	At practical work Lent 1878, in laboratory 1876–99, published on propagation of waves in crystals 1878, and on reflection and refraction in crystals 1880	1872	MT, 5th Wrangler, 1876	pg	FRS. Assistant Director of the Cavendish Laboratory, 1891. First Director of the National Physical Laboratory 1899–1919
Middleton, Henry (4,7)	At practical work Lent 1878, in laboratory 1879–85[125]	Fellow Commoner at St. John's College, Mar. 1879–			Engineer. Inventor of a steam tricycle and electric bicycle
Shaw, William Napier (4,7,9)	At practical work Lent 1878, in laboratory 1876–77 & 1879–99, published an experiment with mercury electrodes 1879	1872	MT, 16th Wrangler & NST, 1st class, 1876	pg	FRS. Assistant Director of the Cavendish Laboratory, 1898–99. Director of the Meteorological Office, 1905–20
Taylor, Sedley (7,9)	'An early attendant of Maxwell's lectures', practical work on soap films	1855	MT, 16th Wrangler, 1859	pg	Fellow of Trinity College. Research in sound and music

PART II

Science

CHAPTER 5

Maxwell and the Science of Colour

MALCOLM LONGAIR

Introduction

James Clerk Maxwell's contributions to electromagnetism, the kinetic theory of gases, thermodynamics and statistical mechanics are rightly recognized as among the greatest contributions to science, providing the essential link between Newtonian physics and the revolutions of relativity and quantum physics that took place in the early decades of the twentieth century. At the same time, Maxwell maintained a deep interest in many other scientific topics, bringing to them his own unique brand of insight, imagination and experimental innovation. His interest in the science of colour remained with him throughout his life, from his early childhood to his last decade. In his studies of colour and colour vision, he lavished the same care and imagination as he did on his epochal contributions to electromagnetism and statistical physics. His endeavours led to the quantitative theory of colour vision.

The Development of Colour Theory from Snell to Young

The story begins in 1621, when Willebrord van Snel van Royen discovered the law of refraction when a light ray crosses the interface between two media. In modern notation, the well-known form of what is now known as *Snell's law* is written

$$n_1 \sin \theta_1 = n_2 \sin \theta_2,$$

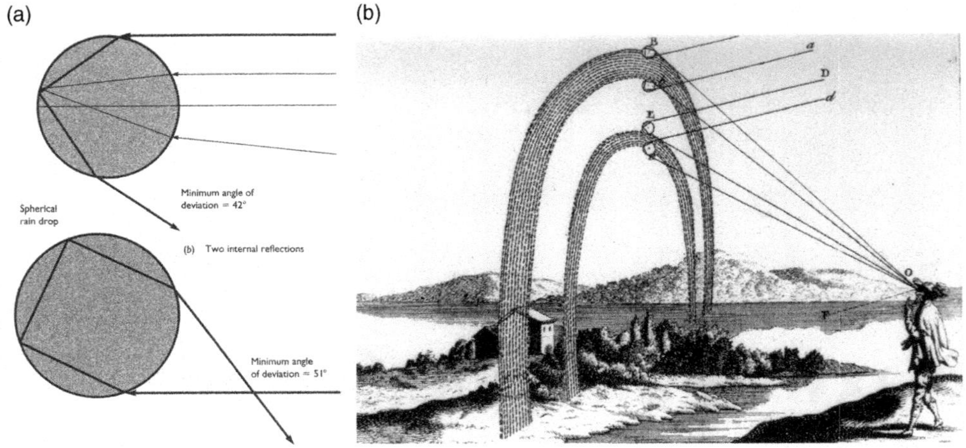

Fig. 5.1 (a) Illustrating the origin of the minimum angles of refraction of light through spherical raindrops. (b) A sketch of the origin of single and double rainbows from Descartes' *Les Météores* of 1637.

where n_1 and n_2 are the refractive indices of the two media and θ_1 and θ_2 are the angles of incidence and refraction. This law is at the heart of optics, lenses, microscopes and telescopes.

One of the early applications of Snell's law was made by René Descartes to account for the formation of rainbows. He studied the internal reflection of light within spherical raindrops and showed that there is a minimum angle of deflection of the light rays amounting to 42° for a single reflection within the raindrop (Fig. 5.1(a) top). This angle corresponds to the condition for total internal reflection at the water–air interface. If there are two internal reflections, the minimum angle is 51° (Fig. 5.1(a) bottom). The origin of the rainbow is associated with the piling up of light rays at the minimum angles of refraction through the raindrop, at the same time accounting for the increased brightness observed within the arc of the primary rainbow. Descartes' theory predicted correctly the observed angles of single and double rainbows with respect to a line from the Sun through the observer's head to the centre of the rainbow. A sketch of the origin of single and double rainbows appeared in Descartes' *Les Météores* of 1637 (Fig. 5.1(b)). There was, however, no understanding of the origin of the colours in the rainbow. This was to be elucidated by Isaac Newton.

In 1665–66, Newton was in his early twenties, and in these years he discovered the binomial theorem, the integral and differential calculus, the theory of colour in optics and unified celestial mechanics and the theory of gravity. These were quite staggering achievements. In the field of optics and colour, the key experiment of these years was his *Experimentum Crucis.* He passed sunlight though a high quality prism and observed that white light was split up into the colours of the spectrum. He then passed this spectrum through a hole which allowed only one of the colours to be selected (Fig. 5.2). On passing this single colour through a second prism, he found that light was not split up into any further colours. In Newton's words:

Light itself is a heterogeneous mixture of differently refrangible rays.[1]

This result gave rise to one of Newton's many disputes with other scientists. The French experimenters could not reproduce the result of the *experimentum crucis*, but found instead that the

(b)

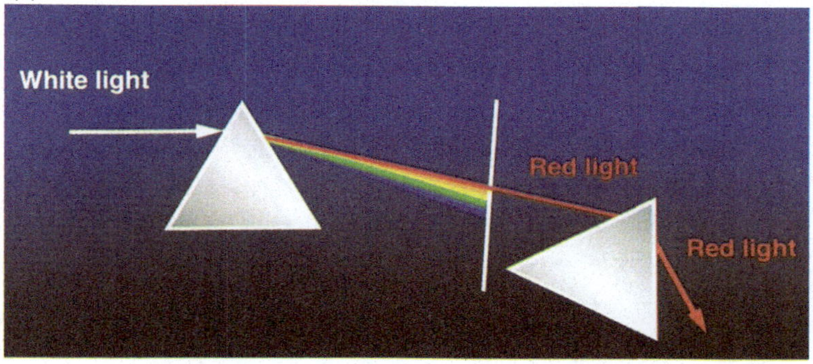

Fig. 5.2 (b) A schematic diagram showing the decomposition of white light into the colours of the spectrum. On passing through the second prism, the colours are not split up into further colours.

Fig. 5.4 An example of a double rainbow showing the primary and secondary bows in which the order of the colours are inverted because of the different paths of the rays through the raindrops, as illustrated in Fig. 5.1(a). (©2000 Kodak.)

(a) (b)

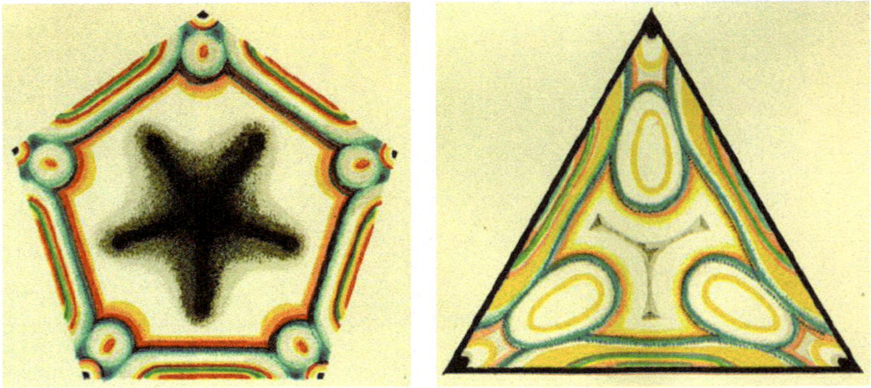

Fig. 5.7 Examples of the coloured images of strained glasses, showing the 'gorgeous entanglements of colours' when polarized light is passed through them (Campbell and Garnett, 1882).

(d)

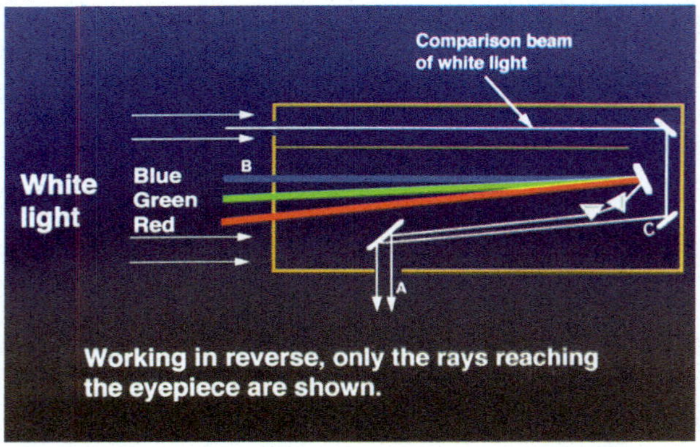

Fig. 5.11 (d) Maxwell's light box working forwards. The blue, green and red regions of the spectrum are focused at A. The amount of each of these lights is precisely varied by changing the width of the entrance slits.

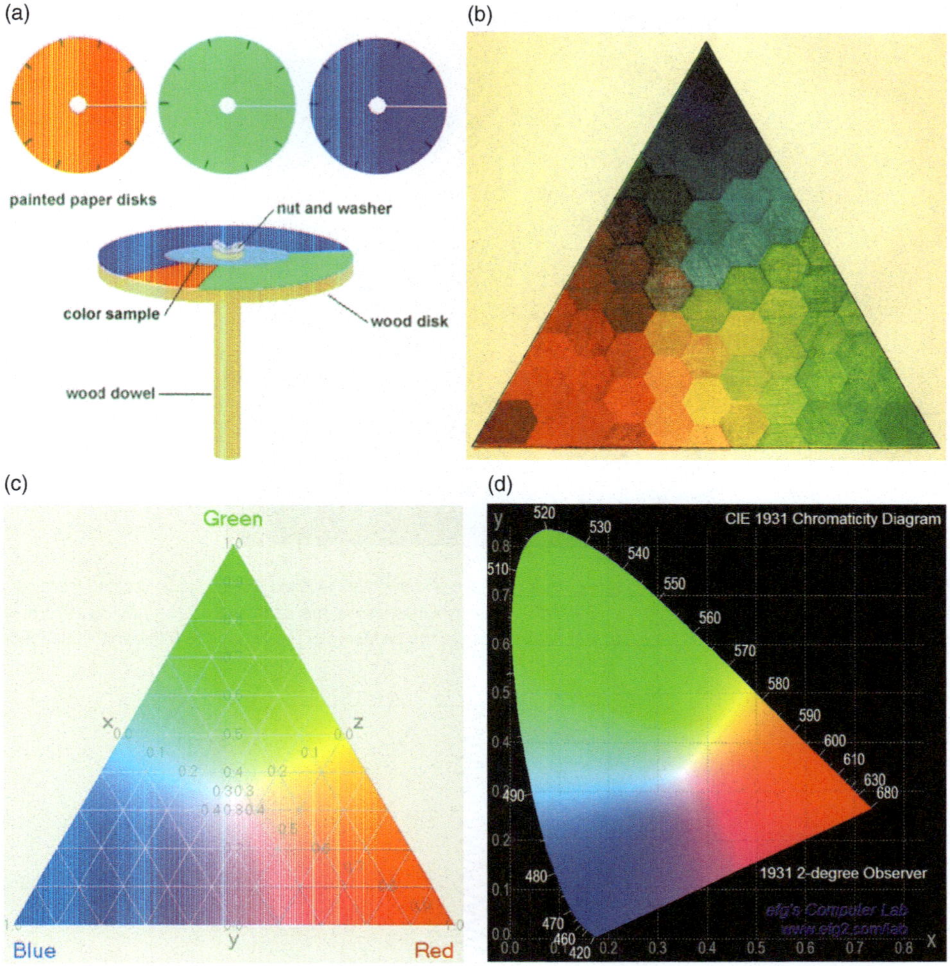

Fig. 5.9 (a) Illustrating the three coloured discs used in Maxwell's colour top (<http://www.handprint.com/HP/WCL/colortop.html>). (b) Maxwell's original version of the colour triangle (Campbell and Garnett 1882). (c) Comparison with a modern version, showing quantitatively the proportions of different primary colours needed to synthesize those within the triangle. The picture is rotated 120° clockwise with respect to (b) (see: <http://www.handprint.com/HP/WCL/colortop.html>). (d) The CIE (Commission Internationale d'Éclairage) diagram. The pure colours in the white light spectrum are shown around the perimeter of the diagram from 420 to 680 nm.

Fig. 5.12 A reconstruction of Maxwell's three-colour image of a brightly coloured tartan ribbon created from copies of the three original plates by Dr. Peter Reid. Intentionally, blemishes on the copy slides have not been removed. (Courtesy of and copyright Dr. Peter Reid. The copy plates are the property of the James Clerk Maxwell Foundation.)

Fig. 5.13 Illustrating how the eye can be fooled in interpreting visual data. As the image is observed further and further from the page, the yellow and white colours seem to leak into the undulations of the wave pattern.

(a)

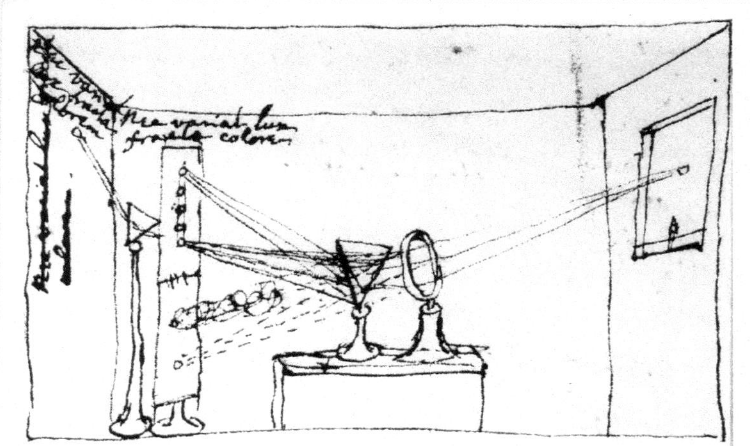

(b)

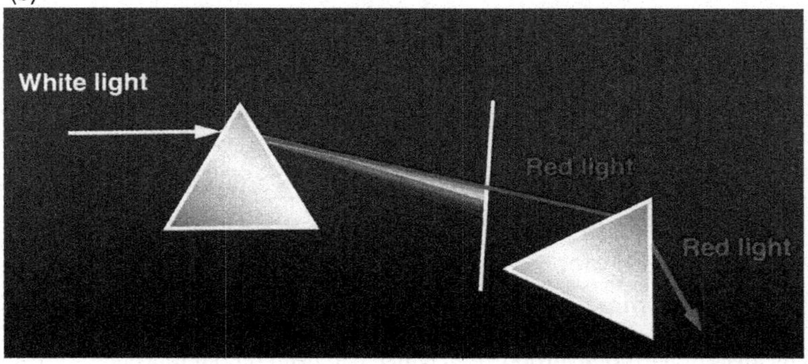

Fig. 5.2 (a) Newton's sketch of his *experimentum crucis.* (b) A schematic diagram showing the decomposition of white light into the colours of the spectrum. On passing through the second prism, the colours are not split up into further colours. Reproduced in colour in the colour plate section.

separate colours could be split up into further colours. It turned out that Newton had found the correct answer because of his use of higher quality prisms than those available to the French scientists, but it was many years before Newton's view prevailed.

Because different colours are refracted through different angles, white light is not focused at a single point by a lens, the phenomenon known as chromatic aberration. To get round this problem, Newton invented the all-reflecting, or Newtonian, telescope in which there is no diffraction of the light rays (Fig. 5.3(a)). He built the telescope himself, including the grinding of the mirrors. The resulting telescope had superior imaging qualities as compared with the Galilean refracting telescope, as he demonstrated by his sketches of the image of a crown observed with the two types of telescope in his paper published in the *Philosophical Transactions of the Royal Society of London* in 1672.

Newton divided the spectrum into seven colours, the famous VIBGYOR sequence of violet, indigo, blue, green, yellow, orange, and red (Fig. 5.3(b)). In his choice of seven colours, he was

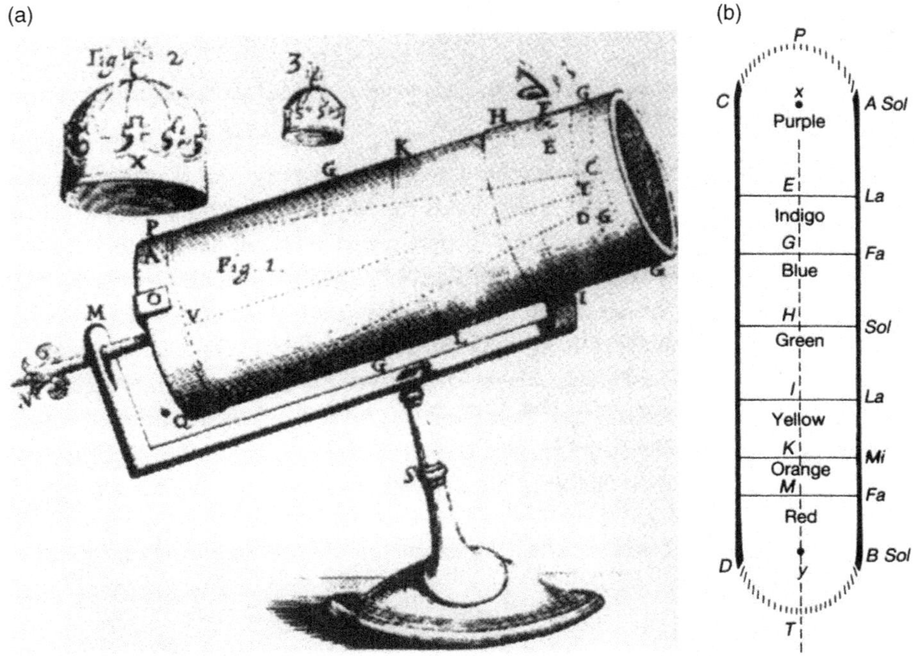

Fig. 5.3 (a) Newton's all reflecting telescope is drawn here (Fig. 1) and two crowns are pictured, the first seen through Newton's telescope (Fig. 2) and the second through a 25-inch refracting telescope (Fig. 3) (Newton 1672). (b) Newton's division of the spectrum of white light into seven colours, with the seven separate tones of the harmonic scale shown to the right of the spectrum (from P. Gouk, 1988. In Fauvel, J. et al. (eds.) *Let Newton Be!* p. 118.)

influenced by ideas of harmonic proportion, which had been set out in Kepler's great treatise *The Harmony of the World* of 1618 and which had influenced all aspects of the arts and sciences in the seventeenth and eighteenth centuries. The treatise contained the discovery of Kepler's third law of planetary motion, which was crucial in Newton's discovery of the law of gravity, and so there is little doubt that Newton knew the treatise well. There are seven different notes in the just harmonic scale, their frequencies all being in the ratios of small whole numbers, and so Newton divided the spectrum into seven colours. Newton's seven colour theory was elaborated by Voltaire in his *Éléments de la Philosophie de Newton* of 1738.

With the understanding that the different colours in the spectrum are refracted by different amounts, in other words that the refractive index is a function of the colour of the rays, the order of the colours in the rainbow could be explained. Figure 5.4 shows the primary and secondary rainbows with the colours inverted in the secondary bow because of the different paths of the light rays through the raindrops. Figure 5.4 also illustrates the increased brightness within the primary rainbow, which is caused by the superposition of the rays of different colours which are diffracted by greater than the minimum angle.

The next major figure in the story is Thomas Young, whose brilliant researches of 1801 not only put the wave theory of light on a firm physical foundation, but also resulted in the theory of three-colour vision. His most famous experiment was the double-slit experiment in which he

Fig. 5.4 An example of a double rainbow showing the primary and secondary bows in which the order of the colours are inverted because of the different paths of the rays through the raindrops, as illustrated in Fig. 5.1(a). Reproduced in colour in the colour plate section. (©2000 Kodak.)

found that, when sunlight is passed through a single narrow slit and then through a pair of similar slits, a characteristic pattern of bright and dark bands is observed on a screen. He interpreted the appearance of this pattern of light rays in terms of the constructive and destructive interference of light waves, a concept which legend has it was inspired by his observations of ripples on the pond in the Paddock at Emmanuel College, Cambridge. This was conclusive evidence in favour of the wave theory of light.

In the same paper, Young also discussed how the eye would perceive lights of different colours. In his picture, the light receptors in the retina of the eye act as resonators which are excited by the incoming light waves. He realized, however, that there was a serious problem with this picture. In his words:

> As it is almost impossible to conceive each sensitive point of the retina to contain an infinite number of particles, each capable of vibrating in perfect unison with every possible undulation, it becomes necessary to suppose the number limited: for instance to the three principal colours, red, yellow and blue . . . and that each of the particles is capable of being put into motion more or less forcibly by undulations differing less or more from perfect unison . . . each sensitive filament of the nerve may consist of three portions, one for each principal colour.[2]

There are several interesting points about Young's proposal. First, he took the three principal colours to be red, yellow and blue. Second, there was no distinction between the mixing of lights and the mixing of pigments. Third, the theory was qualitative rather than quantitative. The theory was to be developed further by Hermann von Helmholtz in 1850. He identified the three types of photo-receptor in the eye as being receptive to blue, green and red light – the theory is commonly referred to as the Young–Helmholtz theory of three-colour vision.

The Young Maxwell

James Clerk Maxwell had an insatiable curiosity about everything. According to the biography by Campbell and Garnett, from his earliest years, he would continually ask, 'Show me how it doos', or 'What's the go o'that?' If he did not receive a satisfactory answer to the latter question, he would ask, 'But what's the particular go o'that?' At home, his father John Maxwell, James and his cousin Jemima, a brilliant artist, were fascinated by the new range of optical toys which became available during the 1830s. These relied upon the phenomenon of the persistence of vision by which the eye and brain preserve the image for about 1/20th of a second. As in a cinematographic film, the appearance of movement is obtained by viewing the images faster than the eye can track the individual frames.

The thaumatrope was invented in 1825 by the London physician, John Paris, who made the toy popular. Two images are printed on either side of a circular card which is then rotated rapidly about a diameter by twisted strings attached at opposite ends of this diameter (Fig. 5.5(a)). Because of the persistence of vision, the images on either side of the disc are observed to be superimposed.

In 1832, Belgian physicist Joseph Plateau and his sons introduced the phenakistoscope, meaning 'spindle viewer'. About a dozen images were drawn on the disc which was then rotated about its axis and the images viewed in a mirror by looking through radial slits cut in the disc (Fig. 5.5(b)). The viewer sees the images very rapidly one after the other, producing the effect of a moving image.

The phenakistoscope was soon overtaken by the zoetrope, which was invented in 1834 by William Horner. Now, the images were painted on a strip which was pasted to the inside of a rotating cylinder. The images were viewed through the slits in the rotating drum (Fig. 5.6(a)).

(a) (b)

Fig. 5.5 (a) An example of a thaumatrope which, when spun, shows the dog chasing the birds. (b) An example of the disc of a phenakistoscope, or 'spindle viewer'. When rotated about its axis and viewed in a mirror through the slits, the moving image of the man riding his scooter is observed.

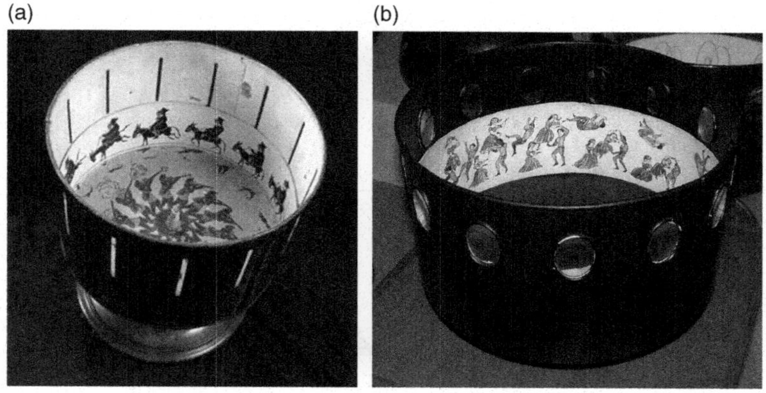

Fig. 5.6 (a) A zoetrope showing the narrow slits through which the image is viewed. (b) Maxwell's improved zoetrope of 1861. By inserting concave lenses instead of slits, the virtual images appear on axis and the field of view is considerably increased (Courtesy of the Cavendish Laboratory, Cambridge).

Zoetropes became popular, a 1905 supplement of the *New York Sunday American and Journal* newspaper including a cardboard cut-out entitled, 'Make Your Own Zoetrope'.[3]

James and Jemima delighted in these scientific toys. As expressed by Campbell and Garnett,

> This was a source of endless amusement to the two cousins, the younger generally contriving, and in part executing, the elder giving life and spirit to the creatures represented ... The cow jumping over the waxing and waning moon, the dog pursuing the rat in and out of his hole, the circus horse, on which the man is jumping through the hoop, have the firmness and truth of touch, the fullness of life, familiar to the many admirers of [Jemima Wedderburn] – the tadpole that wriggles from the egg and changes gradually into a swimming frog; the cog-wheels moved by the pendulum, and acting with the precision of clockwork.

James was not yet 10 when they invented many of their own designs. James remained fond of the zoetrope and in 1861 improved its performance by replacing the slits by concave lenses, as seen in the example in Fig. 5.6(b). As a result, virtual images appeared on the axis of the zoetrope while it rotated, resulting in a much improved image and a wider field of view for the observer. In typical academic fashion, Maxwell then set his invention as a problem in the Cambridge 1869 Tripos examinations.

In 1847, Maxwell was taken by his uncle, John Cay, to visit the optical laboratory of William Nicol, the distinguished optician and inventor of the Nicol prism. Maxwell later recalled that:

> I was taken to see [William Nicol], and so, with the help of 'Brewster's Optics' and a glazier's diamond, I worked at polarisation of light, cutting crystals, tempering glass, etc.

In 1848, he undertook a series of experiments on the chromatic effects of polarized light in doubly refracting materials, crystals and mechanically strained glasses (Fig. 5.7). He remarked on the 'gorgeous entanglements of colours' in strained glasses.

After high school, Maxwell proceeded to Edinburgh University where he attended courses on mathematics by Kelland, natural philosophy by Forbes and logic by William Hamilton. During these years, he read an astonishing range of great and difficult scientific books and he published

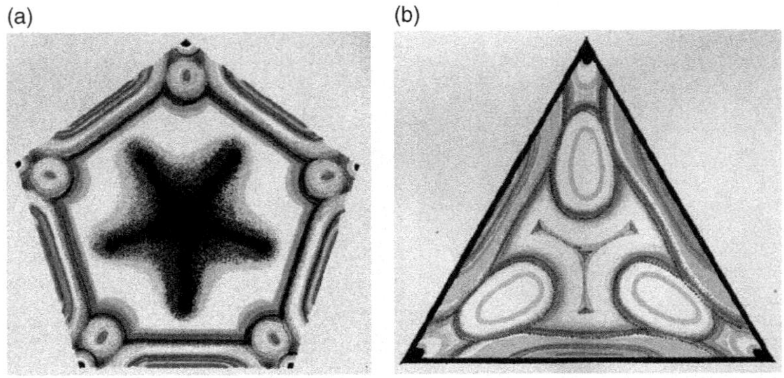

Fig. 5.7 Examples of the coloured images of strained glasses, showing the 'gorgeous entanglements of colours' when polarized light is passed through them (Campbell and Garnett, 1882). Reproduced in colour in the colour plate section.

his first papers 'On the Theory of Rolling Curves' (1849) and 'The Equilibrium of Elastic Solids' (1850) in the *Transactions of the Royal Society of Edinburgh.*

Maxwell went up to Trinity College, Cambridge, in 1850 and was to remain there until his appointment to the Chair of Natural Philosophy at Marischal College, Aberdeen, in 1856. In a letter of 1852, Forbes wrote to William Whewell, Master of Trinity College:

> Pray do not suppose that . . . I am not aware of his exceeding uncouthness, as well Mathematical as in other respects . . . I thought the Society and Drill of Cambridge the only chance of taming him and much advised his going.

In his obituary of Maxwell, his great friend Peter Guthrie Tait wrote:

> . . . he brought to Cambridge in the autumn of 1850, a mass of knowledge which was really immense for so young a man, but in a state of disorder appalling to his methodical private tutor.

Maxwell's Quantitative Theory of Colour Mixing

In 1855, Maxwell was awarded a Fellowship at Trinity College. Among his many interests, he studied the composition of light by means of his colour top, which was central to his first major assault upon the quantitative theory of colour (Fig. 5.8).

Maxwell adopted the Young–Helmholtz theory of three primary lights or colours, which he took to be red, green and blue, following the scheme advocated by Helmholtz. Three coloured paper discs with these colours could be clamped to the wooden disc of the top in such a way as to allow different amounts of the primary colours to be mixed when the top was spun (Fig. 5.9(a)). Note his appreciation of the use of the persistence of vision in accomplishing the mixture of lights with this simple apparatus. A smaller central disc contained the colour sample that was to be matched by adding together different amounts of the primary colours. To eliminate the effect of the different brightnesses of the central disc and the outer colours, different amounts of black could be added to the central disc. Maxwell demonstrated that all colours could be synthesized by different combinations of these three primary lights, red, green and blue. He also distinguished

Fig. 5.8 Maxwell aged 24 with his colour top. Courtesy of the James Clerk Maxwell Foundation.

clearly between the results of mixing lights of different colours and mixing pigments. The former are associated with the addition of lights, or colours, while the latter are associated with the absorption properties of the pigments. For example, he confirmed Helmholtz's discovery that mixing blue and yellow light does not produce green, but rather 'a pinkish tint'.

By 1855, Maxwell had adopted the three-colour receptor theory of Young and Helmholtz and determined experimentally colour equations that quantified how much of each primary colour was necessary to create any particular colour. All colours could then be represented on a colour triangle in which the distance from the corners indicated how much of each primary colour has to be mixed. Figure 5.9(b) shows Maxwell's original version of the colour triangle. Notice that these colour diagrams are projections onto a two-dimensional plane of three-dimensional colour space. He carried out his colour-matching experiments with many independent observers and found little variation between them. These experiments also suggested an explanation for colour blindness in that, if one of the three sets of receptors in the eye were not present, the subject would be unable to discriminate certain colours. Maxwell's pioneering efforts were to be considerably refined over the years, Fig. 5.9(c) shows a modern version of the Maxwell's colour triangle, while Fig. 5.9(d) shows the modern standard CIE (Commission Internationale d'Éclairage) diagram.

Maxwell at Marischal College, Aberdeen

Maxwell accepted the post of Professor of Natural Philosophy at Marischal College, Aberdeen, in 1856, to be closer to his family and the estate at Glenlair. Maxwell's five years at Aberdeen were among the most innovative and important periods of his life. In 1857, he won the prestigious

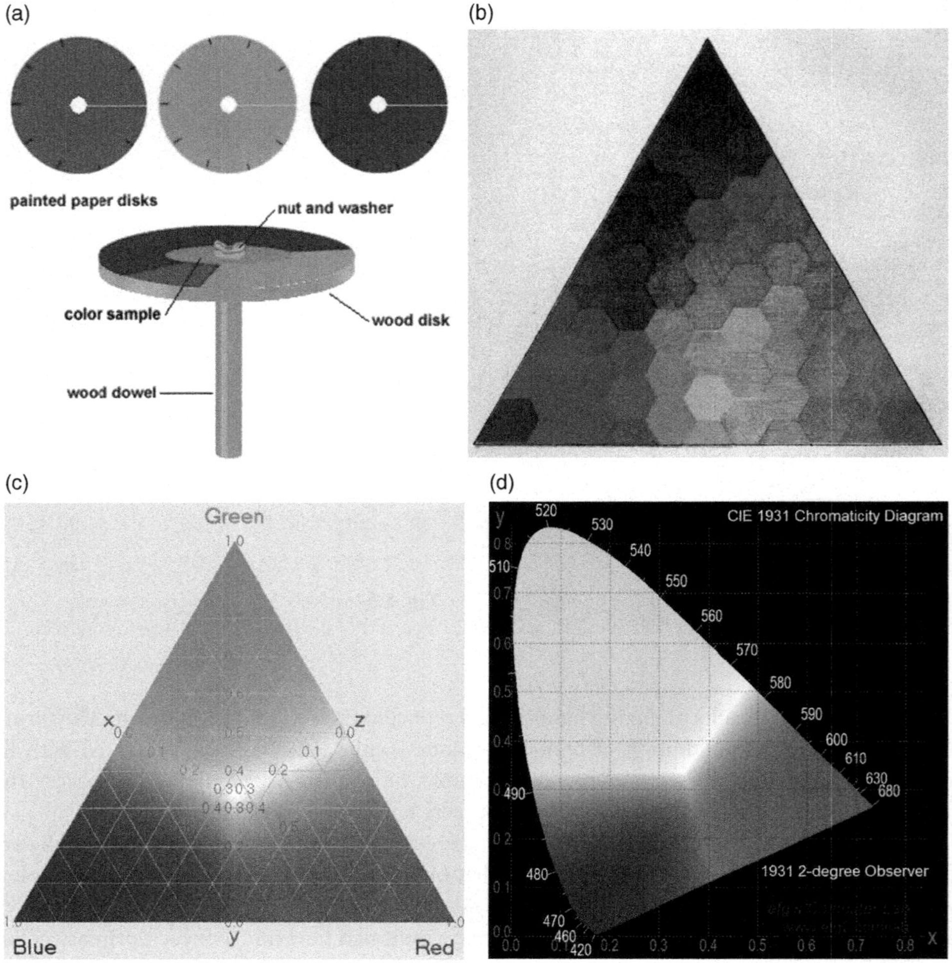

Fig. 5.9 (a) Illustrating the three coloured discs used in Maxwell's colour top (<http://www.handprint.com/HP/WCL/colortop.html>). (b) Maxwell's original version of the colour triangle (Campbell and Garnett 1882). (c) Comparison with a modern version, showing quantitatively the proportions of different primary colours needed to synthesize those within the triangle. The picture is rotated 120° clockwise with respect to (b) (see: <http://www.handprint.com/HP/WCL/colortop.html>). (d) The CIE (Commission Internationale d'Éclairage) diagram. The pure colours in the white light spectrum are shown around the perimeter of the diagram from 420 to 680 nm. Reproduced in colour in the colour plate section.

Adams Prize for his work on 'The Motion of Saturn's Rings'. This was also the period of his papers on 'The Dynamical Top' (1857) and 'The Theory of Colours' (1857, 1859–60). He was also developing his ideas on electromagnetism, which were to reach their culmination over the succeeding five years with the formulation of what we now call Maxwell's equations.

Maxwell realized that he would obtain more precise results on colour mixing if the amount of each colour could be more accurately determined. To this end, he devised a series of 'light boxes'

Fig. 5.10 Maxwell's light box constructed by the firm of Smith and Ramage of Aberdeen. Courtesy of the Cavendish Laboratory.

which enabled different amounts of the three primary lights to be mixed. The perfected light box, constructed by the firm of Smith and Ramage of Aberdeen, was portable and so could be used as a tool to study the colour sensitivities of many subjects (Fig. 5.10).

To understand how the light box works, let us first use it backwards (Fig. 5.11(a)). If white light is shone through the eyepiece at A, on passing through the prisms and being reflected

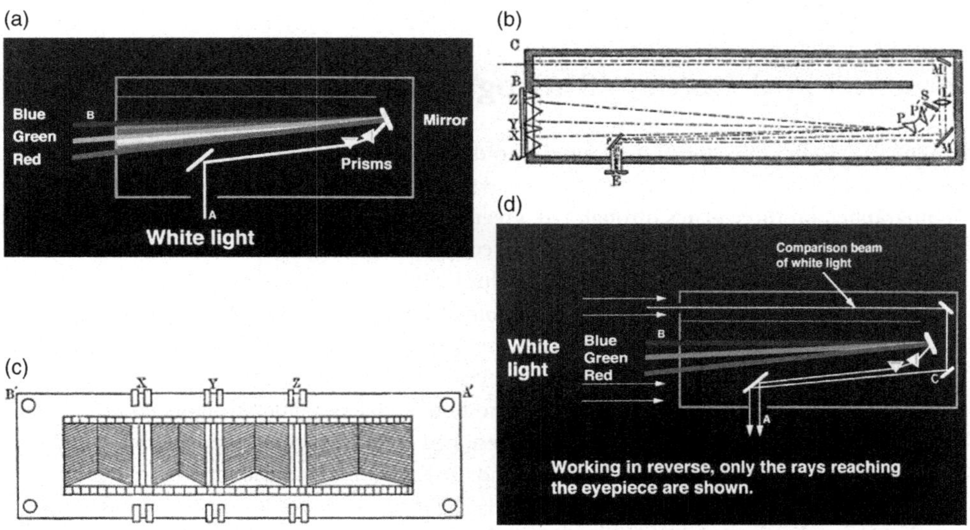

Fig. 5.11 (a) Maxwell's light box working backwards. The diagram illustrates the dispersion of white light when it is shone through the eyepiece onto the reflecting mirror and through the pair of dispersing prisms. (b) The diagram from Maxwell's paper showing the layout of the optics of the colour box. (c) The adjustable slits which allow the precise mixing of different amounts of red, green, and blue lights by measuring precisely the widths of the X, Y and Z slits. (d) Maxwell's light box working forwards. The blue, green and red regions of the spectrum are focused at A. The amount of each of these lights is precisely varied by changing the width of the entrance slits. Reproduced in colour in the colour plate section.

by the mirror, the normal spectrum of colours is produced at B. At B, there are three adjustable slits which allow different amounts of blue, green, and red light to be transmitted (Fig. 5.11(b) and (c)).

If we now shine parallel white light onto the slits at B, only the blue, green, and red colours are combined at A (Fig. 5.11(d)). The amount of each colour could be precisely measured from the width of the slits at B. In addition, a comparison with a white light beam could be viewed at A. To obtain an intense enough light source, bright sunlight was shone onto white paper to provide a white light source.

Maxwell used the light box to study many different aspects of colour vision for large samples of subjects. The experiments gave precise information about the composition of different colours and provided the forerunners of the modern chromaticity, or CIE, diagrams which define how different colours can be synthesized from chosen primary colours. These convey information about the different variables of colour vision, namely, the spectral colour and the degree of saturation. An example of such a diagram is shown in Fig. 5.9(d).

Maxwell was awarded the Rumford Medal of the Royal Society of London in 1860 for his researches in colour vision. He continued using light boxes after he moved to King's College London in 1860. He designed and had built an eight-foot light box which was housed in a large garret which ran the whole length of his house at 8 Palace Gardens Terrace, Kensington, in London. Campbell and Garrett noted that

> When experimenting at the window with the colour-box ... he excited the wonder of his neighbours, who thought him mad to spend so many hours in staring into a coffin.

Maxwell and Colour 'Photography'

In 1861, Maxwell created the first coloured 'photograph'. The story is one of imagination and good fortune. A tartan ribbon with bright colours was photographed by Thomas Sutton, a professional photographer, on three plates through red, green and blue-violet filters. The filters were created by filling a glass trough with differently coloured liquids. Three positive plates, one for each filter, were produced and these were projected through the same filters onto a screen using three 'magic lanterns'. When these images were combined, a remarkably good full-coloured image was observed on the screen (Fig. 5.12). This represented the invention of the additive three-colour separation technique employed in printing, television screens and so on.

Where the luck came in was that, while the blue and green filtered lights could be recorded on the emulsions available at the time, these had essentially zero response in the red filter. A century later, it was shown by R. M. Evans and the Kodak Laboratories that there was an ultraviolet leak in the red filter so that it transmitted light in the 320–400 nm waveband. It turned out that the red dye in the ribbon reflected ultraviolet light and so the ultraviolet light produced the red negative. As a result, when projected, the correct colour balance was obtained. As remarked by Dougal and Forfar, this was an example of 'false colour imaging', far ahead of its time. Thus, Maxwell's remarkable demonstration at the Royal Institution on 17 May 1861 represented the invention of both the three-colour separation technique and false colour imaging.

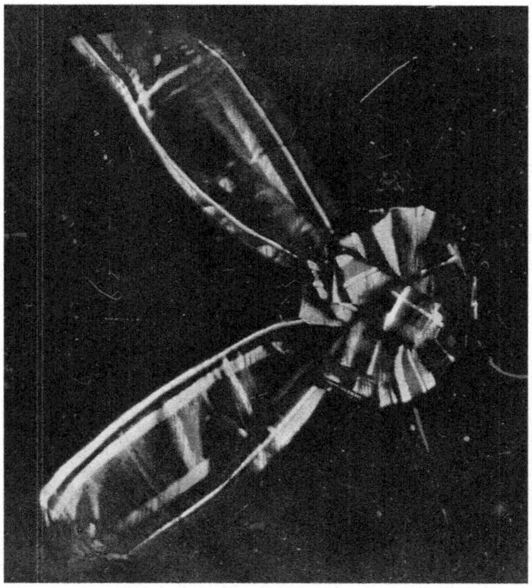

Fig. 5.12 A reconstruction of Maxwell's three-colour image of a brightly coloured tartan ribbon created from copies of the three original plates by Dr. Peter Reid. Intentionally, blemishes on the copy slides have not been removed. Courtesy of and copyright Dr. Peter Reid. The copy plates are the property of the James Clerk Maxwell Foundation. Reproduced in colour in the colour plate section.

Maxwell's Colour Legacy

Maxwell's studies provided the basis for the present understanding of the processes of colour vision and also for the quantitative measurement of colour that underpins a huge variety of modern manufacturing and service industries. The light sensors in the retina are the rods and cones. The rods are very sensitive and can register the arrival of individual photons of light, but have no colour discrimination. The cones are less sensitive but are sensitive to light centred on red, green and blue wavelengths. The important lesson from Maxwell's colour-mixing experiments was that it is only when monochromatic light is used that there is a unique relation between colour and wavelength. Any particular colour can be produced in a number of different ways provided the same total signals are applied to the three different cone types.

Maxwell understood that pigments are in a sense the opposite of lights. They absorb some wavelengths and reflect others. For example, suppose we shine white light onto a pigment which absorbs strongly in the central region of the optical spectrum. The light we see reflected consists of the red and blue ends of the spectrum, resulting in the colour magenta. This understanding of the difference between the mixing of lights and the mixing of pigments also solved the 'green problem'. Blue and yellow pigments absorb the ends of the spectrum and so, when they are mixed, the residual reflected light is centred about green wavelengths.

What we actually perceive is determined by many other factors – is the surface smooth or rough, is it highly reflective or polarizing, and so on? In addition, there are many other physiological and psychological issues, which would take us far beyond the scope of this chapter. Just as one example, the blue cones are rather sparsely distributed on the retina and this can fool the eye and brain if an object is observed with low angular resolution. The brain does an enormous amount of image processing to reconstruct what we perceive. A remarkable example is provided by John Mollon, in which a black wavy line is superimposed on the interface between white and

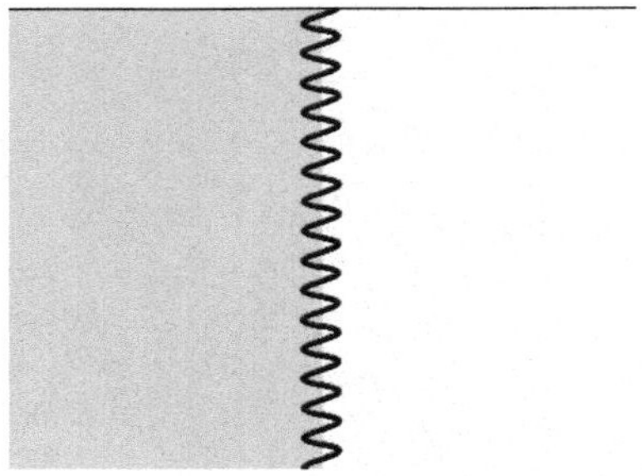

Fig. 5.13 Illustrating how the eye can be fooled in interpreting visual data. As the image is observed further and further from the page, the yellow and white colours seem to leak into the undulations of the wave pattern. Reproduced in colour in the colour plate section.

Fig. 5.14 The statue of Maxwell by Alexander Stoddart in George Street, Edinburgh.

yellow squares (see Fig. 5.13). As the diagram is moved further away from the eye, the yellow colour seems to leak into the white region and the white into the yellow region in the vicinity of black line.

Maxwell continued his studies of colour and optical phenomena during his subsequent career at King's College, London, at his home at Glenlair and at the Cavendish Laboratory, Cambridge. From about 1860 onwards was the period of his development of his deep insights into electromagnetism, the unification of light and electromagnetism, thermodynamics and statistical mechanics, for which he is rightly best remembered. It was characteristic of him, however, that he brought to the quantitative study of colour the same imagination and experimental skill which he displayed in all his other researches. It is remarkable that he was able to maintain his interests over such a vast range of the physical sciences and that he continued to make fundamental contributions to all of them throughout his career.

It is touching that the wonderful statue of Maxwell by Alexander Stoddart in George Street, Edinburgh, inaugurated in 2008, shows Maxwell in his later years contemplating his colour wheel in the company of his dog, Toby (Fig. 5.14). One senses a fondness for his involvement with light and colour through the greater part of his lifetime, from his earliest years at Glenlair, to his twilight years in Cambridge as the first Cavendish Professor of Experimental Physics.

CHAPTER 6

Maxwell and the Rings of Saturn

ANDREW WHITAKER

Maxwell as a Problem Solver

Maxwell's usual approach to his research work in physics was to devote very substantial attention, often over a very long period, to areas that he considered to be of major importance and ripe for intensive study. He would publish a series of substantial papers, usually separated in times of publication by several years, and clearly these papers were a result of sustained and deep reflection.

In the case of his most important work, that on electromagnetism and that on kinetic theory, the measure of his success was that his work was the basis of our steadily evolving understanding of these areas that have dominated much of physics up to the present day. While the other area of science to which he perhaps devoted most attention, colour vision, would naturally be considered less centrally important, it was still a subject of much concern to him, to which he devoted a great deal of thought and effort, both conceptually and experimentally. All three areas are, of course, discussed in detail in the appropriate chapters of this book.

Certainly Maxwell illuminated many other areas of science, and published papers on disparate topics outside his main interests.[1] Examples in this book would include the chapters of Keith Moffatt and John Rowlinson (Chapters 12 and 8 respectively). But it would be wrong to think of these works as problem solving for its own sake. Rather they were areas where perhaps localized questions remained in his view of the edifice of science, questions which he felt should be answered, and which he succeeded in answering.

In discussing Maxwell's work, what may be regarded as labours of 'duty' should not be ignored. Leaving aside the important areas of teaching, both inside and outside academia, and general administration, we should first mention his editing[2] of the electrical researches of Henry Cavendish. Cavendish, who lived from 1731 to 1810, carried out highly important research in many areas of science, including gravitation and the composition of air, but in particular he left 20 packets of material on mathematical and experimental electricity. Cavendish was the great uncle of the Duke of Devonshire, who was Chancellor of the University of Cambridge and provided the funds for the Cavendish Laboratory. As first holder of the Cavendish Chair, Maxwell may have felt something of a moral obligation to edit this material, but it was a mammoth task, taking up much of his time for the last seven years of his life, and entailing much correspondence to elucidate Cavendish's notation; he actually repeated many of the experiments Cavendish carried out with more up-to-date equipment.

Another example of this 'duty' work would be his participation[3,4] in the development of a coherent set of electrical and magnetic units. With Fleeming Jenkin, he wrote a paper for the British Association in 1865, which included a recommendation for a complete set of units. He also did an enormous amount of work on the production of a standard of electrical resistance, also for the British Association. At the end of the 1860s, he put a great deal of experimental effort, yet again on behalf of the British Association, into obtaining a value for the ratio of the electrostatic and electromagnetic unit of electricity; in modern terms, this would be described as checking that the velocity of electromagnetic waves predicted by his own theory of electromagnetism agreed with that found experimentally for light.

Though Maxwell found these labours of 'duty' interesting, it is natural to regret that, particularly in such a short research career, they took away so much time from his personal researches.

In a different category altogether is the work of the present chapter, that of studying the rings of Saturn.[5] This was a task suggested by others, the setters, in fact, of the 1856 Adams Prize of the University of Cambridge. While there is no doubt that Maxwell found the problem fascinating – he writes that[6] the rings are 'the most remarkable bodies in the heavens', except, he suggests, the spiral nebulae – it does not seem that it was an area of study to which he wanted to devote a substantial or continuing part of his scientific life.

Indeed, in their excellent collection of documents concerning Maxwell's work on the rings, Brush, Everitt and Garber[7] (BEG) comment that[8] 'Maxwell had shown no curiosity about astronomical problems as an area of research while either a student at Cambridge or a Fellow of Trinity College'. It is true that, after the award of the Prize, Maxwell spent a very considerable amount of time following up the initial work, partly to complete his task, as he saw it, and to prepare the work for publication, but also because, as we shall see later, this further work stimulated, or possibly was stimulated by, his initial work on kinetic theory. But, as BEG note,[9] once, in his kinetic theory study, he had reached the stage of replacing collisions between gas molecules represented by hard spheres by 'encounters' between molecules represented by centres of force, 'as far as we know, he never returned to the problem of Saturn's rings'. Indeed, they suggest[10] that Maxwell's study of Saturn's rings 'is a case study in the production of first-rate mathematical physics for reasons that lie entirely outside the problem's scientific merit'.

Thus we may take it that Maxwell's work on the rings of Saturn should be regarded as an exercise in problem-solving, rather than work carried out for its own interest. Why Maxwell carried it out, to what extent he may have regarded the work as some kind of 'duty' and, of course, how he carried out the actual work, we shall explore later in this chapter.

John Couch Adams and the Discovery of Neptune

John Adams[11] (Fig. 6.1(a)) was born in 1819, and, though his parents were not well-off, they recognized his great interest in astronomy, and a small inheritance enabled them to send him to Cambridge, where he was a student at St. John's College. In 1843, he graduated as Senior Wrangler, or top of his year, and also won the first and higher of the two yearly Smith's Prizes, probably a better guide than the Senior Wranglership to mathematical creativity. He was also elected to a fellowship at St. John's, though the fellowship would lapse in 1852.

We should also introduce two other characters in the story. George Biddell Airy[12] (Fig. 6.2(a)) was born in 1801. Like Adams, he had been both Senior Wrangler and first Smith's Prizeman, in his case in 1823, and three years later he became Lucasian Professor of Mathematics also at Cambridge. This post was poorly paid, but in 1828 he was appointed to the much better paid posts of Director of the Cambridge Observatory and Plumian Professor of Astronomy and Experimental Philosophy. Then, in 1836, he moved to Greenwich to become Astronomer Royal.

At that date, his positions at Cambridge were taken by James Challis[13] (Fig. 6.2(b)), who was two years younger than Airy, and had also been two years behind him in becoming Senior Wrangler and first Smith's Prizeman. Thus, in the years following Adams' graduation, Airy and Challis were in the two most powerful positions in observational astronomy in Britain.

In 1841, even before his graduation, Adams[14,15] had been struck by the deviations between the path of Uranus predicted using Newton's laws of motion and gravitation, and the actual path of the planet. The French astronomer, Alexis Bouvard, had suggested that these deviations demonstrated the presence of an as yet undiscovered further planet, and Adams determined to locate this planet. From that date, he invested an enormous amount of time in mathematical calculations on

Fig. 6.1 Discoverers of Neptune: (a) John Couch Adams. From Robert Ball, *Great Astronomers* (Pitman, London, 1920). (b) Urbain Le Verrier. From the *Illustrated London News*.

(a) (b)

Fig. 6.2 The main figures in British astronomy: (a) George Biddell Airy (*London Illustrated News* 4 January 1868). (b) James Challis. Courtesy of Institute of Astronomy, Cambridge.

this task. By February 1844, he had determined an approximate location of the proposed planet, and he then asked Challis to request from Airy some more accurate data on the orbit of Uranus in order to make a more accurate prediction. By September 1845, he was convinced that he knew the position, mass and orbit of this new planet.

In September he informed Challis of his work, obviously in the hope that Challis would use the facilities of his observatory to search for the proposed planet, but Challis was clearly unconvinced that the project would be worthwhile and did not do so, though he did mention to Airy that Adams had performed these calculations. Then, in October, Adams twice called on Airy in Greenwich to tell him of his ideas, but he had no appointment and Airy was not at home.

Adams left him a brief report of his results, and it is this report that Sheehan, Kollerstrom and Waff[16] (SKW), in their analysis of the events, call 'the crucial document, the main basis of the British discovery claim'. It gave the form of the orbit of the hypothesized planet. If an astronomer had looked at the indicated position on the appropriate date, SKW say that the planet would have been found about two degrees away.

Airy replied to Adams, asking for clarification as to whether Adams' calculations might explain another anomaly in the orbit of Uranus; it was slightly further from the Sun than Newton's laws would predict. This was particularly important to Airy, because he had put in a considerable effort ascertaining this observational fact. Perhaps if Adams had replied, Airy might have been encouraged to organize a search for the new planet, and indeed he did commence a letter to Airy but he did not complete it. It is not quite clear why he failed to do so; it may be that he was more concerned with some important work he was carrying out on comets, and he was also very busy with his teaching.

Adams had still not published his predictions when a French mathematician, working at the Paris Observatory, Urbain Le Verrier[17] (Fig. 6.1(b)), who had been carrying out the same type of calculations as Adams since June 1845, announced his findings in a paper published in the journal of the French Academy of Sciences on 1 June 1846. He predicted the position at which a new planet would be found at a particular longitude on 1 January 1847.

The journal reached Cambridge later in the month, and Airy would have been one of the very first people to see Le Verrier's paper. Immediately he would have recalled the very similar conclusions that Adams had left at his house the previous year. At this period, relations between Britain and France in astronomy, as in many other fields, were competitive, and Airy must have wondered if he could salvage some glory for Britain from the situation, and perhaps at the same time salve his conscience for his lack of response to Adams.

Merely to announce that Adams had performed the same calculations as Le Verrier, even to suggest that he had done them earlier, might seem a little limp. Much better would be for Cambridge astronomers actually to locate the new planet and then for the work of Adams and this discovery to be announced to the world as a package. So Airy wrote to Le Verrier, asking him the same question that he had asked Adams about the orbit of Uranus, but telling him nothing about the work of Adams. Airy was able to stimulate Challis into carrying out an exhaustive search, commencing on 29 July, in the appropriate area. It has to be admitted, though, that the area surveyed was that indicated by Le Verrier not by Adams, and Adams himself was meanwhile revising his calculations. Again no word of this was communicated to France.

Challis worked exceptionally hard, surveying 3000 heavenly objects in the hope that one of them would turn out to be a planet not a star. Unfortunately he did not have a star map, which would have allowed him to spot any 'intruder' with the minimum of effort. Rather, he had to record the position of each object on two successive nights; stars would remain in the same position in each view, while a planet should move a short distance. Unfortunately practice proved a little more difficult than theory. On 4 and 12 August, he did note an object that would eventually be identified as Neptune, but failed to carry out the required immediate comparison. There was actually another potential clue; Le Verrier had suggested that the planet might have a disc and so be easily recognizable in a telescope. On 29 September Challis did make a highly tentative identification along these lines, but it was already too late.

Le Verrier had actually found it no easier than Adams to persuade his countrymen to search for his proposed planet. On 31 August he published refined predictions and then, on 18 September, he sent his results to Johann Galle of the Berlin Observatory asking him to look for the new planet. The letter did not arrive until 23 September, and on that very evening Galle and his student Heinrich d'Arrest commenced to search for the planet. They had the massive advantage over Challis of having a star map. They were thus able to look directly for a heavenly object not on the map, and it took them a mere half an hour to find what they were looking for, a small blue disc observed within 1° of its predicted location. As a check, they observed the same object the following night, and indeed it had moved slightly through the stellar background; a new planet it decidedly was.

For Airy, it was essentially back to Plan A. He announced that he had received notice of the predicted position from Adams the previous year. Not surprisingly this created something of a furore. The British astronomical community were angry that Airy had not acted on Adams' predictions; the French community was angry that he was trying to obtain British credit for their own discovery!

Adams himself finally published a paper on the topic, which was presented in a meeting of the Royal Astronomical Society on 14 November 1846, and at this meeting Challis and Airy also explained their actions. Adams himself merely reported the facts and made no attempt to claim priority. In this paper he said that:

I mention these dates merely to show that my results were arrived at independently, and previously to the publication of those of M. Le Verrier, and not with the intention of interfering with his just claims to the honours of the discovery; for there is no doubt that his researches were first published to the world, and led to the actual discovery of the planet by Dr. Galle, so that the facts stated above cannot detract, in the slightest degree, from the credit due to M. Le Verrier.

Adams also made no criticism of Airy or Challis. He was to say that:

I could not expect however that practical astronomers, who were already fully occupied with important labours, would feel as much confidence in the results of my investigations, as I myself did.

Airy and Challis, it may be said, were much more willing to claim credit for Adams. Their position, and it is probably true that it has held sway in the English-speaking world, was that Adams and Le Verrier carried out the same work and deserved equal amounts of credit. Indeed, it was possible to argue that Adams' initial conclusions came before those of Le Verrier, and it was Adams who should be awarded the prime place of honour.

And, one might think rather cheekily, Challis suggested a name for the new plant – Oceanus. Galle, who one think had more right to offer a name, suggested Janus. Le Verrier had originally suggested the name of Neptune, but soon changed his mind and decided the planet should be called – Le Verrier! In this desire he was backed up by the Director of his Observatory, the well-known astronomer François Arago,[18] and French almanacs actually printed the name of the new planet as Leverrier, while, presumably as a sop to the British, they renamed Uranus as Herschel, after its discoverer Sir William Herschel. (To be fair, the latter was not a new suggestion.) However the Russian astronomer, Friedrich von Struve, argued strongly for the name of Neptune, which was soon universally accepted.

In the eyes of the French, Le Verrier deserved all the credit. Arago wrote that:

In the eyes of all impartial men, this discovery will remain one of the most magnificent triumphs of theoretical astronomy, one of the glories of the Académie and one of the most beautiful distinctions of our country.

An amusing French cartoon[19] (Fig. 6.3) shows Adams looking though his telescope for Neptune, conspicuously in totally the wrong direction, before finding it – in the work of Le Verrier! (It may be noted, though, that, even in his homeland, Le Verrier[17] was not the most popular of men. A contemporary Frenchman is reported to have written that: 'I do not know whether M. Le Verrier is actually the most detestable man in France, but I am quite certain that he is the most detested'. He was Director of the Paris Observatory from 1854 to 1870, but in that year his arrogance and unpopularity were such that he was removed and replaced by Charle-Eugène Delaunay. He only returned to the post when Delaunay accidentally drowned in 1873, from which date Le Verrier was again Director, this time under a steering committee, until his own death in 1877.)

In Britain, of course, things were totally different. It is true that Le Verrier was given the premier award of the Royal Society, the Copley Medal[20] for 1846: 'For his investigations relative to the disturbances of Uranus by which he proved the existence and predicted the place of the new Planet; the Council considering such prediction confirmed as it was by the immediate discovery of the Planet to be one of the proudest triumphs of modern analysis applied to the Newtonian Theory

Fig. 6.3 A French cartoon deriding Adams' achievements. In the first picture and wearing a dunce's cap, he is searching for Neptune, conspicuously in totally the wrong direction! In the second, still wearing the cap he finds it – in the writings of Le Verrier!

of Gravitation'. Adams was awarded the same medal two years later: 'For his investigations relative to the disturbances of Uranus and his application of the inverse problem of perturbations thereto'. No mention of Neptune for Adams!

However, for many in Britain, particularly those with strong Cambridge connections, Adams was to be regarded as perhaps the greatest mathematician since Newton. He was offered, but turned down, a knighthood as early as 1847. And it was in that spirit that members of St. John's College endowed the Adams Prize, which was approved by the Senate in 1848. Originally it was restricted to Cambridge graduates, and was worth £130 – equivalent to perhaps £13,000 today. (Today it is open to any mathematician working in England, usually under 40, and is worth around £14,000, though a third of this is paid to the institution of the candidate for research expenses.) By comparison, the published salary for Maxwell's position in Aberdeen was £350.[21]

The purpose of this lengthy preamble on the genesis of the prize is to make the point that when, in the mid 1850s, Maxwell worked on his entry for the Adams Prize, the prize itself was by no means a remembrance of a hero of yesteryear, and certainly not merely an opportunity for a private competition among Cambridge men, but rather a continuation of the attempt to promote British and particularly Cambridge mathematics and astronomy. Though Adams himself took no part in the administration of the prize, both Challis and Airy were central in the task, as was William Thomson, later Lord Kelvin, who was Belfast-born and Glasgow-based, but retained strong links with Cambridge. All these not only assisted in choosing the topic and judging the work, but did their best to improve the prize-winning essay further and then to publicize it. Following final publication of Maxwell's work, it was Airy[22] who produced a highly laudatory review; BEG comment[23] that this review paid particular attention to refuting any suggestion that previous workers, particularly the American Benjamin Peirce, had pre-empted Maxwell in some of his conclusions.

BEG remark[24] that it is 'ironic' that Challis, who they say was 'generally seen as having frustrated Adams' work on Neptune' was in a position to control the administration of the prize named after him. From the point of view of those supporting the traditional idea that Challis and Airy were culpably and totally responsible for the failure to follow up Adams' ideas, one would agree. If, though, one is more cynical and feels that, at least to an extent, the two 'took a hit' in accepting the full blame so as to achieve glory for Adams and Cambridge, it is entirely natural that they should be determined to push home whatever advantage had been gained.

Before we move onto Maxwell, we briefly review the later lives of our principals. Adams[11,12,19] had a long and successful career and, at least in Britain, was highly respected for his retiring and amiable personality as well as for his astronomical advances. Not being ordained, he had to leave his Fellowship at St. John's College in 1852, but he became a Fellow of Pembroke College in the following year, retaining this fellowship until his death in 1892. He was briefly Professor of Mathematics at St. Andrews, but then became Lowndean Professor of Astronomy and Geometry at Cambridge in 1855, and then succeeded Challis as Director of the Cambridge Observatory in 1861, though he was able to arrange for the actual observing to be done by others, concentrating on theoretical work himself.

He made many contributions to astronomy, of which we mention only his demonstration that, in his theory of the orbit of the Moon, the famous French mathematician, Pierre-Simon Laplace, had left out some terms which should not have been neglected. By including them, Adams was able to decrease the disagreement between theory and observation by a half, the remainder now known to be due to the interaction between the Moon and the oceans on Earth. It will be readily understood that his correction of Laplace by no means improved the view of Adams in France.

Airy[12] gained many important positions, including becoming President of the Royal Society from 1871 to 1873. However he was generally felt to be more of an administrator than a scientist, to be snobbish and to be dictatorial with his staff at the Royal Observatory. To physics students, he is known for the Airy disc and pattern,[25] the diffraction pattern resulting from a uniformly illuminated circular aperture.

Challis[13] retained the Plumian Professorship until his death in 1892, but he was an ordained clergyman, and in his latter years he was more interested in theology, or the relationship between theology and science, than science itself, and his later writings were of extremely questionable value. Eggen[13] comments that: 'At a later time, or under less amiable circumstances, he would have been branded a charlatan'. The reputation of neither Airy nor Challis perhaps ever recovered from the Neptune affair.

Le Verrier[17] continued to be revered in France for his research though, as has been said, very much disliked for his personality. Apart from the discovery of Neptune, his most interesting work was the similar idea of attributing the anomalies in the orbit of Mercury either to the existence of a planet that he called Vulcan, which had to be nearer to the Sun than Mercury, or alternatively to an asteroid belt, so near the Sun that it is invisible. Search for such objects was unsuccessful, and in the following century, Einstein was to explain the anomalies as effects of general relativity.

How is the Neptune controversy viewed today? Investigation was hampered for 30 years or so by the fact that the Adams–Airy documents went missing in the mid 1960s. It was suspected that Olin Eggen,[26] who had worked as Chief Assistant to the Astronomer Royal from 1956 to 1961, and had legitimately borrowed the material during that period, had taken it with him when he moved to the United States, then to Australia, and finally to Chile, where he was Director of the Cerro Tolelo Inter-American Observatory from 1977 until his death in 1998. After his death, the material was indeed discovered and returned to Greenwich, along with two large tea chests of other material he had likewise 'borrowed'. Dennis Rawlins,[27] the fiercest critic of the British lionizing of Adams, has pointed out the convenience for this process of the documents being lost. However, it seems that Eggen intended eventually to write a book on the discovery of Neptune, and, like any author, recognized the merits of having all the material at hand, no matter who else might want it! It will be noted that he wrote the accounts of both Airy and Challis cited here.[12,13]

When this material was examined, it became clear that the various notes, including the crucial one exchanged between Adams and Airy, certainly existed, but the meaning of the documents was equivocal. A date on one document seemed suspicious, and the versions published in the 1840s had been subtly edited. With the aid of a Royal Society grant, Nicholas Kollerstrom prepared a website[28] giving the full information and, with William Sheehan and Craig B. Waff, he also published a paper (SKW)[16] provocatively called 'The case of the pilfered planet: Did the British steal Neptune?' However their strictures on the British are comparatively mild. They write that: 'We can affirm that Adams does not deserve equal credit with Le Verrier for the discovery of Neptune. That credit belongs only to the person who succeeded both in predicting the planet's place and in convincing astronomers to search for it. The achievement was Le Verrier's alone.'

The approach of Rawlins,[27] admittedly a maverick of his profession at odds with many of the established figures, is far more aggressive. He points out that Adams' predictions varied substantially between his visits to Challis and Airy in mid 1845 and his final publication a month after the discovery of the new planet. His thesis is that, at least until Galle's work, Adams was labouring to improve his predictions, though not especially successfully. Only after the discovery was he able to finalize his work. The suggestion is that dates were written on some of the central documents to fit in with the re-writing of history, according to which re-writing Adams made the correct predictions in 1845.

Although Rawlins considers his later work showed Adams to be 'in the front rank of history's mathematical astronomers', he believes that in the discovery of Neptune:

> his rôle was actually nil, and his behaviour has always been inexplicably murky. Thus Airy and-or Challis are not the villains but the (flawed) saints of the affair, in the sense that in order to semi-create and defend that needed great-Brit-mathematician rôle-model, they kept silent about or even obscured Adams' limitations – at terrible costs to their own reputations in history.

It is indeed possible to make much of the fact that, as mentioned before, Challis' search actually used Le Verrier's predictions rather than those of Adams. However we may also note[16] that the prediction of neither man was actually correct. Both greatly overestimated the distance of the planet from the Sun, and there was a measure of lucky timing that prediction agreed with observation.

The History of the Rings of Saturn up to 1855[29]

We now sketch the history of investigation of the rings of Saturn. Galileo was the first to observe structure around Saturn in 1610, but he interpreted it as an indication that Saturn consisted of three stars in contact with each other. The situation is indeed confusing, for when the Earth is well out of the plane of the rings, the general structure is obvious, but when it is in the plane of the rings, the rings themselves may be seen as a comparatively dark line across the planet and a comparatively light line across the rest of the sky, or even be invisible. Of course this tells us that the rings must be exceptionally thin.

It was Christian Huygens, in 1656, with a much improved telescope, who actually identified the structure as a ring surrounding the planet, with a diameter somewhat more than twice that of

Fig. 6.4 A view of the rings of Saturn (from Gribbin, John and Goodwin, Simon (1998). *Empire of the Sun: Planets and Moons of the Solar System*. Constable, London.)

Saturn (Fig. 6.4). Soon afterwards, Robert Hooke observed that ring and planet cast shadows on each other. This led Huygens and Edmund Halley to suggest that Saturn rotated in the plane of the rings and then, towards the end of the eighteenth century, by observation of dark spots on the surface of Saturn and others on the ring, William Herschel confirmed that the ring and the planet rotated practically together. In fact the planet rotated on its axis in 10 hours 29 minutes, while the rotation of the ring took about 3 minutes longer.

Huygens had hypothesized that the ring was a single solid structure, but in 1675 Jean Cassini observed that the ring was split into two concentric rings by a dark band, the inner ring being the brighter. The internal and external diameters of the outer ring were 150,000 and 170,000 miles, those of the inner ring were 114,000 and 147,000 miles, while the diameter of Saturn itself was 75,500 miles. By the mid 1850s, incidentally, the thickness of the rings system was estimated as less than 100 miles, while the mass of the rings was thought to be around 1 percent of that of the planet.

It was the nineteenth century before it was observed (or at least claimed by some) that the outer of these two rings was itself divided into two concentric rings, and indeed that, when the line of sight was inclined to the plane of the rings at a substantial angle, a series of dark elliptic curves could be seen, indicating that each ring is further broken up into still thinner rings.

There was considerable discussion on the nature of the rings. A popular view was that they were composed of a very large number of small rock-like satellites, though these views had little theoretical backing. Then, in 1787, Laplace applied some rigorous theoretical analysis. First he showed that a solid stationary ring would collapse under the gravitational attraction of Saturn. The stress would be so great that we can conceive of no material capable of sustaining it.

Laplace then considered the possibility of a rotating ring, for which the gravitational attraction would provide the necessary centripetal force (or, to express matters more physically, the gravitational attraction would be balanced by the centrifugal force). He was able to check that the calculated speed of the outer portion of the ring system agreed with the value determined by Herschel.

However, there were still two difficulties. First, for a solid ring, necessarily the angular speed of each part must be the same. But if one balances the magnitudes of the gravitational attraction and the centrifugal force for each part of the ring, one finds that, for each distance from the centre of the planet, a different angular speed is required. Clearly there is a contradiction, and the result is that the ring would inevitably break up. Second, the motion is unstable; any disturbance, even the most minute, from the usual orbit, such as will inevitably be caused by minor changes in the gravitational background, will result in ever-increasing divergence from the stable arrangement, and eventually the ring would come into contact with the planet.

Laplace concluded that, to allow different angular speeds, the rings must consist of a great number of concentric rings each rotating at the appropriate angular speed. To avoid the instability, he deduced that the rings were not homogeneous; their centres of gravity were displaced by some distance from their geometrical centres. Laplace concluded that: 'The several rings that surround the globe of Saturn are, consequently, made up of irregular solids of unequal size at different points in their circumferences.'

In his review of Maxwell's paper, Airy[22] remarked that: 'Probably no competent mathematician has ever read [Laplace's] paper without remarking the want of ground for the last conclusion.' Maxwell's own conclusion was surely correct, as Airy pointed out: 'If the rings were solid and uniform, their motion would be unstable, and they would be destroyed. But they are not destroyed, and their motion is stable: therefore they are either not uniform or not solid.'

Laplace also concluded that, for the attraction of the ring to neutralize its tendency to split up, it is necessary that its density should be $\frac{10}{13}$ that of the planet. As we shall see, numerous other factors demanded an enormously lower density for the ring.

It is interesting to consider these views of Laplace in the light of his own nebular theory of the origin of the solar system. In his proposal, material is spun off from the cooling, contracting and rotating Sun, and a ring of vapour is produced in what is now called an accretion disc. This may survive in the unlikely event that there is no irregularity, but much more likely is that there will be condensation of the disc material into a planet. Similarly a planet will have its own disc, which may condense into a moon. In other cases, though, the disc may remain. In the case of Saturn, of course, we must allow for both the ring and moons.

In the few years around the middle of the nineteenth century, there was a revival of interest, both observational and theoretical, in the rings of Saturn, which was to lead to the setting of the Adams Prize in 1855.

The main observational advance was that several astronomers – William Bond and his son George in America, and William Dawes in England – practically simultaneously observed an inner dark ring. This ring is partially transparent – the planet can be seen through it, and, since the planet is seen in all positions without refraction, it would seem that it cannot consist of a transparent gas or liquid. It could be a partially transparent solid ring or it could consist of unconnected particles.

Several interesting theoretical papers were also published. George Bond argued that a system of solid rings as proposed by Laplace would be unstable, but a fluid ring could solve all the problems faced by Laplace. Benjamin Peirce agreed that solids rings were not possible; he proposed that the ring was a stream of fluid, but this itself had to be stabilized by the gravitational influence of the moons of Saturn. Daniel Kirkwood regarded this dark ring as a proof of the truth of the nebular hypothesis.

Also very important in encouraging the Adams Prize examiners to make the rings of Saturn their prescribed topic, were various suggestions that the nature and dimensions of the rings were changing, although there was no general agreement on this. First, the claimed subdivisions between rings, existence of which we already said was controversial, were claimed by some astronomers to appear and disappear. More significantly, Otto von Struve, son of Friedrich who we met earlier, reviewed measurements of the rings from their first discovery, and claimed that the inner edge of the inner bright ring was decreasing in radius, while the total breadth of the ring system was increasing.

Another interesting paper was published in 1849 by Edward Rouche. He discussed the stability of a satellite under the tidal action of its planet. He showed that, if the planet and satellite have equal densities, the satellite will break up if it comes within 2.44 times the radius of the planet, and he noted that this distance is a little further than the *external* radius of the ring of Saturn.

It is tempting to view Maxwell's own work as, at least to an extent, a response to these papers which had been published in the previous few years, or, to put things the other way round, some of these papers as pre-empting some of Maxwell's conclusions. This would be misleading. To start with the work of Roche, this was published in an obscure journal, the *Academie des Sciences et Lettres de Montpellier*, and was not well known until after Maxwell's work was published. Maxwell did not know of the work of Bond and Peirce when he wrote his original prize entry, and, having been alerted to it by Challis, mentioned it only briefly in a footnote in the published edition. Though he felt the conclusions were interesting, he did not feel that the arguments were strong. And, as already mentioned, Airy[22] was extremely keen to rule out the possibility that Maxwell owed anything to Bond or Peirce.

As we shall see in the following section, the suggestion of von Struve that the form of the rings was changing in time was quite important in the initial suggestion of the topic for the Adams Prize. However, it became less prominent as the further discussions of the examiners progressed, and also in Maxwell's actual analysis.

Maxwell was, of course, very familiar with the work of Laplace, though, as we have seen, he did not accept all its conclusions. Indeed it is more accurate to think of Maxwell's work as a detailed study of the topic from first principles.

The Adams Prize of 1856

In the year under consideration,[30] it was actually the fourth time a topic for the Adams Prize had been set. It may be remarked that the topic was announced in March 1855, the essay had to be sent in to the examiners by 16 December 1856, and the prize was actually awarded in 1857, so there is often confusion in stating the year of the particular prize!

The conditions of the prize as it was set up were that it could be in pure mathematics, in astronomy, or in other branches of natural philosophy. However, as we shall see, in the early years there was a total bias to astronomy, presumably because Challis was in charge of the organization. In 1849, the topic was on the perturbations of Uranus and Neptune caused by the near-commensurability of their motions. The one entrant, Robert Peirson, was awarded the prize. In 1851, the topic was on the perturbations of the moon caused by Venus; the single entry was not of sufficient merit to be awarded the prize. Then, in 1853, the subject was Biela's Comet and there were no entries. It may be remarked that Airy was highly involved in setting the topic in these years.

Thus, for the topic to be set in 1855, Challis aimed for a subject that would attract more attempts, so he wished to pick a topic not requiring long astronomical calculations although, as it turned out, he was unsuccessful in both aims; Maxwell's attempt certainly had extremely long astronomical calculations! Challis's fellow examiners, as well as the Vice-Chancellor, Chairman *ex officio* but not actually participating in the task, were to be William Thomson, later Lord Kelvin, and Stephen Parkinson,[31] who was a Fellow of St. John's College, so that he was effectively the representative of the members of the college who had provided the finance.

Thomson and Parkinson had been rivals for the Senior Wranglership in 1845, and famously Parkinson had won, despite Thomson clearly having much more ability and justifiably going on to win the first Smith's Prize ahead of Parkinson. Parkinson was actually a very capable mathematician, becoming a Fellow of the Royal Society in 1870. He was a Fellow of St. John's from 1845, and President of the College from 1865 until his death in 1889. At St. John's, the Head of College is the Master, and the President deputizes for him when necessary and also acts as senior Fellow. I am grateful to Dr A.M. Nicholls, Librarian and former President of St. John's, for providing information both on this matter and on the present role of the College in the Adams Prize – it provides two adjudicators alongside two from the Faculty of Mathematics and one from the Royal Society.

In February 1855, Challis wrote to Thomson suggesting a number of topics. Two of them were very close to the topics set in 1849 and 1851, and another was on the aberration of light. The fourth was on the rings of Saturn. Challis made it clear that one of his motivations for suggesting this topic was the work of von Struve, with which he said he agreed. He felt the central point of any attempt would be to decide whether the perturbations of the moons of Saturn, the planets and the Sun could produce such a change in the form of the rings. He also suggested that it would be interesting to consider the effect of a resisting medium, as the motion of the planet about the Sun was coupled with the motion of the Solar System itself.

Thomson suggested three other possible topics, all rejected by Challis. The first concerned zodiacal light, and the perturbations of Mercury and Venus; it was rejected because Challis felt it would require long calculations. The second was the question of mechanical stability, rejected for being too general. The third was the elasticity of solids; Challis rejected this not only because it was rather remote from the usual concerns of Cambridge mathematics and might, in any case, require experiments, but also because he wished the topic not to be totally removed from astronomy – the latter point interesting since, as has been said, the range of topics allowed was much wider, and the topic in previous years had all been exclusively astronomical.

Thomson had by this time in the correspondence become interested in the rings of Saturn, so it was easy to settle on that topic. Candidates would be asked to consider the following possibilities: (a) the rings are rigid; (b) they are fluid 'or in part aeriform'; or (c) they consist of incoherent masses. In each case, candidates should aim to discover whether the conditions of stability are satisfied, and whether the bright rings and the recently discovered dark ring could be explained. From their analysis, candidates were asked to ascertain if any change of form of the rings could be explained. It will be seen that the question of von Struve's claim of change of form had moved from being the motivation for setting the topic to being tacked on at the end. Actually, while Thomson had changed the title from 'The Stability of the Motion of Saturn's Rings' to 'The Motion of Saturn's Rings', on the grounds that the motion might not be stable, Maxwell changed it back in his own work, certainly showing he stressed stability over change.

Candidates had a year and nine months to do the necessary work. It was an extremely busy period for Maxwell. For the first part of the period, from the end of March 1855, his thoughts must have been very much dominated by his efforts concerning his first work on electromagnetism. In April 1856, his father died, and, apart from the emotional strain, he had many duties at the family estate. Also, in February 1856, he had applied for the vacant Chair of Natural Philosophy at Marischal University in Aberdeen, and had to spend time requesting testimonials. He was appointed to the position and moved to Aberdeen in November; naturally he had to present an Inaugural Lecture as well as to prepare material for, and commence, teaching.[32]

Nevertheless, during this time, he found time to put a very great effort into preparing his essay for the Adams Prize. During this period, there does not seem to be much reference to this in his letters, at least those that have reached us. One reason for this was that, in a sense, the work was an examination. He could certainly not discuss the topic with those who could be most expected to have interesting ideas, Challis and Thomson, as they were his examiners. He did, in fact, write some letters of length to Thomson[33] in this period. As a newly appointed professor it was natural for him to seek advice from the more experienced man. He mentioned a variety of physical topics, and included a request for a testimonial, which was speedily provided, but he did not mention Saturn.

We do have two letters he wrote to others in which Saturn was mentioned. On 4 July 1856 he wrote[34] to Richard Litchfield, a College friend, saying that he was giving 'a portion of time to Saturn's Rings, which I find a stiff subject but curious'. At that time he was clearly working on the fluid ring.

Then, on 14 October 1856, he wrote[35] to Cecil Monroe, who had been elected to a fellowship at Trinity College Cambridge. At this time, Maxwell's thoughts were certainly not wholly on Saturn. He mentions that he is writing his Inaugural Lecture, which he describes as 'a solemn address or manifesto to the natural philosophers of the North', and his first mention of physics relates to a problem in magnetic induction. However, he then turns to Saturn's Rings, and writes down a fourth-order equation in time that he has reached, again for the fluid case. He states the form of the two large roots but demands that Monroe finds him the small ones. Monroe apparently was not able to help, and Maxwell does not state the roots in the prize essay.

The essay[36] was presumably handed in on the deadline, because the day after this deadline, on the 17 December 1856, the Vice-Chancellor passed it to Challis,[37] who had read it and had thought 'very favourably of it' by 31 December when he wrote to Thomson[38] to see if the latter was ready to receive it. Maxwell's submission was the only one, so Challis's wish for greater quantity of submissions was not borne out, though high quality was definitely there. Thomson and Challis both made comments on the manuscript, but all three examiners approved it, and Maxwell was awarded the Adams Prize[38] for 1856 in June 1857.

However, Maxwell was by no means finished. Challis and Thomson had raised issues on the work and had made some criticisms, and, even apart from that, he wished, as he saw it, to complete his work. Only now could he communicate with Thomson and Challis, and he wrote several letters, most of them very extensive, on Saturn, two to Thomson[39] in August 1857, and two to Thomson[40] and one to Challis[41] in November 1857. He was checking up on their criticisms – some of which were certainly justified, as he had made some mistakes – and reporting on his progress. He also mentioned his work on Saturn to several other correspondents. Clearly he was working very hard at the problem despite his professorial duties. By January 1858, when he wrote to Thomson,[42] his remarks on Saturn were to report a machine he was having made to illustrate a feature of its behaviour, which will be discussed shortly.

Then, in September 1858, he wrote to George Stokes[43] saying that he was finishing his essay on Saturn's Rings, but needed some information from Stokes on viscosity. Also in September 1858, he wrote to publishers MacMillan and Company[44] discussing how the figures should be arranged in his forthcoming book, and eventually it was published[5] the following year.

It is rather a peculiar result of the foregoing that nearly all the letters in which he describes himself as working hard on Saturn were written *after* he had been awarded the prize. It is not surprising that an excellent author[45] may come to the conclusion that the submission was due in December 1857 not 1856. Even the magisterial Campbell and Garnett[46] are, to say the least, a little confusing on the matter!

Maxwell and a Solid Ring System

We now turn to Maxwell's actual publication. We use the word 'publication' rather than 'essay' because, as has been said, there were major changes between the prize essay[36] and the published version of the work.[5] Rather than dwell on these, we shall just discuss the final version.

We must also stress that the work is highly mathematical. Any hopes that Challis had of receiving solutions which did not consist of 'long mathematical calculations' certainly did not come to fruition. We shall only be able to hint at the methods and discuss the results.

Yet, it must be said that the mathematics used was really not advanced. It was based around potential theory, and used linear differential equations, Taylor series and Fourier analysis. An able undergraduate of today should have the knowledge and ability (though certainly not the patience!) to work through the material. The same would apply to an undergraduate of the mid nineteenth century, with the possible exception that Fourier analysis was perhaps not as well-known and appreciated as it is today. (It is noteworthy that Maxwell proves some basic formulae in Fourier analysis[47] which would certainly be assumed in any publication today.)

Indeed, it is tempting to think of the problem rather as a Senate House problem (or Cambridge examination problem) writ enormously large. It postulated a physical situation and the student was expected to use a selection of well-known but carefully chosen techniques to find the desired solution. In the Senate House, of course, the solution was known, at least to the examiner; in the case of Saturn, it was merely hoped that a solution existed.

Of course, nothing should be taken away from Maxwell's achievement. It is one thing to say that standard methods are used, an entirely different thing to decide on the best possible course through exceptionally complicated and difficult mathematical developments, and to follow this course with immense skill, accuracy and attention to detail. It was a bravura performance.

We first study Maxwell's treatment of the rigid or solid ring system. It will be remembered that Laplace had shown that the ring system could not be both solid and uniform, with which Maxwell would agree, and Laplace had also claimed that it was 'therefore' solid and non-uniform, which was clearly unmotivated and with which Maxwell would *not* agree.

Maxwell commences his analysis by selecting a set of suitable coordinates; the distance between the centres of the ring and the planet; the angle between the line joining these centres and a fixed direction in the ring; and the angle between this line and a fixed direction in space. This may seem fairly arbitrary, but BEG[48] stress that it demonstrated his 'sureness of touch'. Choosing the right set of variables is essential in obtaining fundamental equations which are as simple as possible and which may be used to analyse the situation as readily as possible.

He then determines the potential of the system in terms of these coordinates, and is able to write down the equation of motion of ring and planet. It is then relatively straightforward to obtain conditions of uniform motion. As Maxwell says: '[I]f they were exactly fulfilled, the uniform motion would go on for ever if not disturbed'. But there will always be disturbances, however small, and so we need to find out whether, if the motion is slightly disturbed, the subsequent motion will be periodic about the original position or whether it will increase indefinitely, leading to collision between the inside of the ring and the planet. We must explore whether the motion is dynamically stable or unstable.

To do this, Maxwell obtains the rates of change of the potential with respect to the relevant coordinates, and expands these expressions out in terms of the powers of these coordinates using Taylor's theorem. He now demands that the types of motion predicted for these quantities must be periodic rather than increasing. (Any term could conceivably decrease, but Maxwell's analysis shows that any term that decreases must be produced together with one that increases, so both may be ruled out.) It is fairly straightforward to decide this from the nature – positive or negative – of the first terms in the expansion.

He then tests various possible forms of ring system to see whether periodic terms may be predicted. First, he examines the case of a uniform ring, and it is straightforward to show that periodic terms will definitely not be predicted. This is not surprising; it merely confirms Laplace's result from many years before.

Next, he considers a ring thicker at one side than the other, and again shows that no periodic solution may be obtained, so this form of ring cannot be reconciled with the conditions of stability.

However, he now considers a uniform ring loaded with a heavy particle at a point of its circumference. In this case, it is found that periodic motion and thus stability is possible. However, if that is the good news, there is plenty of bad! Maxwell is able to express his results in terms of a parameter f, where f is the ratio of the displacement of the centre of gravity of the ring to the radius.[49] Different mathematical conditions demand both that f is less than 0.8279 and greater than 0.815865.

To put things slightly differently,[50] we must have for this case a uniform ring loaded at a single point with a heavy satellite whose weight lies between 4.43 and 4.81 times the weight of the ring. But simple observation tells us that the rings do not have so grotesque a structure, and if they did, they would collapse under gravitation, so this mathematical oddity should not divert us from the fact that Maxwell has shown with mathematical rigour that Laplace's statement that the rings of Saturn consist of a solid non-uniform ring is incorrect.

Let us sum up the situation in Maxwell's own words, from an extremely short summary of his work[51] published only in 1862, though originally presented in 1858. His results are

> 1st. That a uniform ring cannot have a permanent motion; 2nd. That it is possible, by loading one side of the ring, to produce stability of motion, but that this loading must be very great, compared with the whole mass of the rest of the ring, being as 82 to 18; and 3rd. That this loading must not only be very great, but very nicely adjusted; because, if it were less than .81, or more than .83 of the whole, the motion would be unstable.

In his published account, he adds[52] that another powerful argument against solidity is 'the tendency of the fluid or loose parts of the ring to accumulate at the thicker parts, and thus to destroy that nice adjustment of the load on which stability depends'. And he added that

> When we consider the immense size of the rings, and their comparative thinness, the absurdity of treating them as rigid bodies becomes self-evident. An iron ring of such a size would be not only

plastic but semifluid under the forces which it would experience, and we have no reason to believe these rings to be artificially strengthened with any material unknown on this earth.

Finally in this section we note that Maxwell gives an 'intuitive [though long and complicated] argument to show why, in the anomalous case, the ring might be stable'. In this case, the effect of the planet on the ring may be expressed as the sum of a couple that rotates the ring, and a force through the centre of gravity. In the very special circumstances uncovered by Maxwell, the effect of couple and force may cancel out leading to stability, but, as we have seen, this requires an extremely delicate balance of factors.

Maxwell and a Fluid Ring System

The second possibility that candidates had been asked to consider was the fairly obvious idea that the rings were fluid. Actually, writing to Thomson while arranging the terms of the prize, Challis[53] commented that: 'It seems to me that the treatment of the question may be very nearly the same whether the Rings be fluid and aeriform, or be supposed to consist of numerous small and unconnected masses'. We shall indeed see that Maxwell's methods of investigation, that of analysing the motion in terms of waves, were broadly the same in both cases.

Maxwell was at first very interested in the possibility of the rings being fluid, but in a letter[54] to Thomson on November 1857, he said that: 'I have abolished my off hand theory of the attractions of a thin fluid filament affected by waves, long compared with the diameter of the filament, and this because the *short* waves are the only dangerous ones ... The liquid continuous ring is doomed.'

We shall say a little more about waves in the following section. Here we note that Maxwell finds that long waves around the fluid ring will be stable provided the density of Saturn is at least 42.5 times that of the ring. But Laplace had shown that, since the inner and outer parts of the ring must have the same angular speed, the density of Saturn should not be more than 1.3 times that of the ring. Since there is a contradiction, already instability is assured.

Even more striking is the fate of short waves. Let us imagine that as an aspect of this wave motion, fluid has moved to a region which has therefore become thick. Adjacent regions ahead and behind have become thin. It is now time for the excess fluid to return from the region which is temporarily thick.

However, there is a crucial aspect to the motion. Let us consider the fluid that should now move forwards. When a portion of the ring moves forward, it is moving faster and so the centrifugal force on it increases. The previous balance between gravitational attraction and centrifugal force is broken, and the portion of ring moves outwards. However, here it is on a path with a longer circumference and its angular speed decreases. Compared to adjacent parts of the ring, it is going backwards in its path round Saturn. Thus the expected movement forwards is stymied. Similarly, the portion that should move backwards relative to surrounding portions of the ring, is forced inwards, and here has a greater angular speed and catches up the nearby parts of the ring.

As Maxwell[55] writes: 'It appears therefore that the motion of a particle is either outwards and backwards or inwards and forwards'.

Thus, in the case of the wave, the expected motion of fluid from the temporarily thick region is inhibited. The result is that the irregularities increase without limit in the part of the ring where they originate, and the ring falls apart. In fact, as we shall see in the following section, clumping occurs.

It appears that Maxwell was as surprised by this result as the reader of today. To Thomson, he[54] wrote that: 'This effect is very remarkable as showing the destructive effect of an apparently conservative force'. In his published work,[55] he wrote that: 'It may seem paradoxical that a tangential force, acting *towards* a position of equilibrium, should produce instability . . . but it is easy to trace the destructive tendency of the apparently conservative force'. Of course what may seem rather an unlikely account to the reader was made unavoidable for Maxwell by several pages of intricate algebra.

Maxwell[51] summed up the possibility of a fluid ring as follows: '[W]e find that, in the case of a fluid ring, waves would be generated which would break it up into portions, the number of which would depend on the mass of Saturn directly and on that of the ring inversely'.

Maxwell and a Fragmented Ring System

Almost by default, Maxwell had ended up with the only apparent possibility being that the each ring consists of an enormous number of unconnected masses, each of which might be solid or fluid.

As Maxwell[51] pointed out, there was actually fairly direct evidence that the recently discovered inner dark ring was formed in this way. For the planet could be observed somewhat dimly through the ring, but also it was seen in its true position, not refracted as it would have been if it were seen through a fluid. As Maxwell said, this indicated that 'the light [had] passed through interstices between the separate particles composing the ring'. It seemed not at all unlikely that the outer rings had a similar but much denser constitution so that no light could penetrate them.

Now, of course, Maxwell had to show that such a system of rings was possible physically, and to establish any conditions on its formation. To make at least a good start on a complicated problem, he took a system of equal satellites equally spaced around the ring, and examined the possibility[56] of waves travelling round the ring in a stable way. As in the previous parts of the work, Fourier methods were essential for obtaining the appropriate equations. He calculated forces between different parts of the ring–planet system.

At the outset, it may be mentioned that any individual satellite may undertake a motion from its equilibrium position in any one of three directions, which BEG[57] call radial, tangential and normal. The third is out of the plane. It would immediately seem that the resulting gravitational force on the satellite will be back to the plane in this case, and the motion will be a stable oscillation about the plane. Such is, indeed, the case, and we shall not need to discuss this type of motion any further.

Tangential and radial displacements are in the plane. It would again seem that, in the case of a radial displacement, the resultant gravitational attraction will be back to the ring. This again will assure stability. However, at first sight, any tangential displacement would seem to be unstable. For, under such a displacement, the satellite will be closer to one of its nearest neighbours, further away from the other one. The resultant force will be towards the closer of its neighbours, and, from this argument, one would predict collisions between satellites and extremely rapid dissolution of the ring.

Yet this does not happen. Before discussing the reason physically, we shall sketch Maxwell's mathematics. He has to couple together the radial and tangential motions – or to put it another way, he finds that nature does the coupling, and he ends up with a biquadratic equation for the rate of oscillation of the wave – such an equation has four solutions, grouped in two pairs.

As Maxwell[58] says: 'When the four roots of the equation are all real, the motion of each satellite is compounded of four different oscillations of different amplitudes and periods, and the motion of the whole ring consists of four series of undulations, travelling round the ring with different velocities. When any of these roots are impossible [i.e. mathematically complex], the motion is no longer oscillatory, but tends to the rapid destruction of he ring.'

For each of the four types of wave, the number of undulations may be any whole number from one up to half the number of satellites. Since the danger of destruction of the ring increases with the number of undulations, Maxwell puts the number of undulations equal to half the number of satellites, and obtains[59] the condition of stability as

$$M_S > 0.4352\ \mu^2\ M_R$$

where M_S and M_R are the mass of Saturn and the rings respectively, and μ is the number of satellites. This is in agreement with Maxwell's statement at the end of the previous section. If one were to take the number of satellites to be 50, we can see that the mass of the ring would be roughly 0.1% that of Saturn.

We now discuss briefly the form of the different types of wave. The first two may be called fast waves. They are practically identical and travel around the ring in opposite directions, one in the direction of the rotation of the ring, the other against it. For an observer on the ring, each wave takes the same time, about 50 hours, to move round the circle. From a stationary point, both waves take about 10 hours to move round the circle, the first type being faster than the second. In each case the satellites follow an elliptical path about their mean position, the mean position itself, of course, travelling round with the ring. (See Fig. 6.5.)

The other two types of waves may be called slow. They travel in opposite directions at about 1% of the speed of the first two. Again the satellites themselves move in elliptic orbits about their mean positions, but in this case their motion is most meaningfully described, from a stationary position, by saying that they move in circular orbits about Saturn, but with the radius of the circle increasing and then decreasing very slowly over many revolutions of the rings.

Let us now return to the rather surprising stability of the tangential displacements. As has been implied in the account of Maxwell's mathematics, this is a result of the coupling of tangential and radial displacements, and indeed the effect is just the same as that described for fluid rings in the

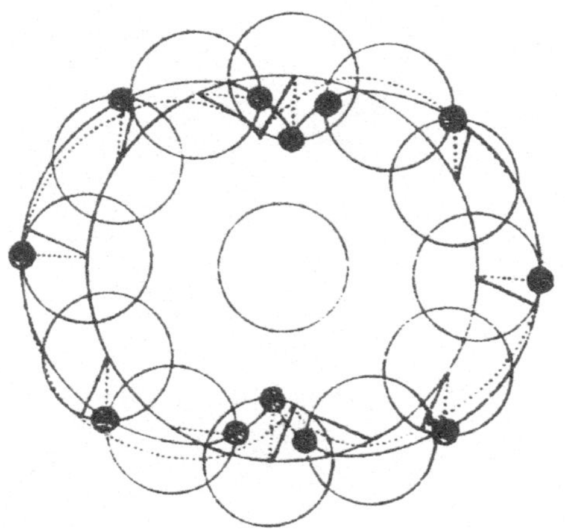

Fig. 6.5 A representation of one of Maxwell's fast waves. The satellite travels in an elliptical (practically circular) path round its mean position, which itself travels round the ring. (From SP, Volume 1, Plate V, which follows p. 376.)

previous section. While in the previous section it acted against the stability of a continuous fluid ring, causing clumping into satellites, in this section it prevents the breaking up of the satellites. Again, if any satellite moves ahead of its normal position, it is forced outwards by the radial effect and thus loses angular speed, and the reverse occurs for the any satellite falling behind.

It may be mentioned that Maxwell's work lends no support to the nebular theory of Laplace. He finds no tendency for the satellites to conglomerate into one larger mass.

As mentioned above, by early 1858, Maxwell had arranged for a model of his theory of a ring of 36 satellites to be constructed by Ramage of Messrs. Smith and Ramage, instrument makers of Aberdeen. In the letter[42] to Thomson sent in January of that year, he described it as follows:

It consists of two wheels on the parallel parts of a cranked axle. There are 36 little cranks ranged round the circumference which all move parallel to the axle-crank and to each other. Each carries a satellite placed excentrically on the part of the crank which sticks through the wheel, so that when the axle turns and the wheel moves, every satellite describes a small circle. The position of the satellite in this circle is arbitrary, and its arm may be adjusted on the crank at pleasure, so that by properly arranging these arms the ring of satellites may be thrown into waves of any length which travel round the rings, the wheel being fixed.

Thus by fixing the crank and moving the wheel, the absolute motion of the fast waves could be displayed. (See Fig. 6.6.)

As Maxwell said, the other waves were not so interesting and could not be displayed by this apparatus.

To sum up, Maxwell's work suggested that a ring consisting of a large number of masses travelling independently was feasible mechanically. Since no other of the suggested forms of ring was similarly feasible, the only conclusion must be that the ring did indeed consist of such independent masses.

Fig. 6.6 Maxwell's machine for displaying the fast waves for the system of 36 satellites. The original machine is now in the Cavendish Laboratory.

Maxwell and Systems of Many Rings

Next, Maxwell had to recognize that there were several rings rather than a single one, and so it was necessary to consider interactions between different rings. He was able to set up equations for either ring much as in the previous section, and found that rather than four types of wave, there are now eight. Mathematically, of course, Maxwell had to deal with equations of the eighth degree, not a pleasant task, but he managed effectively by treating interaction terms between the two rings as small.

The details of the waveforms do not need to be discussed, but the significant feature of the analysis is what happens if, for certain ratios of the radii, a wave of one type in one ring comes into resonance with a wave of a different type in the other ring. The wave on each ring will split into two, one steadily decreasing in magnitude, but the other steadily increasing so that it will grow uncontrollably, leading to collapse of the rings.

As Maxwell[60] says, if the rings are concentric and quite close, it is almost certain that some pairs will be at the right distance for instability. When they break up, their particles will fly off in all directions and some must collide with satellites in other rings. In principle, one would have to predict eventual destruction of the ring system, or at the very least, great changes.

But, as he also says, the discussion has been restricted to satellites of equal size distributed at equal distances around each ring, and the results my not apply to more general cases. He had at that time no method for treating satellites of irregular shape, size, and mass, spaced irregularly around any ring.

Maxwell[61] was also able to make some general comments about the long-term future of the rings. He suggested that, even without any consideration of viscosity, there may be a tendency for the exterior rings to move further away from Saturn and the interior ones to move closer. However, if the interior rings were approaching the planet, but there was no motion of the exterior ones, viscosity must be involved, and it was to consider such an effect that he had approached Stokes for the required data and information.

As to stability, Maxwell[61] concluded that, provided the satellites were arranged in series of narrow rings, while it could not be guaranteed that destruction, as discussed earlier in this section, could never take place, it would almost certainly be very slow.

Aftermath

As has been mentioned before, soon after the publication of the published version of Maxwell's work, Airy[22] presented a highly laudatory account of Maxwell's work.

He obviously felt that his first task was to remove any suspicion that Maxwell had take advantage of the work of any of his predecessors. Having criticized Laplace's conclusion that the ring must be an irregular solid, he discusses the work of George Bond and that of Peirce. He first briefly mentions Bond's ideas that there must be a very large number of rings, and that a fluid symmetrical ring may be stable. He actually criticizes this in the words of Peirce, who suggests that a fluid ring without gravitational support must be unstable, but that it could be stabilized by the effect of the moons of Saturn. Airy comments that:

This paper [that of Peirce], it must be remarked, contains no symbolic investigations. Perhaps many mathematicians will think that a very complete mathematical treatment will be required to establish the two laws, that the attraction of *Saturn* will not support a fluid ring, and that the attraction of satellites will tend to support it.

He then tackles a later paper of Peirce, but, though admitting that the early part of this work is 'skilfully based on the method of potentials', he criticizes the rest of the paper as relying on 'general reasoning', and suggests that his mention of 'foreign support' [by the moons of Saturn] as unmotivated. He concludes by saying that:

The author, however, has spoken so briefly on the subject that it is possible we have mistaken his meaning. An expectation was held out that this paper would be continued, but, so far as we are aware, no continuation has yet appeared.

Having disposed of any 'opposition', Airy then gives a fairly straightforward account of Maxwell's work, and ends with the statement that:

The abstract which we have given will, we think, fully justify the opinion that the theory of *Saturn's rings* is now placed on a footing totally different from any that it has occupied before, and that the essay which we have abstracted is one of the most remarkable contributions to mechanical astronomy that has appeared for many years.

Maxwell, meanwhile, was moving on from his published book of 1859. He recognized a major defect of the theory was that it was restricted to the case where the satellites were arranged regularly around the ring. Indeed, he remarked that he was unable to analyse an irregular array of such satellites with the tools available to him at the time. This was to change, at least to some extent, with the arrival of a moderately mature kinetic theory. Maxwell was, of course, one of the main producers of this theory, as described by Garber in this book, and there are two possible narratives for this event. In the first, Maxwell realized that the problems of his work on Saturn could be tackled with such a theory and decided to work on it. In the other, he saw in the work of Clausius the tool that he needed and decided to develop it, which he did with great skill.

BEG[62] have helpfully provided a number of drafts or sets of notes that Maxwell wrote at that time and that are in the collection of his manuscripts at Cambridge University. The first would appear to be a draft of a paper specifically commencing to apply the methods of kinetic theory to the rings of Saturn. It starts by describing the results of his study of the rings of Saturn, and then discusses, very much in outline, how the ideas of kinetic theory – he mentions the work of Clausius and his own – could be applied to treat such problems. Collisions between satellites had to be handled, while the gravitational effect of Saturn still had to keep them in the rings. Maxwell recognized it as an extremely complicated problem.

The subsequent documents become distinctly more mathematical. One later document begins with the words: 'We assume the Rings to consist of independent portions of solid matter of small dimensions compared with the thickness of the rings (100 miles or less) revolving around Saturn in orbits very nearly circular . . .' and it continues to justify its mathematics by references to the rings.

However, the documents become less and less to do with Saturn, and more and more to be exercises in kinetic theory. At a certain point, it seems, Maxwell effectively turned his back on Saturn and on mathematical astronomy, never to return.

Why did he put in the enormous effort that the Adams Prize task involved, at a time when he had a major interest in electromagnetism, and had important tasks to undertake in connection with his Aberdeen chair? We mentioned the quite substantial amount of prize money, but of course it does not seem so large balanced against the amount of work entailed over a number of years! His father's comments on the Aberdeen position and the salary involved – 'I believe there is some salary, but fees and pupils, I think, cannot be plenty. But if the *postie* be gotten, and prove not good, it can be given up' – do not suggest that money would have been a major consideration.

Maxwell probably relished the challenge of tackling a complicated problem. BEG make the point that, at the age of 26, Maxwell had not yet established a reputation. They comment that:

> This bold synthetic work [on electromagnetism] was largely incomprehensible to his contemporaries, who could not understand his insistence on mathematizing Michael Faraday's sophisticated and essentially geometrical ideas on electric and magnetic forces . . . A solid piece of work of mathematical physics solved effectively yet imaginatively was necessary to consolidate a rather shaky career . . . The Adams Prize Essay was the perfect opportunity for demonstrating mathematical facility along with just the right amount of physical intuition.

While this is certainly true, we should also see Maxwell's work as a component of the continuing struggle to reassert and strengthen the position of Cambridge astronomy in the wake of the Adams affair. Challis and Thomson were not just judging and, to some extent, correcting and improving Maxwell's initial work as prize examiners, they were, by choice of prize topic and their fine-tuning, helping to create a major achievement for Cambridge. Airy was on hand to dispel any doubt about the originality of the work, and to commend it wholeheartedly. And when G. A. Hirn published work similar to that of Maxwell over a decade later, it was again Airy who pointed out that Maxwell had priority.[63]

Maxwell may well have felt not only a desire to be awarded the prize, but also some obligation to support his Cambridge colleagues by creating a substantial piece of research in the astronomical area. To this extent, his work on Saturn may have been as much an effort of 'duty' as that on the research of Cavendish and the work for the British Association on units.

And so to Saturn itself. While, as ever, a range of opinions continued to be expressed on the constitution of the rings in the years after Maxwell's work, by roughly 1880 it could be said that it was agreed both that Maxwell's conclusions were correct, and that his were the authoritative arguments that had demonstrated them convincingly. (Airy may not have been unbiased, but he was still right!)

In 1895 the well-known astronomer James Keeler,[64] Director of the Allegheny Observatory of the University of Pittsburgh, used a spectrograph attached to refracting telescope to show that different parts of the ring reflected light with different Doppler shifts, because of their different speeds of motion round the planet. This was regarded as experimental confirmation of Maxwell's idea that the rings consisted of small objects each rotating at its own speed.

With the Voyager missions of the 1980s, backed up by NASA's Cassini probe in 2009, a fairly full understanding[65] of the rings was obtained. The particles themselves are now known to consist mostly of ice with some traces of rocky material, and their sizes range from micrometres up to metres. The total mass of the rings is thought to be something like 10^{-7} times that of Saturn, a much lower number than those discussed in the nineteenth century! The thickness of the rings may vary between 10 metres and 1 kilometre.

The general structure of the rings may be described as nine continuous rings and three discontinuous ones. The rings are known as A, B, C . . . in order of discovery, so C is the inner dark ring discovered in the early 1850s.

However, it should by no means be thought that each ring is uniform in constitution. Quite the contrary, each has an enormous amount of structure, with many maxima and minima of density and brightness. They have *gaps* where the particle density drops sharply practically to zero. Some of these gaps are known to have been caused by the presence of moons, while others are the result of orbital resonances with moons. Some resonances, on the other hand, are stable and lead to build up of density. Indeed, as well as gaps there are *ringlets*, where density is high.

Many of these gaps and ringlets are named after those who investigated the rings, and it is pleasant to note that in the middle of the C ring is the *Maxwell gap*, while in the middle of this gap is the thin *Maxwell ringlet*.

CHAPTER 7

Maxwell's Kinetic Theory 1859–70

ELIZABETH GARBER

When James Clerk Maxwell died in 1879, his scientific reputation rested largely on being the foremost 'molecular scientist' of his age.[1] Today it would seem clear that, important as this work undoubtedly was, his contributions to electromagnetism, described in this book by Daniel Siegel in Chapter 10, were decidedly more significant, but at the time of his death these were not generally and fully appreciated. Indeed it would take the remainder of the century, and the work of the extremely able group of scientists usually called the Maxwellians and described in this book by Chen-Pang Yeang in Chapter 11, before the full implications of his theory of electromagnetism were understood. In contrast, the importance of his work on kinetic theory was clear almost immediately, even though the subject continued to be developed by Maxwell himself and others for many years afterwards; some examples of this later work are discussed by John Rowlinson and Andrew Whitaker in the following Chapters 8 and 9 in this volume.

When Maxwell began his work on kinetic theory, there was no general consensus among chemists and natural philosophers [physicists] about the nature either of heat or of matter. In the eighteenth and the first half of the nineteenth centuries, there had been two main theories of heat. The first described it as a substance, caloric, and the caloric theory of gases had led to an early theory of specific heats, while the second theory described it as a form of motion. The theory of thermodynamics, which was obtained by Rudolf Clausius, William Thomson (later Lord Kelvin), and Macquorn Rankine in 1850–51, served to rule out the idea of caloric, and showed that heat was a form of energy. However, it still left undecided the precise nature of heat, and also the structure of matter at the atomic or molecular level. The development of kinetic theory effectively answered these two questions

Nineteenth-century natural philosophers had carried out very few investigations of the inner structure of molecules, their sizes, or their distances apart in the three states of matter, though early in the nineteenth century Thomas Young and John J. Waterston, using rather primitive kinetic theories, had separately estimated their size. Maxwell followed the same path, the initial goal of his theoretical and experimental work on gases being to estimate the size of molecules.

In his earliest work, Maxwell established the foundations of a theory of gases that led to consequences which could be experimentally investigated. His work in the 1860s, first theoretical and later experimental, created a foundation for later explorations of all the transport properties of gases – viscosity, diffusion and heat conduction. As early as 1859, he used the ratio of the specific heats of a gas to deduce a value for its viscosity. However, he quickly abandoned his first theory, which treated molecules as elastic spheres, because his own experiments on the coefficients of viscosity led to a value of viscosity which differed from that deduced from his theory. He turned instead to a model of molecules as centres of force, though unfortunately this model led to the same theoretical results as his earlier theory. Despite these problems, his theoretical and experimental papers on gases established his reputation as both a theorist and experimenter, and the types of mechanical models of molecules that Maxwell developed dominated work in molecular physics until the first decade of the twentieth century.

Maxwell's Work in Kinetic Theory: First Paper of 1859–60

Maxwell's kinetic theory was not the first. As already mentioned, the German physicist, Rudolf Julius Emmanuel Clausius (Fig. 7.1), had established his reputation in 1850 with his seminal work in thermodynamics, and later in the same decade he developed a theory of gases. He published two papers, one in 1857, the second in 1859, in which he concentrated on three aspects of the theory: a model of the molecule, examination of a gas composed of such molecules, and the distribution of kinetic energy amongst their various modes of motion. Clausius then focused on the kinetic energy of the molecules of a gas, and a topic he had explored in an earlier paper, the link between this kinetic energy and heat. With this work he tried to bridge the gap between statements in thermodynamics on the nature of heat, and his mathematical description of heat as the motion of the molecules of a gas. Josiah Willard Gibbs noted that Clausius' objective was less the development of a theory of heat and more to study 'the nature of molecular phenomena whose laws have thermodynamics as its sensible expression'.[2]

Thus, Clausius' intention in exploring the theory of gases was to understand the sizes and structures of molecules. In his paper of 1857,[3] he derived the now well-known expression connecting the pressure and volume of a gas with the mean square free velocity of its molecules. He distinguished atoms from molecules, and he demonstrated that molecules were complex bodies,

Fig. 7.1 Rudolf Clausius – he made crucial contributions to both thermodynamics and kinetic theory.

subject to rotary as well as translational motions. He also introduced two assumptions that were essential for Maxwell's own later work, that molecules were not necessarily elastic spheres, and that, according to the laws of probability, we can assume that when molecules collide there are as many molecules whose angles of reflection fall within a certain interval 'as there are molecules whose angle of incidence fall within the same limits'. Clausius' arguments were mechanical not statistical, in the sense that, while he recognized that in nature gas molecules had a distribution of speeds, he was content to work with a mean value. He was able to deduce the mean speeds of the molecules of common gases, and he also introduced the notion of the mean free path.

While the analysis in Clausius' first paper led to known gas laws, some of his conclusions seemed to lead to results contrary to known observations, but he was able to address this problem in his second paper.[4] The most telling of the arguments against his theory was the observation that, given the molecular speeds Clausius was indicating, it was difficult to understand why it took minutes, sometimes hours, for gases to diffuse across a room. In answering this important observation, Clausius introduced the idea of a 'sphere of action' of a molecule, which was half the distance between the centres of mass of two colliding molecules. If molecules approached too closely they were driven apart by forces within the molecules. Molecules would travel, on average, a distance called the 'mean free path' between undergoing such collisions, but Clausius could not deduce the length of the mean free path, and so he did not develop a way of estimating the size of a molecule.

Maxwell's first paper, initially presented in 1859 but published in 1860, resulted from his reading translations of Clausius' two papers on gases in the *Philosophical Magazine*. While Clausius' fundamental aim in studying the properties of gases was to explore the nature of heat, Maxwell used kinetic theory to study the properties of gases in their own right. The goal of this paper was primarily to develop a theory to ascertain the mean free path of a molecule. He treated the problem as an exercise in dynamics, and Maxwell considered intermolecular collisions far more comprehensively than Clausius had done. He did this by assuming that the motion of a large number of molecules could be dealt with using probabilities, and thus, by introducing statistical methods into physics, he was able to describe the state of the system at any moment in time. Indeed his statistical description of molecular motions was the key to the later development of the theory, carried out almost entirely by himself and Ludwig Boltzmann.

His matter of fact introduction and easy manipulation of probabilistic methods suggest that he was already familiar with them, and indeed he had been introduced to them as a student at the University of Edinburgh, where probability was included in the philosophy course taught by Sir William Hamilton. The Edinburgh Professor of Natural Philosophy, James David Forbes, also used such arguments when criticizing a paper on the probability that stars seemingly close together when seen through a telescope were necessarily binaries. Forbes' argument was itself criticized by the Cambridge don, Robert Leslie Ellis, who had already published on the deficiencies of earlier derivations of the normal law of probability, and while a student at Cambridge Maxwell himself was exposed to the ensuing arguments on the foundations of probability theory in the pages of the *Philosophical Magazine.*

He had also had to pay attention to such problems while in training for the Tripos examination at Cambridge, and in June 1852 such a problem actually appeared on the examination at Peterhouse.[5] Also in the early 1850s, a controversy over probability had appeared in an issue of the *Cambridge and Dublin Mathematical Journal*, and we can assume Maxwell was aware of it because this was an issue in which he also had a paper published. Most conclusively, he had dealt with the motions of myriads of particles in his prizewinning essay of 1859 on Saturn's rings.

After analysing the motions of solid and liquid rings, he turned to rings consisting of small, separate particles. In this last case he noted that mechanics was useless in following the motions of such satellites as 'we can no longer trace the mathematical laws of their motions with any distinctness'.[6] (Maxwell's work on Saturn's rings is discussed by Andrew Whitaker in Chapter 6 of this book.)

It is in this context that he performed his first work on kinetic theory in 1859. Maxwell was well aware of his debt to Clausius, as he explained in a letter to William Thomson in 1871, but Maxwell turned the problem around from Clausius' approach, and asked what measurable properties of gases could be used to find their, as yet unknown, unmeasurable properties, such as size and mass.

In 1859, he sent George Gabriel Stokes an outline of his first paper;[7] it had been presented to the British Association in that year, and was to be published in the *Philosophical Magazine* in the following year. This paper included an expression for the mean free path and a sketch of his theory of viscosity. His goal was to develop a theory of the transport coefficients of gases and to use these to explore the size of molecules. Crucially, he chose to describe the state of the whole system in probabilistic terms at a particular point in time. This was to be the key to the later extremely significant development of his theory, in which he obtained his important distribution function, and in which he also connected molecular motions to the measurable transport properties of gases. In addition, he used Clausius' concept of the mean free path to deduce the transport coefficients and to express the pressure of a gas on the walls of its container.

Before he could do all this, Maxwell had to establish the laws of motion for the collisions of the elastic spheres. In 1859, he laid out his plan to attain this goal. While there was no method available to measure the mean free path, certain phenomena, such as the viscosity, heat conduction, and diffusion of gases, 'seem to indicate the possibility of determining accurately the mean length of path which a particle describes between two successive collisions'.[8]

Maxwell began by considering the motions and collisions of perfectly elastic spheres, and he demonstrated that all directions in space are equally likely as the paths for those spheres. He also argued that, although the velocities of individual molecules might change, if the gas was in equilibrium the numbers of molecules whose velocities were within certain limits were constant. He then established expressions for the number of particles whose velocity components lay between u and $u + du$, v and $v + dv$, and w and $w + dw$; let us call this range A. He assumed a uniform distribution of the molecules throughout the volume of the gas, and also that the probabilities of the velocity components of a molecule lying between u and $u + du$, v and $v + dv$, and w and $w + dw$ are independent. These probabilities may be written as $f(u)\,du$, $f(v)\,dv$, and $f(w)\,dw$. Then the number of molecules in range A, dN_A is given by:

$$dN_A = N\,f(u)f(v)f(w)\,du\,dv\,dw$$

where N is the total number of molecules in the gas. Since the probability of the molecule having any particular velocity depends only on the magnitude of that velocity, which we may call c, and not on its direction, we may write:

$$f(u)f(v)f(w) = \varphi(u^2 + v^2 + w^2) = \varphi(c^2)$$

Maxwell took these conditions as defining the required distribution function,

$$dN_A = N\,\varphi(u^2 + v^2 + w^2)\,du\,dv\,dw$$

He was then able to deduce that the number of molecules with speeds lying between c and $c + dc$ was

$$N\frac{4}{\alpha^3\sqrt{\pi}}c^2\exp\left(-\frac{c^2}{\alpha^2}\right)\,dc$$

where α is a constant. Maxwell stressed the important conclusion, which is that the speeds of the particles are distributed 'according to the same formula as the errors are distributed in the theory of errors'. He had made the crucial breakthrough from the dynamics of Clausius to the statistical approach which would be followed by himself and Boltzmann.

He proceeded to deduce expressions for the mean speed and the mean square speed. The distribution function Maxwell derived was valid for a group of infinitely many elastic spheres uniformly distributed through the total volume occupied by the gas, provided no forces act on the spheres and the velocity components of the molecules are independent of each other. (He later abandoned this last assumption.) It is important to stress that the function calculated was independent of time, and hence only valid for a system already in thermal equilibrium. Deriving the distribution function was only the first step in his discussion of the transport properties of gases, so Maxwell did not explore the foundations of the distribution function, but rather used it to investigate the physical consequences of the elastic sphere model of the molecule. Then, in conjunction with the concept of the mean free path, he used kinetic theory to investigate the transport phenomena of gases and develop expressions for them.

However, Maxwell found his first derivations of the distribution function unsatisfactory. An initial worry was that when he related his expression for viscosity to the number of collisions between molecules per second, he found, surprisingly, that the coefficient of viscosity was independent of the density of the gas. In his letter to Stokes, Maxwell noted that this result was 'curious' and it was 'certainly very unexpected that the friction should be as great in a rare gas as in a dense gas'. However he was able to come up with an explanation; he argued that in rare gases the mean free paths were so long that 'frictional actions extend to greater distances'.

He also enclosed an estimate of the mean free paths using the experimental results of Thomas Graham on the rates of flow of gases through fine tubes, and Stokes' own work on the decay of the amplitudes of pendulums. Maxwell concluded that they were in 'satisfactory agreement'.[7] There were, though, some surprises. If friction was assumed to be proportional to density, then Graham's results were correct, but in his own theory friction differed from Graham's results 'very much'. This was 'a curious result'. There seemed also to be another more important flaw in his argument, in that his theory of gases seemed to be incompatible with the second law of thermodynamics; when his theory was extended to the case of a gas under external forces it appeared to lead to perpetual motion. However, it later emerged this was a mistake on his part and Maxwell corrected his paper; he also concluded that a gas in mechanical equilibrium under gravity was also internally in equilibrium.

An equally important result was that the coefficient of viscosity must vary as the square root of the temperature. This result, though, depended on his use of the elastic sphere model for the molecules of the gas, and in 1865 his own experiments on viscosity were to convince him to abandon this model.

The most controversial aspect of his theory of 1860 was his derivation of the coefficient of conductivity, his derivation actually being little more than a sketch of such a theory. He expressed the kinetic energy of translation of a particle as $\frac{1}{2}\beta m \langle c^2 \rangle$, where $\langle c^2 \rangle$ is the mean square velocity, m is its mass, and β is the ratio of the total kinetic energy of the molecule, including any rotational and vibrational energy, to its translational energy. The latter was an expression that had been used by Clausius to obtain the rate of transfer of heat across a plane. Maxwell then developed an expression for the transfer of energy across a unit area, in unit time and in energy units,

$$JH = -\frac{1}{3}\frac{d}{dx}\left\{\frac{1}{2}\beta m{<}c^2{>}\; N\; {<}c{>}\; \ell\right\}$$

where H is the heat transferred across the plane, and J is the mechanical equivalent of heat, so JH is the mechanical energy transferred across the plane, ℓ is the mean free path of the molecules of the gas, and $<c>$ is the mean speed of the molecules.

Using the expression, $\frac{1}{2}mN\ {<}c^2{>} = RT$, Maxwell obtained an expression for the energy crossing a plane,

$$JH = -\frac{3}{4}\beta P\ell\ {<}c{>}\frac{1}{T}\frac{dT}{dx}$$

where P is the pressure.

In the above equations, the mean velocity and the mean square velocity were taken to be those deduced from the equilibrium velocity distribution function. From this expression, Maxwell deduced the coefficient of thermal conductivity, K, to be

$$K = \frac{3}{4}\beta P\ell\ {<}c{>}\frac{1}{T},$$

From this, and using his own estimate for the mean free path, Maxwell obtained a numerical value for the conductivity of air. However, he focused most of his discussion on diffusion rather than viscosity or conductivity, as diffusion was the only transport phenomenon for which there was any experimental data; diffusion of gases had already been investigated experimentally by Thomas Graham and others. Graham's Law, that the rate of flow of was inversely proportional to the square root of the density, was valid for the diffusion of one gas into another, and also for the case of a gas filtering through thin plates. But his law did not apply to gases flowing through long tubes.

Maxwell investigated the case of diffusion of gases through a tube and gave an estimate for the mean free path of the gaseous molecules. He confined himself to discussing the volume of two gases that diffuse in opposite directions through a tube, closed at each end by spherical vessels. Stokes had developed theories of viscosity and elasticity in the 1840s, and had performed experiments on the damping of pendulums, experiments that Maxwell was very well aware of; in fact he used data from Stokes' paper to estimate the mean free path of gaseous molecules. It thus becomes very clear why Maxwell sent a fairly detailed account of his theory to Stokes in May 1859.

With his early work on kinetic theory, Maxwell developed the first mathematical discussion of the transport properties of gases, and introduced probabilistic arguments into the body of physics. Then, in 1865, he turned to experiment to measure the coefficient of viscosity for air. Maxwell used a pendulum made up of three horizontal glass discs clamped perpendicularly to a vertically hung wire, and the wire and the attached discs were rotated about the vertical axis under various pressures and temperatures. Stationary discs between the moving ones kept potential draughts to a minimum. The inertia of the system remained constant, while the surface area of the pendulum could be varied by separating the glass discs along the vertical wire, or clamping them together. He heated his London house, where the experiments were performed, to various temperatures, and collected data on the coefficient of viscosity at these different temperatures, the apparatus being protected from air currents by a wooden box.

In this way he confirmed his prediction, which had initially seemed 'curious', that viscosity was independent of the pressure of the gas. The results of these experiments became one of the most significant pieces of evidence in favour of his kinetic theory, and led to his Bakerian Lecture on the subject at the Royal Society in 1866.[9]

Maxwell's Second Kinetic Theory Paper of 1866

However, all was not well with this theory. Maxwell's theoretical prediction was that the coefficient of viscosity should vary as the square root of temperature. On the other hand, his own experiments convinced him that actually it varied linearly with temperature. In addition, the ratio of specific heat at constant volume to that at constant pressure, deduced from his theory, did not agree with experiment. Added to this was his unsatisfactory derivation of the coefficient of conductivity in which he used the mean velocity and mean square velocity from the equilibrium distribution function. We shall shortly see why this cannot be correct. His own experiments on the viscosity of air gave him decisive evidence that he should abandon his first molecular model, and he now turned to molecular models based on centres of force.

In addition, he had to face Clausius' criticism of his first paper. Clausius saw Maxwell's 1860 kinetic theory paper as a serious attempt to solve the problem of gaseous behaviour, and he studied Maxwell's paper in detail, pointing out several mistakes and omissions. His criticism rested on Maxwell's incorrect use of the equilibrium distribution function when analysing transport phenomena. Clausius then developed his own mechanical alternative to Maxwell's probabilistic argument, using the example of the conduction of heat along the x-axis, taking into account the additional kinetic energy associated with the motion of the molecules in the direction of the phenomena, usually along that axis.

The problem he faced was developing integral expressions for the three transport properties of mass, momentum and kinetic energy or heat. In the case of transporting mass, he considered the velocity of the molecules along the x-axis, and included a 'proportion', I, of the molecules, between given limits, moving with an average velocity, at an angle to the x-axis. In the case of momentum, the integral contained the square of the molecule's average velocity. For heat conduction, gas molecules possessed additional momentum in the direction of the temperature gradient. This integral included the cube of the average velocity. He concluded that this led to a distribution of molecules that could not be spherically symmetric. Hence he argued, as has been mentioned above, that Maxwell's use of his own distribution function was invalid.

Clausius' method of obtaining averages was unwieldy and he neglected terms in the second and higher degrees. This led to results that seemed no different from a spherical distribution, but his paper was important even if his mathematical approach did not work well. He clearly established the physical problems in calculating the transport coefficients, and he developed a method for quantifying the case for heat conduction that Maxwell would use later. Clausius also assumed that the spread in velocities about the mean was the same as the spread occurring when the gas was in equilibrium, plus that due to a gradient in the gas.[10] The additional momentum of the molecules in, say, the x-direction, was directly proportional to the mean free path of the molecules in equilibrium. Since the mean lengths of the actual paths of the molecules could be written as a series of terms in increasing powers of the mean free path in equilibrium, the velocities of the molecules depended directly on the mean free path in equilibrium. The actual path travelled also depended upon the angle of projection of the molecule with respect to the x-axis. All averaging was over the cosine of the angle of the molecular path with the x-axis, or the mean free path, of the molecules in equilibrium.

Reacting to this paper, Maxwell recognized the contribution of Clausius to kinetic theory and he fully acknowledged the 'gross mistakes' Clausius had noted in his own work. He then reworked his theory in a paper which was published in 1867.[11] He discarded the notion of the mean free path and introduced a new model of gas molecules as centres of force, subject to mutual repulsions. He also generalized his theory.

He was able to produce what he felt was a more rigorous proof of his distribution function. He began by considering a single 'encounter' between two centre-of-force molecules. Two molecules with masses m_1 and m_2 and speeds c_1 and c_2 approach each other. Using centre-of-mass coordinates, Maxwell examined their deviation from their original straight-line paths. During the encounter, the velocity of the common centre of gravity is unchanged, and the velocities of the two molecules are turned through an angle of 2θ with respect to their common centre of gravity. The angle θ depended on the relative velocity between the two molecules, and b, the distance of closest approach of the centres of gravity of the molecules. Maxwell argued that the velocity components of the molecules after such an encounter depended upon their initial velocities, the angle between their centres at impact, and their distances apart, and he then applied statistical methods to the analysis of the encounters of the molecules. The number of encounters between molecules whose initial velocities were c_1 and c_2 depended on the number of molecules with those velocities, the limits put on the variations in those velocities, the relative velocities between the molecules, and the angle through which the molecules are turned by their encounters, as well as the distance of closest approach.

He denoted the velocities after such an encounter as c_1^* and c_2^*. Maxwell then considered the possible collisions between two molecules whose initial velocities were c_1' and c_2' and whose final velocities are also c_1^* and c_2^*. When the numbers of collisions in which the velocities change from either starting point and reach the same final velocities are equal, Maxwell was able to argue that the gas is in equilibrium, and

$$f_1(c_1)f(c_2) = f(c_1')f(c_2')$$

where $f_1(c_1)$ is the number of molecules in a unit volume of gas whose velocity lies between c_1 and $c_1 + dc_1$, $f_2(c_2)$ is the number of molecules per unit volume of gas whose velocity lies between c_2 and $c_2 + dc_2$, and so on. In all these encounters energy is conserved:

$$m_1c_1^2 + m_2c_2^2 = m_1c_1'^2 + m_2c_2'^2$$

and momentum is also conserved.

He found that the number of molecules is given by,

$$f_1(c_1) = A_1\exp\left(\frac{-c^2}{\alpha^2}\right)$$

where

$$A_1 = \frac{N}{\alpha^3\pi^{3/2}}$$

The number of molecules whose velocity components lie between u_1 and $u_1 + du_1$ and so on is:

$$dN = N\frac{\exp\left[-(u^2 + v^2 + w^2)\right]}{\alpha^3\pi^{3/2}}du\,dv\,dw$$

This final expression for dN was the same as the one Maxwell had derived in his 1860 paper. The condition he imposed upon the system in 1860, that the number of molecules with a particular x-component of velocity was independent of the number with a particular y-component and so on, was not applied explicitly in this later derivation.

However, Maxwell implied that the average velocity components must equal one another because: 'If the velocities of the molecules moving in different directions were independent of one another, then the pressure at any point of the gas need not be the same in all directions, because the pressure between two portions of the gas separated by a plane need not be perpendicular to that plane'.[12]

Despite this new derivation of the distribution function being more general than his earlier one, it was still only valid only for a gas whose molecules were uniformly distributed throughout the space. Further improvements in his method of derivation were only developed in the 1870s, after he was able to fathom Boltzmann's exhaustive investigations of the distribution law that were published between 1868 and 1871. However, to develop expressions for the transport coefficients, Maxwell needed to investigate the distribution function in specific physical circumstances, to find expressions for the various transport coefficients where uniform distribution could not be assumed.

In this later development, there was a unity of construction and concept that emerged from his use of Clausius' method, and he focused on developing a more generalized, accurate theory of the transport phenomena. He considered the rate of change of a property, Q, of a molecule that could change during collisions with other molecules. Q could be related to mass, momentum or kinetic energy, depending on whether diffusion, viscosity or heat conduction was being considered. The derivation of the expression for the rate of change of Q in time was incorporated into his calculation of the velocity distribution. The rate of change of Q, for the molecules whose initial velocities lie between c_1 and $c_1 + dc_1$ was

$$\frac{\delta Q}{\delta t}dN_1 = (Q' - Q)\,V_{12}b\,db\,d\phi\,dN_1dN_2$$

where Q is the value of a property Q before the encounter of two molecules, Q' is the value of the property after the encounter, V_{12} is the relative velocity of the two molecules, b is the distance of closest approach for the two molecules, ϕ is the angle between c_1 and c_2, dN_1 is the number of molecules per unit volume whose velocities lie between c_1 and $c_1 + dc_1$, and dN_2 is the number of molecules per unit volume whose velocities lie between c_2 and $c_2 + dc_2$.

This equation was central for Maxwell's second transfer theory. To find the motions of the molecules became an exercise in orbital mechanics. His results were posited in terms of the distance, b, between parallel asymptotes of the molecules' paths before and after their interactions, their initial velocity, V_{12}, θ the angle between the lines of the closest approach, and ϕ the angle determining the plane of V_{12}. There was one difficulty: the exchange of energy between the translational motion of the molecules, and the motions within the molecules themselves. Maxwell made the assumption that the effect of latter was negligible.

This was a fundamental reworking of Maxwell's theory. Integrating with respect to b, N_1 and N_2, the total rate of change could be deduced for any quantity Q, providing expressions for diffusion, viscosity or thermal conductivity. To complete these integrations, Maxwell had to be more specific and discuss the number of molecules of the two types whose velocities lie within certain given limits. He had to choose the relevant form for Q, and then proceed with the integration. For example, to study diffusion, Q must be put equal to $m_1 u_1$.

In the final form of the equation for rate of change of Q, V entered the integral in the form $V_{12}^{(n-5)/(n-1)}$, where n is an integer. The definition of n is that the force law between interacting molecules is of the form k/r^n, k being a constant and r the distance between the molecules. Maxwell also pointed out that from experiments on the viscosity of gases, which he conducted, $n = 5$, and therefore V_{12} would disappear from the expressions, and they will be capable of immediate integration.

Using his new method, Maxwell derived expressions for all the transport coefficients of a gas. At each stage, he was careful to see that the equations on the molecular level were compatible with the hydrodynamic equations of continuity and conditions for the equilibrium in fluids. The only derivation that changed was that for thermal conductivity. He then expressed his new coefficient of thermal conductivity in terms of the coefficient of viscosity,

$$K = \frac{5}{3(\gamma - 1)} \frac{P}{\sigma\theta} \frac{\mu}{\rho}$$

where μ is the coefficient of viscosity, γ is the ratio of the specific heats of the gas (c_P/c_V, where c_P and c_V are the specific heats at constant pressure and volume respectively), ρ is the density of the gas and σ the diameter of the molecules of the gas. From this expression, Maxwell obtained a value for the mean free path using his own experimentally determined value for the coefficient of viscosity.

Considerable space in this paper was taken up by a theory of diffusion, which had been the most closely studied transport property of gases from the experimental point of view. Maxwell again analysed the diffusion of gases down long, thin tubes, and then their diffusion through a hole. In the latter case, the gas leaked through the hole as if no other gases were present. In addition, when the pressure on either side of the hole was equal, the rate at which the gases diffused was given by Graham's Law of Diffusion. Yet Maxwell could only obtain the conditions he wanted provided no energy was transported into or out of a stratum. If, say, the temperature was constant, there was no heat conduction.

He also worried that his gas theory appeared to be incompatible with the second law of thermodynamics. Neither he, nor the referee of his paper, William Thomson, was able to solve the problem. In fact the issue remained unresolved until Maxwell solved the thermodynamic problem after Francis Guthrie spotted the dilemma for Maxwell's gas theory while reading Maxwell's thermodynamics text, *Theory of Heat.*

These first six years of Maxwell's interest in gas theory were his most fruitful. After 1866 he concentrated on finishing his treatise on electricity and magnetism, and it was not until 1873 that he again became interested in gas theory. In the intervening years, Boltzmann's first papers on gas theory and the distribution function had appeared, and Maxwell gained a deeper insight into gaseous behaviour. Until that time, Clausius continued to influence Maxwell. Specifically, he was the first to express the equipartition theorem.

The Equipartition Theorem

In the middle of the nineteenth century, all molecular models were mechanical, and they ranged from those based on centres of force or using ball bearings, to the mathematically elegant and mechanically stable configurations of interlocking fluid vortex rings. Vortex rings and later more sophisticated models were used to explain the chemical as well as physical properties of matter, and conversely chemical evidence was used to justify the structures of molecular models.

Clausius was the first physicist to use molecules to develop a theory of the three states of matter. His molecules had inner structure, and the components could vibrate and rotate with respect to the centre of mass of the molecule. The total kinetic energy of the molecule was shared equally by three motions, the translational motion of the molecule and two internal motions, rotation and vibration. He used evidence from thermodynamics to investigate the properties of a gas made up of such molecules, and deduced that, for a gas in equilibrium, the ratio of the energy of its translational motion to its total energy was a constant. He also determined the value of this constant for air, and found that the kinetic energy of translational motion did not equal that of the internal motions of the molecules. However, the kinetic theory of gases indicated that 'all the molecules [of different gases] in thermodynamical equilibrium, on the average, have the same vis viva [effectively kinetic energy] of translational motion'.[13] He showed that, although the total kinetic energy of a molecule did not divide itself equally among its different types of motions, the average translational energy per molecule was the same for each gas.

This was a partial statement, without proof, of the equipartition of energy. In his subsequent work, Clausius only applied the equipartition law to linear motion and never extended it to the internal motions of the molecules. He merely deduced the total kinetic energy taken up by internal motions from experiments devised by others, and thus he was never troubled, as Maxwell was, by the difference between a theoretically derived ratio for the specific heats and that deduced from experiments. Clausius, though, did deduce that for a gas in equilibrium the ratio of the energy of translation to the total kinetic energy was a constant.

Maxwell's first gas theory was independent of any need to use experiment to define the energy ratios in the various types of motion of the gas molecules. His general derivation of the principal of the equipartition of energy was a necessary deduction from his assumptions, although it would lead to more problems for his theory than it solved. In his paper of 1860, where his molecules were elastic spheres, the gas laws were obeyed and he expressed the pressure, P, as

$$P = \left(\frac{1}{3}\right) mN{<}c^2{>} = KT/V$$

where N is the number of molecules in a unit volume and K is a constant. (Actually K is NVk, where k is today's Boltzmann's constant.)

For equal volumes of different gases, at the same temperature and pressure, the total kinetic energy of translational motion per unit volume was the same. This statement, by itself, proves nothing about the average translational kinetic energy of molecules of each gas, and Maxwell demonstrated the equipartition of energy by a separate argument. He considered elastic collisions between two masses and the changes in kinetic energy of each mass associated with the collisions. He then constructed and examined the change in the difference of the kinetic energies of the particles before and after the collision. At each impact, the difference in the average kinetic energies is reduced by a constant ratio, until the difference no longer exists and their average kinetic energies become equal. Thus, even if particles of the different gases start with different average kinetic energies, eventually these become equal. This result can be extended to any number of different kinds of particles colliding elastically with each other. This statement of the equipartition theorem is still incomplete and equivalent to that of Clausius, as it deals only with translational energies.

To find the ratio between the rotational and translational kinetic energies, Maxwell examined collisions between perfectly elastic bodies of arbitrary shape. In the case of irregularly shaped bodies colliding with one another, the impact will, in general, change both the translational and the rotational energies of the masses until 'some relation is established between them'. He assumed that the nature of the distribution of rotational kinetic energy about the three principal axes of the bodies was the same as that of the distribution of translational energy. He then examined the change in total kinetic energy for one body during an elastic collision, and expressed the velocity of the body after impact as a function of its own initial velocity and that of the colliding body. He was only interested in the relationship between the average energies of the colliding molecules.

After many collisions, a situation is reached in which the average translational energies of the molecules are equal, as he had previously demonstrated. Since the total energy was conserved during every elastic collision, the difference in total kinetic energies before and after a collision was zero. Maxwell assumed that this relationship held when using average values for the velocities. Yet all directions for the velocities and positions in space of the molecules were equally probable. Therefore, in the expression linking the average total kinetic energies before and after a collision, the terms involving position and direction drop away. Conservation of energy holds for the average square velocities only if the total translational kinetic energy of the molecules is equal to the energy associated with the rotational velocities of the molecules.

In the final state of the gas, the average translational kinetic energy of the molecules is the same along each axis and equal to the average rotational kinetic energy of the molecule about each of the three principal axes of that molecule. Maxwell then made his first, general statement of the Equipartition theorem: '[A]dding the vires vive [effectively the kinetic energies] with respect to the other axes, we find that the whole vis viva of translation is equal to that of rotation in each system of particles, and is also the same for different systems.'[14] This is true for all colliding elastic bodies of any shape, but restricted to systems of such bodies for which the distribution of rotational and translational velocities follows the law established by Maxwell. Thus, the equipartition theorem is a consequence of Maxwell's distribution law.

In today's terms, Maxwell was suggesting that his gas molecule has six degrees of freedom, by which we mean six ways of storing energy, three translational, one corresponding to motion along each axis, and three rotational, again one corresponding to each axis. The average energy associated with each degree of freedom is the same at any particular temperature.

For a general number of degrees of freedom equal to n, the quantity β used by both Clausius and Maxwell, and equal to the ration of the average total energy of the molecule to the average translational energy, is equal to $n/3$. The ratio of c_P to c_V, always called γ, is greater than 1 because of the work done by the gas expanding if it is heated at constant pressure. In fact it is equal to $\frac{n+2}{n}$. So

$$\beta = \frac{2}{3}\frac{1}{\gamma - 1} \quad \text{or} \quad \gamma = \frac{3\beta + 2}{2}$$

So, for $n = 6$, the values predicted are $\beta = 2$; $\gamma = 4/3 = 1.333$.

There was, though, an alternative idea. Maxwell usually regarded spherical molecules as smooth, so that they were not set into rotation by collisions. Thus, for that assumption, $\beta = 1$, and $\gamma = 1.667$.

However, the value deduced from experiment, using the experimentally known ratio for the specific heats, agreed with neither set of values. For air, the experimental value of γ used by Maxwell was 1.408, and this leads to a value of β of 1.634. Clearly there was a fundamental difference between experiment and either theory. This appeared to indicate that colliding molecules did not behave as elastic rotating bodies of any shape. Maxwell noted that 'The result of the dynamical theory being at variance with experiment overturns the whole hypothesis, however satisfactory other results may be'.[15]

Maxwell was not content to leave this gap in the structure of his theory and searched for a more satisfactory theoretical model for the molecule. He replaced the molecular model by turning to Clausius' suggestion of treating molecules as centres of force, but Maxwell still deduced the same distribution function using a centre-of-force model as before, and the mean translational energy was also the same. Some flexibility appeared to be possible; with this model the proportion of kinetic energy stored in the internal motions of the molecules could be guided by the experimental value for the ratio of the specific heats. However, the fraction of the total energy carried as rotational energy depended on the nature of the molecules themselves, and, what seemed worse, this fraction changed with each encounter.

Centre-of-force molecules move either singly, or in clusters, and the aggregate molecules moving together have a surface that is not spherical. In addition, the rotary motion of the system will store up a certain proportion of the whole kinetic energy. Assuming this to be the case, Maxwell could account for the value of the specific heats being larger than that deduced from the elastic sphere model, and the experimentally deduced specific heats ratio was the key to exploring the nature of the molecule. Indeed, after 1860, his papers on gases were part of a search for a satisfactory molecular model, although he did not entirely discard the spherical ball. The problem of the internal motions of molecules continued to occupy Maxwell. In a letter to a life-long friend, H. R. Droop, in 1862, Maxwell noted that while most of the known, gaseous phenomena supported the spherical model, 'still one phenomenon goes against the theory – the relation between the specific heat at constant pressure and that at constant volume, which is, in air, 1.408 while it ought according to theory to be 1.333'.[16]

The experiments Maxwell conducted in 1865 on the viscosity of air gave him the evidence he needed to finally discard the elastic sphere model. The conclusion he drew from these experiments was that: 'If, as the experiments of Graham and those of this paper show, the viscosity is as the first power of the absolute temperature, then, in the dynamical theory, which is framed to explain the fact, we must assume that the force between the molecules is proportional inversely to the fifth power of the distance between them'.[7] However, a gas of centre-of-force molecules led to the same velocity distribution law as elastic sphere molecules. And, from this distribution law, he deduced that the mean translational energy of all molecules was equal, which he noted was an important result for the theory as 'it is independent of the nature of the action between the molecules, as are all the other results relating to the final distribution of velocities'.[17]

The above result was only valid for the translational motion of the molecules. The fraction of the total energy of each molecule carried as rotational energy depended upon the nature of the molecules themselves, and this fraction changed with each encounter. But, on the average, in equilibrium, this fraction was a constant, as had been demonstrated by Clausius. Because this paper was a complete reworking of his earlier kinetic theory, Maxwell again derived the ratio of the two specific heats. To do this he considered the total energy of translation of the gas as a body, and the energy of agitation due to the random motions of the molecules, which was equally divided, on average, between all molecules. In addition, he finally took into account the internal motions of the molecules. He used the energy of the molecules, due to their random motions and their own internal structure, that is the total energy in the unit volume of the medium, to find the ratio of the specific heats of a gas. However he obtained exactly the same values for β and γ as in 1860.

Concluding Remarks

In his 1860 paper, Maxwell presented kinetic theory as a hypothesis, and this tone continued in his later paper of 1866 in which he discussed as many ideas on the structure of molecules as could throw light on the theory of gases. Whether he believed any of these models actually represented reality is another question, which he avoided addressing.

Yet, as time passed, his general confidence in the ideas grew, and the emphasis of his later papers shifted from setting up and justifying the theory of gases, to choosing between theories. In addition, increasing amounts of experimental evidence allowed Maxwell to return to the topic and express his doubts about the validity of the equipartition theorem, and allowed him to turn to a theory of molecular models that he hoped might avoid the difficulty. At no time did Maxwell express any doubt that a kinetic theory of gases was correct, for even his first elastic sphere molecule yielded results that explained many of the known, previously isolated, properties of gases.

It should be mentioned that the opinions of chemists were important for the acceptance of kinetic theory, since the results of the theory accounted for many known chemical properties of gases. However, amongst the chemists there was a more basic opposition to kinetic theory, because the molecular theory of matter, although used in chemical theories, was not considered a reality. Maxwell, finally, presented his case to the Chemical Society in 1873, where he used as much chemical evidence as possible to support molecular theory.

The equipartition theorem and its significance for kinetic theory bothered Maxwell even at its inception, and he avoided its complications in much the same way as Clausius, by taking into account the complexity of the molecule, indicated by chemical evidence, and he later rationalized these actions with his own experiments. At this stage, Maxwell's concept of the molecule was as a collection of centres of force, bound together by internal forces, which interacted as a group with other groups of the same type.

History has shown that Maxwell was quite right to regard the failure of the equipartition theory as exceptionally significant. Today, we would work in terms of the atomicity of the molecule. A monatomic molecule, such as argon, exhibits only translational motion, so n, the number of degrees of freedom is equal to 3, β should be 1, and γ should be $5/3$. These are the values obtained by experiment, and also agree with those for Maxwell's smooth spherical molecule.

However, let us now consider a diatomic molecule with the line joining the molecules along the x-axis. There should be two modes of rotation, one about the y-axis and the second about the z-axis. (Rotation about the x-axis is not considered because this is a rotation only of the atoms, not of the molecule.) But there is also a mode of vibration along the x-axis. And while, as we have taken for granted, each mode of translation or rotation corresponds to a single degree of freedom, because only kinetic energy is present, any mode of vibration corresponds to two degrees of freedom, because there are two stores of energy: potential and kinetic. Thus the total number of degrees of freedom, n, is predicted to be 7, the sum of 3 for the translational degrees of freedom,

2 for the rotational ones, and 2 for the single vibrational mode. β and γ are predicted to be $7/3$ and $9/7$ respectively.

We may say that this is a definitive conclusion of nineteenth-century physics. It is not surprising or concerning that the values do not agree with any of Maxwell's models – there is a marked difference between his models and our own. But it is extremely important that they do not agree with experiment. As has already been stated, the value of γ for air, which consists largely of the diatomic molecules of oxygen and nitrogen, is 1.408 at usual temperatures, which is very close to $7/5$. This suggests that for diatomic molecules, n is 5 rather than 7, and it is often said that the vibration of the molecule is 'frozen out' at usual temperatures. However a genuine understanding of this 'freezing out' had to wait for the coming of quantum theory in the twentieth century.

So, in his concern about equipartition, Maxwell touched on the aspect of molecular physics that would demonstrate the inadequacy of nineteenth-century physics. As with his concern for Maxwell's demon, discussed later in this book, he showed his ability not just to produce theories of the greatest importance, but also to home in on those parts of the accepted theories of his day that demonstrated significant gaps in scientific understanding, and would lead to important discoveries and advances long after his own death.

CHAPTER 8

Maxwell and the Theory of Liquids

JOHN S. ROWLINSON

The curious incident of the dog in the night-time.
The dog did nothing in the night-time.
That is the curious incident.
'Silver Blade', from *The Memoirs of Sherlock Holmes.*

To paraphrase Conan Doyle, we might now write: 'the curious incident of Maxwell's contribution to the theory of liquids – but Maxwell made no contribution to the theory of liquids – that is the curious incident.' The burden of this chapter is to explain why Maxwell, who was deeply interested in liquids, did so little to advance their theory.

Most simple substances can exist in three states: gas, liquid and solid. The mechanical and thermal properties of liquids and solids are easier to measure than those of gases, and by the early nineteenth century the classical subjects that we now call hydrodynamics and the elasticity of solids were well explored. It was more difficult to arrange the information so obtained into ordered or rational mental structures. The efforts made by the French school of applied mathematics, which arose mainly from the military and engineering works at the time, fell into two broad classes. Laplace and his disciples – Poisson, Biot, and, with reservations, Navier – maintained that matter was formed of small, invisible but real *molécules* and that the mechanics of liquids and solids should be explained in terms of forces acting between such particles. Laplace had one great success with this approach when he explained in such terms the mysterious capillary rise of liquids in narrow tubes. He set out his theory, which we accept today as broadly correct, at the end of the fourth volume of his *Mécanique Céleste*, in 1806.[1] He tackled the problem of the interaction of a great number of particles, N, by considering one of them in interaction with (N-1) others that were distributed in a random way. This device allowed the effect of these (N-1) neighbours on the particle under consideration to be estimated by averaging their positions over the whole available space, so producing a uniform field, proportional to the square of the number density of the particles. This assumption now goes by the name of a 'mean-field approximation' and was to be revived later in the century by Athanase Dupré,[2] and, with greater success, by J.D. van der Waals. A different approach was taken by those, many of them engineers, who distrusted reliance on hypothetical particles. They did not usually deny that matter was made of particles, but they thought that their needs were better met by ordering their experimental results into theoretical structures in which matter was treated as a continuum with contact forces between small contiguous portions of matter. This group, led by Fourier, comprised such men as Clapeyron, Lamé and, again with some reservations, Cauchy. They were followed by George Green in Britain.

When, by the 1840s, the French way of doing theoretical physics had taken hold in Cambridge, it was the continuum school which was taken as the model. Laplace, who had died in 1827, was then in low repute, partly because he had tried to hold to the view that light, as well as matter, was particulate. The successful wave theory of light, Faraday's experiments in electricity and magnetism, and a theory of heat that evolved into classical thermodynamics, all convinced the leaders of the day that non-molecular physics was the way forward. In his Presidential Address to the British Association in 1845 John Herschel said:

> The time seems to be approaching when a merely mechanical view of nature will become impossible – when the notion of accounting for *all* the phaenomena of nature, and even of mere physics, by simple attractions and repulsions fixedly and unchangeably inherent in material centres (granting any conceivable system of Boscovichian alternations), will be deemed untenable.[3]

Maxwell was raised at Edinburgh and Cambridge in this prevailing 'continuum' tradition. He took Faraday's theoretical speculations seriously and his immediate mentors were George Stokes and William Thomson. The latter was an admirer of Fourier's work and, until the 1870s, he was distrustful of molecular methods for tackling physical problems. In 1861 he, in Glasgow, and his friend Peter Guthrie Tait, in Edinburgh, embarked on the writing of a major work which aimed to provide a foundation for the understanding of physics, their *Treatise on Natural Philosophy*. There was trouble from the beginning, since Tait believed in atoms and Thomson then did not. In 1861 Tait wrote to Thomas Andrews in Belfast: 'Thomson is dead against the existence of atoms; I though not a violent partisan yet find them useful in explanation'.[4] A few days earlier he had written to Thomson: 'It is amusing to see how definitely you go into the ease of conception and treatment of the continuous uniform medium in which atoms (or at all events matter) are supposed to float'.[4] Maxwell followed Faraday and Thomson in his electrical work, and Stokes and Thomson in his early paper 'On the Equilibrium of Elastic Solids'.[5]

The comfort of this consensus was disturbed by a group of outsiders who had not had the benefit of the formal mathematical instruction of the French Grandes Écoles or of Cambridge University. Working independently of their academic leaders, and of each other, they embarked on brave attempts to interpret the properties of gases in terms of rapidly moving particles that transferred momentum to the walls of the containing vessels and exerted negligible forces on each other except for rare elastic collisions; that is, these four men developed a kinetic theory of gases. In order of seniority they were John Herapath (1790–1868) a journalist and enthusiast for railways; John James Waterston (1811–83) a Scots engineer who taught naval cadets in India; James Joule (1818–89) of a Salford brewing family, a man who had yet to establish his reputation for his thermodynamic work; and August Krönig (1822–79) who taught in a Berlin Technical College. Their ideas were basically sound but their enthusiasm often outran their technical competence. Their work from the 1820s to the 1850s did, however, wake up the leaders of the day and make some of them wonder if there was more to learn about such molecular interpretations of physical properties. Two men took up the challenge, Maxwell and Rudolf Clausius (1822–88), who put the innovators' ideas on the behaviour of gases on a sounder mathematical basis. The result was that the theory of gases became detached from that of liquids and solids where the methods of these 'kineticists' could not be used. Some years later, in his *Theory of Heat*, Maxwell wrote:

> The mathematical difficulties arising in the investigation of the motion of molecules are so great that it is not to be wondered at that most of the numerical results are confined to the phenomena of gases. The general character, however, of the explanation of many other phenomena by the molecular theory has been pointed out by Clausius and others.[6]

The continuum school soon lost any control of the theory of gases when Maxwell accepted the real existence of atoms and molecules, and Clausius came out of his shell and also acknowledged their existence. He had long believed in them, but when he was doing his pioneering work in classical thermodynamics he had suppressed any reference to them, believing correctly that his thermodynamic work should rest entirely on a macroscopic description of matter and not be compromised by dubious statements on its ultimate atomic structure. This was a trap that Macquorn Rankine had fallen into when he made his generally sound statements on classical thermodynamics depend on his ideas of an atom as a vortex in an ether, so making his influence on the field much less than it might have been.

In the hands of Maxwell, Clausius and later Boltzmann, the kinetic theory of gases became an acknowledged part of physics by the 1870s, but no extension to liquids or solids seemed possible without an assumption of a mean-field approximation of the kind used by Laplace, and in 1873 also by van der Waals. Lamé, writing in 1852, said that imitation of Newton's theory of gravitation was not then the way forward. He accepted that attractions between particles might in time provide an explanation of 'terrestrial phenomena' but 'to presuppose [the existence] of this unknown principle . . . is to hold back, perhaps for a long time, the epoch of its discovery'.[7] Thirty years later, Thomson came to accept atoms and molecules, and even to make estimates of their sizes, but he was as convinced as Lamé had been that the molecular approach was still not the best way of tackling liquids and solids. For solids, the discussion between the atomic school of 'elasticians' and those who eschewed atomic hypotheses centred around the number of independent elastic constants needed to characterize a crystal of given symmetry. The atomic school used what are called 'Cauchy's relations' to reduce the number of independent elastic constants; thus, for example, they reduced the three independent constants of a crystal with cubic symmetry to two, by adding the additional assumption that the cohesion of the crystal arises from forces acting between the centres of each of the constituent particles. Thomson wrote of this reduction:

> Under Properties of Matter, we shall see that an untenable theory (Boscovich's), falsely worked out by mathematicians, has led to relations among the coefficients of elasticity which experiment has proved to be false.[8]

The argument over the elasticity of crystals ran on into the twentieth century, and its resolution came only when much more was known about their molecular structure.

The impasse in the theory of liquids was broken by the supporters of the atomic school making what Maxwell thought was an arbitrary assumption, that some of the molecular properties of a gas could be taken to have the same values in a liquid, and, presumably, in a solid also. Did, for example, Maxwell's expression for the distribution of molecular velocities in a gas hold also in a liquid? To Maxwell, there seemed at first to be no good reason for assuming that this was so.

He himself had guessed that, at a given temperature, the speed of molecules in a liquid might be less than that in the vapour with which the liquid was in equilibrium. This guess appeared in the first edition of his monograph on *Heat*,[6] and it survived in all later editions up to the last in 1891, 12 years after his death. Nevertheless, Rankine[9] and Clausius[10] made the assumption that the speeds were the same in liquids and gases, with the corollary that the 'kinetic' contribution to the heat capacity was the same in both states of matter. Rankine rather obscurely called this part the 'true' specific heat. Clausius, in a sentence that he italicized, wrote: '*The quantity of heat actually present in a body depends only on its temperature, and not on the arrangements of its component particles.*' In a letter to Tait, Maxwell noted sarcastically that 'the condition of energy – in the privacy of bodies – is known only to R, C and Co'.[11] Sixteen years later, when he was reviewing Tait's book on *Thermodynamics*, Maxwell called Clausius's statement 'the most important doctrine, if true, in molecular science'.[12] But was it true?

Maxwell had to face this problem in public when he undertook to review for *Nature* an obscure thesis, in Dutch, by a hitherto unknown schoolmaster. This was the thesis 'On the Continuity of the Gaseous and Liquid States'[13] that J. D. van der Waals had submitted to the University at Leiden in June, 1873. Here he had boldly adopted Laplace's mean-field approach and the stance of Rankine and Clausius of assuming that the mean translational kinetic energy of the molecules is a measure of the temperature, indeed: 'This may be considered to be our definition of temperature', and so that 'a molecule of water and a molecule of steam at 0 °C have the same velocity of progressive motion'.[14] His assumption is, however, an evasion of the problem, since he does not show that this definition of temperature necessarily coincides with that of the absolute scale furnished by the second law of thermodynamics. Maxwell could not accept an assumption as brave as this on so fundamental a point. He wrote

> At present we know the conditions of thermal equilibrium only in the case of gases . . . With respect to substances of more complicated constitution, we know, as yet, nothing of the physical condition on which their temperature depends although the researches of Boltzmann on this subject are likely to result in some valuable discoveries . . . M. Van der Waals seems, therefore, somewhat hasty in assuming that the temperature of a substance is in every case measured by the energy of agitation of its individual molecules.[13]

Maxwell did, however, add one important result to van der Waals's theory, whose equation of state correctly predicted that there was a critical temperature above which there was no distinction between liquid and gas; for each pair of values of pressure and temperature there is, above that temperature, only one possible solution of his equation of state which yields the volume appropriate to these values of pressure and temperature (see Fig. 8.1). Below this critical temperature there are three real solutions, the largest of which is related to the volume of the gas or vapour, and the smallest of which is related to the volume of the liquid. (The third or intermediate solution was recognized as a physically unstable state in which the compressibility of the fluid was negative and which was therefore a solution that could be discarded.) Van der Waals, however, although he could identify the branches of the isotherms on which the volumes of the coexisting liquid and gas lay, could not calculate the volumes themselves. An additional constraint was needed for this to be done. It is the constraint that we now identify with the requirement that coexisting phases must have not only the same pressures and temperatures, but also the same chemical potentials. In the days before Gibbs's great paper on thermodynamics had been published and understood, this condition could be shown to be satisfied only by requiring that the work done in going from one phase to the other be the same whether the isotherm followed the continuous looped line of

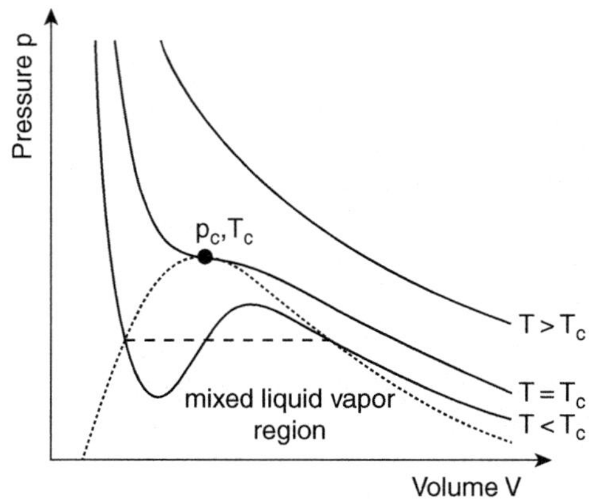

Fig. 8.1 Three isotherms for a fluid. The top isotherm is for a temperature above that of the critical point and shows volume as a monotonic function of pressure. The second is at the temperature of the critical point and has zero slope and a point of inflection at the point itself. The lowest is for a temperature below the critical where simple theoretical curves, such as that of the van der Waals equation, give three values of the volume for a range of values of the pressure. Experiment, however, shows a line with three distinct segments: van der Waals-type curves in the liquid region to the left of the figure and in the gas region to the right, joined by a horizontal line connecting the volumes of the coexistent liquid and gas phases. Maxwell showed that the position of this line on a theoretical curve is found by placing it so that, in the mixed liquid vapour region, the areas of the theoretical curve above and below the straight line are equal. The dotted line is then the locus of the coexistent liquid and gaseous phases.

van der Waals's equation or the horizontal line that represented the experimental facts. Maxwell showed that this condition was satisfied if the areas of the loops above and below the horizontal line were equal, a condition that came to be called Maxwell's 'equal-area rule'.

Maxwell struggled to find this result. As early as 1871, he was asking James Thomson: 'The next difficulty is: What determines the true boiling temperature which is found to be so constant?' He thought at first that evaporation of the liquid started at the pressure of the maximum of the loop, a proposal to which Thomson objected, saying that 'There is just one intermediate point of pressure . . . at which the liquid and its gas can be present together in contact with one another; and this is the boiling, or rather the evaporating or condensing pressure, for the temperature to which the curve belongs'. Neither he nor Maxwell knew at first how to find this intermediate pressure. Maxwell sent an unsound argument to Thomson in a letter of 24 July 1871, but he arrived at the correct result three years later, first in a postcard to P. G. Tait and more publicly in a lecture before the Chemical Society in 1875, in which he set out his belief in the atomic constitution of matter and the evidence from the kinetic theory of gases on which it was mainly founded.[15]

Maxwell wrote his textbook on *The Theory of Heat* while he was in semi-retirement at Glenlair, in Scotland, between his appointments at King's College, London, and the Cavendish Laboratory at Cambridge. He wrote his review of van der Waals's thesis soon after his arrival in Cambridge in 1871 as the first Cavendish Professor. He was therefore probably not able to keep up with what Boltzmann was doing, for he notoriously found it difficult to follow Boltzmann's proofs, 'so long and roundabout, that it requires no ordinary exercise both of penetration and patience to find out if they are proofs at all'.[16] In fact, Boltzmann had taken, from 1871 onwards, the first tentative steps that led from the simplicities of the kinetic theory of gases to the more abstract, and ultimately more powerful, methods that we recognize today as statistical mechanics. Conservation laws played an important role in nineteenth-century physics, and the most important of these was that of conservation of energy. In a theory of gases, it was possible to make a start by ascribing a definite energy to each individual molecule and to assume that energy was exchanged between molecules at two-body collisions. One consequence of such a scheme, as both Maxwell and Boltzmann showed, was the equipartition of energy between different modes of molecular motion, such as translational energy, rotational energy and energy of internal vibrations of the constituent atoms within a molecule. No such assumption of ascribing the energy unambiguously to independent molecules was possible, however, in a dense gas or in a liquid where each molecule is close to and exerts forces on the neighbouring molecules. The total energy of an assembly of molecules could be specified, but there was no unique way of dividing it between the constituent molecules. It was this impasse that Boltzmann first broke in his papers of 1871–72, by the device of considering a group of molecules as a single, but larger, super-molecule.[17] In the first paper of this group, he had already established the identity of the velocity distribution in a gas and in an assembly of interacting molecules: 'The mean kinetic energy is thus the same for all atoms in the interacting molecules. The thermal state which the many interacting atoms assume, we will call the state of equal temperature; thus at equal temperatures the mean kinetic energy is equal; this can be taken to be an important measure of the temperature.' Maxwell seems not to have persevered in any study of Boltzmann's long and difficult papers of these years, and so was at first unconvinced by his arguments. His conversion seems to have come in the years between his review of van der Waals's thesis in 1874 and his review of H. W. Watson's book on kinetic theory in 1877.[16] In the latter review, he noted that Watson had excluded all intermolecular forces except for those acting between colliding pairs of molecules. He now believed that this restriction is unnecessary 'and indeed it is easy to remove it' which he did by Boltzmann's device of treating a group of interacting molecules as one polymolecular system. His justification of Boltzmann's result came in one of his last papers; it was of a length of which Boltzmann might have been proud.[18] Boltzmann was impressed by this paper and wrote an abstract of it for the *Beiblätter* of the *Annalen der Physik*, and persuaded the editors of the *Philosophical Magazine* that his abstract was of sufficient importance to the subject to justify an English translation.[18] Maxwell died before this was published.

It is pointless to speculate what the future of statistical mechanics might have been if Maxwell had survived to the beginning of the twentieth century. Suffice it to say that the way that Boltzmann and he had opened up was not to prove the best way forward. The clumsy device of considering groups of two, three, four and so on molecules had one important success; it yielded an expression for first-order correction of the perfect-gas law when the product pV was expanded in powers of $(1/V)$. This equation related rigorously a quantity measurable in the laboratory, which we now call the second virial coefficient, to the forces acting between a pair of colliding molecules. Maxwell derived it in 1876 when reviewing for publication a lecture of Thomas Andrews to the Royal Society.[19] Unfortunately, he never published it, and it did not come into general knowledge and thus into general use until it was derived again by Boltzmann in 1896.[20] The route taken by Boltzmann and Maxwell was followed again by James Jeans in his *Dynamical Theory of Gases* of 1904.[21] In the 1920s and 1930s it was found that this so-called virial expansion was a useful way of studying the equation of state of gases, since each successive group of two, three, four molecules could be rigorously related to the corresponding term in an expansion of the pressure of a gas in the second, third, fourth powers of its density. But the mathematical difficulties of calculating these terms rapidly limited the degree to which the expansion could be taken, and doubts about the convergence of such a series made it useless for any discussion of liquids. The second and, occasionally, the third coefficients were often measured in the twentieth century and were the foundation of much of what we came to know about intermolecular forces. A more radical advance, however, was needed before the theory of liquids could make any progress.

Hitherto the centre of attention of molecular theory had been, not unnaturally, the molecule itself, but there was another option. Instead of considering a single ensemble of interacting molecules with a fixed total energy, but with the attendant difficulty of deciding where this energy was residing, one could consider an assembly of such ensembles of interacting molecules. Each ensemble is now a macroscopic system and there is no difficulty in assigning an energy to it. The assembly of these ensembles is now a gross system at equilibrium, in which some energy may flow from one ensemble to another when all are contact with each other and so at the same temperature. By magnifying the scale of operations, one has bypassed the difficulties of the single ensemble that plagued the early practitioners of statistical mechanics in the closing years of the nineteenth century. The single ensemble is called a micro-canonical ensemble, and for it one attempts to calculate the entropy as a function of the number of molecules, the volume to which they are confined, and the total energy of the ensemble; that is one seeks to evaluate the entropy S as a function of N, V and U, or, alternatively, U as a function of N, V and S. The larger entity, the assembly of these ensembles, is called a canonical ensemble, and for it one seeks to calculate the Helmholtz free energy, F, as a function of N, V and T. This change of perspective to the larger system was made first by Willard Gibbs in a book of 1902, where he also went a step further and introduced the grand canonical ensemble, in which the volume is still fixed but the number of molecules and the energy in each micro-ensemble can both now fluctuate, driven only by the need to keep the temperature and chemical potential constant.[22] These more sophisticated ensembles were to prove the way forward in developing a theory of liquids, a key early player being again a Dutch physicist, L. S. Ornstein, whose Leiden thesis of 1908 was one of the first in which Gibbs's methods were applied to liquids.[23]

The use of a canonical ensemble solves at once the difficulty that faced Rankine, Clausius, Maxwell and van der Waals of finding the distribution of molecular velocities in a liquid. The energy of the canonical ensemble is written in Hamiltonian form when the translational kinetic term is $\Sigma(\boldsymbol{p}_i)^2/2m_i$, where m_i and $\boldsymbol{p}_i$ are the mass and linear momentum of molecule i, and the sum is taken over all molecules. This sum is independent of the other terms in the energy which relate to molecular rotations, vibrations and the potential of the intermolecular forces, and so the integration over the linear momenta in the canonical partition function can be made at once whatever the intermolecular energy is. The classical distribution of molecular speeds is therefore the same in all three phases of matter, as Rankine, Clausius and van der Waals had supposed.

The term 'theory of liquids' is not generally taken today to include classical thermodynamics, but the two subjects were more closely connected in the nineteenth century, and thermodynamics was also a field in which Gibbs's and Maxwell's work overlapped. Gibbs had re-laid the foundations of classical thermodynamics in his monumental paper of 1876–78.[24] Maxwell received a copy of the first part of this early in February 1876 and was at once bowled over by its rigour and elegance. Within weeks he was writing to Tait: 'Have you Willard Gibbs on the Equilibrium of Heterogeneous Substances? If, not, read him'.[25] Both Maxwell and Gibbs independently formally solved the problem of finding the conditions that determine the equilibrium between the phases of a system with more than one component. Maxwell had generalized Gibbs's basic primitive surface[26] for a single substance, $U(V, S, M)$, to a hypersurface, $U(V, S, M_1, M_2, M_3 \ldots M_n)$, where M_i is the mass of component i, and there are n components. He introduced what he called the 'reactions', $r_i = (\partial U/\partial M_i)$, for $i < n$, where the differentiation is at constant values of V, S and all M other than M_i. He showed that the condition of equilibrium was the equality in each phase of pressure, temperature, the $(n - 1)$ reactions and of the function $(U - TS + pV - \Sigma\, M_i r_i)$, where the sum is taken over the $(n - 1)$ molecules. He never solved, however, the problem of determining the form of his hypersurface, writing 'But the difficulty is to see how to form the surface for a mixture when those of the constituents are given'.[27] All this he had achieved and reported in the postscript of a letter to George Stokes on 3 August 1875, some months before he received Gibbs's paper. Meanwhile, Gibbs had proposed the equivalent, but more symmetrical, condition of the equality of the n chemical potentials, $\mu_i = (\partial U/\partial N_i)$ for all i, where N_i is the 'amount of substance i', measured usually in moles, and the differentiation is under the same restrictions as Maxwell had used.[24] The two schemes differ trivially; Maxwell's reactions are related to Gibbs's potentials by:

$$r_i = \mu_i - \mu_n \text{ for } i < n, \text{ and } r_n = \mu_n.$$

Maxwell recognized at once the superiority of Gibbs's scheme, because of its greater symmetry and because the appropriate surface could readily be formed, at least for a binary mixture, by starting not with the primitive $U(V, S)$ surfaces but from that for the Helmholtz free energy $F(V, x_1)$, where x_1 is the mole fraction of the first component. Maxwell acknowledged that 'The idea of the potential of a substance is, I believe, due to Prof. Gibbs'. The paper[28] in which Maxwell acknowledged the superiority of Gibbs's scheme was lost from sight for many years, which was unfortunate, for it contained one innovation that would have improved the teaching and understanding of thermodynamics had it become more widely known. For many years students were, and indeed, still are, taught to distinguish between *extensive* and *intensive* thermodynamic properties; the former are proportional in magnitude to the size of the system and the latter are independent of the size. Maxwell, however, distinguished between *magnitudes* and *intensities*. The latter are pressure, temperature and Gibbs's potentials, all of which are uniform throughout a system at equilibrium, while the former can change from phase to phase. The thermodynamic description of a system is in terms, each with the dimensions of energy, formed of pairs of products of an intensity and of a magnitude, as, for example, pV, TS and μN. Thus pressure and mole fraction are both called intensive properties, but they are of very different kinds; only the former is one of Maxwell's *intensities*, the latter is a *magnitude*. Maxwell's distinction was rarely made explicit in thermodynamic writing until Robert Griffiths and John Wheeler re-introduced it in 1970 under the names *densities* (= *magnitudes*) and *fields* (= *intensities*).[29]

One further interest that Maxwell and Gibbs had in common was the study of the surface tension of liquids, or the subject of capillarity as it was then usually called. Here Gibbs was a real innovator and Maxwell's principal contribution was a fine review of the field that he wrote for *Encyclopaedia Britannica*.[30]

The Cambridge school of theoretical physics was surprisingly slow to take advantage of Gibbs's innovations in statistical mechanics. One could not expect Jeans in 1904 to have mastered the difficulties of Gibbs's book of two years earlier, but he was still following the Boltzmann–Maxwell line in his second edition of 1916.[21] Between the two World Wars, the leading writer on statistical mechanics at Cambridge was R. H. Fowler, whose monograph appeared first in 1929, with a second edition in 1936 and a closely related book aimed also at chemists in 1939.[31] Fowler wrote that there are two possible foundations for statistical mechanics, 'the Gibbsian ensemble and the general conservative dynamical system'; he adopted the latter. He therefore discussed only gases and idealized solids in which the state of each molecule can be described independently by supposing that its neighbours remain fixed on their lattice sites. For an imperfect gas, he considers quantitatively only the interaction of molecules two at a time, noting that the chance that they are at a separation where their mutual energy is ε is proportional to the 'Boltzmann factor', exp $(-\varepsilon/kT)$, which was a step that was no advance over what Maxwell and Boltzmann had achieved. He does not discuss liquids. By 1939, liquids could no longer be ignored, but he and his co-author, E. A. Guggenheim, evade the real difficulties by assuming that a liquid could be discussed as a pseudo-solid (since the densities were similar at low temperatures) in which a mean-field assumption again allows the molecule under consideration to have an identifiable energy from its interaction with the group of surrounding molecules. It was a model introduced by their colleague, John Lennard-Jones. They make no mention of Ornstein or others who were taking a less antique view of the problems. One hopes that, had Maxwell survived longer, the Cambridge school might have done better than this.

CHAPTER 9

Maxwell's Famous (or Infamous) Demon

ANDREW WHITAKER

A View from the 1950s

For physics students in the central decades of the twentieth century, the seemingly obligatory reference to Maxwell's demon in textbooks on thermodynamics[1,2] must have seemed little more than a nod to history. By this time, the two different ways of discussing heat, work and energy had diverged into effectively separate disciplines.

Of these, classical thermodynamics concerns itself with macroscopic phenomena as observed and described by humans; in the laws of thermodynamics it forbids the type of event that never had been, and, it was confidently predicted, never would be, observed. While the first law says that heat is a form of energy, and that conservation of energy is never violated, the second law says, for example, that you cannot transform all the heat from a hot object into useful work; rather some heat must be deposited at a lower temperature. In a different statement of the law, you do not observe heat, of its own accord, flowing from a cooler to a hotter object.

The second discipline of statistical thermodynamics, or statistical mechanics, studies the system in terms of the behaviour of the atoms that build up the macroscopic objects; the first law again expresses conservation of energy, but, when we turn to the second law, this approach tells us that those events that are ruled out categorically in classical thermodynamics can, in principle, occur, but only with a probability so low that one might have to wait many times the lifetime of the universe to observe them.

Statistical thermodynamics has the advantage of dealing with detailed microscopic models of the system, but classical thermodynamics has the balancing advantage of complete generality, its working being completely independent of any – perhaps temporary – ideas of atomic physics. Thus the two disciplines complement each other in a useful way. The inclusion of

Maxwell's demon into the discussion may have been somewhat instructive, or even possibly a little amusing, but it would not have been seen by 1950s students as adding anything in any sense important.

Such students would have been surprised to learn that in the previous century the demon had actually played an important role in clarifying the distinction just described. They would probably have been even more surprised to find out that, earlier in the twentieth century, the demon had also been crucial in stimulating the initial development of information theory, and thus the early theory of computation.

And as the twentieth century moved on, they would have been astonished to find that the now definitely elderly Maxwell's demon was to play an important part in obtaining a deeper understanding of information theory, and in particular of clarifying the possibility of achieving reversible computation; thus it would elucidate the distinctly modern 'hot topic' of quantum computation. Certainly it must be interesting to return to the days when the demon was youthful, and to investigate exactly why Maxwell announced the idea to the world.

Thermodynamics, Kinetic Theory and the Background of the Demon

Two of the most celebrated theories of the nineteenth century formed the intellectual climate in which Maxwell felt it necessary to promote the idea of his demon. The first was (classical) thermodynamics.[3] The two immediate stimuli to the creation of this theory were the seemingly rather contradictory ideas of Sadi Carnot and James Joule.

Carnot's famous work of 1824 developed what is known as the *Carnot cycle*. He showed that a heat engine that is most effective in obtaining mechanical effort from a given supply of fuel must be *reversible*; there must be no losses, for example, due to friction, and no flow of heat between objects at different temperatures. Also implicit in the concept of the cycle is that, as well as a source of heat at high temperature in the form of the fuel, there must be a sink of heat at a lower temperature.

Carnot was working in the caloric theory of heat, in which heat is a so-called state variable – for a fixed mass of gas, for example, the amount of heat in the gas can be written as a function of pressure and volume. In this theory, heat was always conserved, so the amount of heat gained from the hot reservoir must be equal to that lost to the cold reservoir. (The analogy was with a waterwheel, in which water moves from high pressure to low pressure, and mechanical work is done, but the amount of water is clearly unchanged; we definitely don't imagine that water is transformed into work!)

During the 1840s, William Thomson, later Lord Kelvin, and his brother James, (Figs. 9.1(a) and (b)) who was himself to become Professor of Engineering, first at Queen's College Belfast and later at the University of Glasgow, thought deeply about these matters.[4,5] While hugely impressed by Carnot's work, they felt that it left one major dilemma. In the case of an engine, Carnot says that when heat moves from high to low temperature, work is produced. Suppose, though, that hot and cold objects are placed adjacent to each other so that heat passes directly from high temperature to low by conduction. In this case, no work is done, and so it seems that the work that might have been produced has somehow been lost. This may be described as destruction of the (potential) work. The brothers found this difficult to accept because they felt that only God could create

(a) (b) (c)

Fig. 9.1 William Thomson (later Lord Kelvin), his brother James Thomson, and Peter Guthrie Tait were long-serving friends of Maxwell, with whom he liked to discuss his ideas, often in a humorous way. (a) William Thomson. (b) James Thomson. (c) Peter Tait.

or destroy. Destruction implied the *irreversibility* of the universe. What was destroyed was gone for good.

Carnot's ideas also seemed to be in conflict with those of James Joule, which in other ways the Thomson brothers also felt had much to offer. Joule believed, quite contrary to the idea of caloric, that heat was a form of energy and could be turned into work and vice versa. Thus he said that in the steam engine (or the Carnot cycle) part of the heat from the high temperature reservoir is transformed to become the desired mechanical work, and the remainder is deposited still as heat at the lower temperature. From this point of view, heat was not conserved and could certainly not be a state variable.

Since he regarded heat as a form of energy, Joule is considered one of the founders of the law of conservation of energy. However, as Thomas Kuhn[6] has pointed out, discovery of this law is a prize example of 'simultaneous discovery'; at least 12 scientists and engineers have been given the title of discoverer of this principle, with Julius Mayer and Hermann von Helmholtz probably Joule's greatest rivals.

By 1849, the Thomson brothers felt that the conflict between Carnot, dissipation and irreversibility on the one hand, and Joule, conservation and reversibility on the other, was practically impossible to resolve, and William wrote an important paper stressing this point.[7] In fact, though, his ideas were very soon trumped by those of Rudolf Clausius,[8] a young man of 28, who showed how to combine the best of Carnot and Joule. Clausius followed Joule by renouncing caloric and considering heat to be a form of energy – his was a dynamical theory of heat. However the energy that was seemingly lost, in conduction for example, is not actually destroyed. Conservation of energy is upheld, and in fact becomes the first law of thermodynamics. The energy that seems to be destroyed is merely 'lost to man irrecoverably' or *dissipated* as heat; it cannot be transformed back to useful work, and this is the second law of thermodynamics.

In today's physics education, conservation of energy is at the very centre, practically the most important *motif*, in the whole of physics, while the second law is studied much later, and appears as an awkward and probably rather confusing addition; Stephen Brush[9] amusingly makes the point that historically things were quite the other way round:

Here we see vividly how the Second Law of Thermodynamics, already born [in the work of Carnot], cannot be christened until it has dragged its brother the First Law out of the womb.

It should be mentioned that, as well as Clausius, William Thomson was effectively overtaken in these few months by Macquorn Rankine, an engineer, who was later to become Professor of Civil Engineering in Glasgow University; James Thomson was his successor after Rankine's untimely death in 1872. Rankine's work addressed much the same points as that of Clausius, but his work was restricted to a very special molecular vortex model, and so he sadly gained little credit for his ideas.[10]

While Clausius was the first to announce the new ideas, he admitted that he made great use of the analysis of Thomson, and indeed he knew of Carnot's work only through the writing of Thomson. As Crosbie Smith[11] has pointed out, perhaps one reason why Clausius moved ahead of Thomson was that Clausius's interests were narrower than those of Thomson, being mainly restricted to the theory of heat engines.

In any case, Thomson caught up with Clausius extremely quickly, and then established the subject of thermodynamics on a vastly wider basis than Clausius would probably have been capable of, putting the quantity of energy, actually, as we have just seen, a new arrival on the scene, at the very centre of physics, where modern teaching has kept it. He realized that the second law had profound consequences.[12]

[W]e find that the end of this world as a habitation for man, or for any living creature or plant at present existing in it, is *mechanically inevitable* . . .

It was Clausius, though, who, through the early 1850s inched towards the concept of entropy. This is the quantity that, according to the second law, never decreases, remaining constant in reversible changes and increasing in irreversible ones. It thus makes clearer the fact of irreversibility and an arrow of time. Entropy is linked with the *disorder* of the system. To say entropy increases is to say that the system becomes less ordered. Clausius also took the final step to understanding the so-called heat death.[13]

The more the universe approaches this limiting condition in which the entropy is a maximum, the more do the occasions of further changes diminish: and supposing this condition to be at last completely obtained, no further change could evermore take place, and the universe would be in a state of unchanging death.

Heat death would play a part in some of the discussions concerning, and attitudes towards, Maxwell's demon.

As well as thermodynamics, the other nineteenth-century physical theory that led to the invention of the demon was kinetic theory,[14] but the general acceptance of this theory was itself due to the arrival of thermodynamics. In kinetic theory the molecules of a gas move at high speeds; their collisions with the walls of their container give rise to the gas pressure.

Daniel Bernoulli had produced a basic, but serviceable, kinetic theory as early as 1738, but this work was soon forgotten. It is well-known that, in the first half of the nineteenth century, John Herapath and John James Waterston separately produced articles on kinetic theory significant enough that their rejection by the Royal Society 'was to retard the progress of molecular physics by a decade or two'.[15] The lack of judgement of the Royal Society is doubtless shocking, but it must be admitted that Herapath and Waterston took advantage of several other means of publication, including the *British Association Reports*. Unfortunately their publications in these places drew no more interest or support than their submissions to the Royal Society.

It seems that the fundamental idea of kinetic theory was generally regarded at this time as ridiculous. Brush[16] points out that, while many caloric theorists were atomists, 'they attributed the expansive force of a gas to the repulsion of its atoms rather than to their free motion'. In this context, it is easier to understand the report of Baden Powell,[17] the referee of Waterston's most important paper of 1845, who admits that the paper 'exhibits much skill and many remarkable accordances with the general facts', but is still inclined to reject it because the fundamental idea seems 'entirely hypothetical'.

What changed matters was, of course, the coming of thermodynamics, a set of ideas that included the conservation of energy, and the acceptance that heat was a type of energy consisting of some form of motion. It was not immediately clear that this necessarily implied the truth of a kinetic theory attributing heat to the kinetic energy of atomic translational motion. In fact, as late as 1853, scientists such as Clausius, Thomson and John Tyndall were quite non-committal about what kind of motion heat was.[18]

Both Thomson and Rankine came to prefer the idea of the vortex atom, in which the kinetic energy corresponded to rotational motion of the atom, and, for Thomson this would be the start of the 40 years of work on atoms and molecules described by Elizabeth Garber.[19] As she says, Thomson's work was a search for a dynamical theory of the interaction of matter and the ether, with the ultimate goal of uniting all known phenomena, and it was not in the end particularly successful. At least in the 1860s and 1870s, though, Thomson was reasonably open to the idea that the vortex atom might be the basis of kinetic theory,[20] in the sense that the atoms would have both rotational and translational kinetic energy.

Certainly it was the case that the construction of a kinetic theory was now seen to be reasonable and timely, and it was not at all surprising that Joule and Clausius, two of those contributing greatly to thermodynamics, should have been among those taking this step. Joule was the first, with a paper read in 1848 and published in 1851; perhaps finding the subsequent development of the theory too mathematical, he never published in the area again.[21] August Karl Krönig published a well-read paper in 1856, but, like all the work so far, he was able to obtain the ideal gas law and relate absolute temperature to the kinetic energy of the molecules, but little more.

It was Clausius, in a paper of 1857, who was the undoubted founder of modern kinetic theory. He moved beyond the work of Joule and Krönig by taking into account rotational and vibrational motion of the molecule as well as its translational motion, and he attempted to relate the energy involved in the various forms of motion. In a later paper, he introduced the idea of the mean free path of a molecule, the mean distance it would travel before colliding with another molecule, and later still applied the idea to an attempt at a theory of heat conduction, diffusion and viscosity.

It should be mentioned that, by 1856 or 1857, the reception of the theory of kinetic energy was almost uniformly favourable because, unlike the earlier attempt of the unfortunate Waterston, it coupled Waterston's empirical success with newfound confidence in its basic assumptions.

Before turning to Maxwell, I shall note one further point about these pioneers of modern kinetic theory, Joule and Krönig in particular. They were broadly content with the simplest possible assumptions. They would consider a cubical box, with 1/3 of the molecules moving along each of the three axes – often called the Joule division, and they would assume each of the molecules had the same speed. Some scientists actually attempted to justify the latter point by suggesting that, in the case of an assumed initial non-uniform distribution of speeds, collisions would serve to equalize the speeds; this was probably more of a wish than a genuine argument.[22] Any modern student of elementary kinetic theory will realize precisely why they used simple assumptions: in this way a considerable number of important results may be produced with the absolute minimum of effort!

The far more mathematically sophisticated Clausius was well aware of the limitations of such an approach. He was able to make excellent progress with the situation where molecules travel in all possible directions. He was also aware that molecules had a wide range of speeds; for example, this was his explanation for evaporation of a liquid – even below the boiling point, some molecules were travelling fast enough to escape from the surface.[23] Yet, in his papers[24] he was always inclined to replace any range of molecular properties by the corresponding average value, making the calculations more straightforward, but also sometimes making the results less meaningful.

And now at last we reach Maxwell. Born in 1831, Maxwell was too young to have taken part in the resolution of thermodynamics. William Thomson was only seven years older but, by the mid 1850s, when Maxwell's scientific work may be said to have started, Thomson already had enormous scientific achievements to his name. However, it may not be too churlish to suggest that, by that time, and despite the 50 years of extremely hard work to come, Thomson's most important work in pure science, the work for which he is remembered, was over, as distinct, of course, from his technological successes, his important educational advances and his contributions to scientific affairs and politics.

In contrast, Maxwell's career for the next quarter-century until his untimely death was, of course, to remain centred on pure science, though he was certainly also devoted to education, both for the student and the working man. By the end of the 1850s, he was extremely well placed to take further some of the arguments of Clausius on kinetic theory, and for the remainder of Maxwell's life, he and Clausius each made major contributions to the subject and laid the basis of the kinetic theory that we know today. Maxwell's part in particular is described by Elizabeth Garber in Chapter 7 of this volume.[25]

Here we dwell only on one of Maxwell's major contributions, his explicit introduction of statistical methods into kinetic theory by his production, in 1860, of the law of distribution of speeds, today always known as the Maxwell distribution.

Clausius and Maxwell: Mechanics and Statistics

From one point of view, Maxwell's introduction of the speed distribution was just an extremely important advance within the reasonably narrow field of kinetic theory. However, from a different perspective, it may be seen as far more than that – the instigation of an entirely new field of research, based on the statistical distribution of properties of an ensemble of particles.

In 1889, the great American expert on thermodynamics, J. Willard Gibbs,[26] wrote in his obituary of Clausius:

> In reading Clausius, we seem to be reading mechanics; in reading Maxwell, and in most of Boltzmann's most valuable work, we seem rather to be reading in the theory of probabilities.

To that comment, it should be added that the early Boltzmann would have been identified with Clausius rather than Maxwell. It was Maxwell's influence on Boltzmann that caused a drastic change of approach and led to Boltzmann becoming a follower of Maxwell's ideas, and to his indeed highly valuable work. In fact, while Maxwell may be said to have introduced statistical methods, it should be admitted that he did not produce a genuine statistical mechanics; the credit for that must go to Boltzmann.

Let us for the moment consider the general approach of Clausius, and it is fair to describe it as one based on mechanics. As early as his great 1850 paper, in which he introduced thermodynamics, it seemed that he intended to go beyond the abstract identification of heat as motion, to its definite representation as a form of molecular motion.[27] Though, as has been mentioned, he was not immediately sure how to follow this idea up, he soon became convinced that kinetic theory was the correct approach, and Brush describes kinetic theory as 'one of the most typical and comprehensive applications of the *mechanistic* approach in science' [my italics].[28]

We have already seen that Clausius made extremely limited use of the distribution of molecular speeds. Garber has stated that he made only one application of the actual Maxwell distribution; this was in an 1874 calculation of the mechanical expression for the heat added to a gas.[29] Even in his later work, in which he attempted to improve the van der Waals equation of state by studying the effect of variation of the intermolecular forces with temperature, he rather degenerated into empirical curve-fitting, because of his failure to take advantage of the developments of Maxwell and Boltzmann.

One important contribution of Clausius was the development of the virial theorem in 1870; this relates the mean value of the kinetic energy of a system of particles to a function of the rectangular components of a particle and of the force acting on the particle, summed over all particles.[30] Today, the virial theorem plays an important role in determining the pressure of the system given any particular force law. However, Clausius did not use it for this purpose, but in furtherance of his scheme for finding what he thought of as the true mechanical foundation for thermodynamics, in particular the second law.

In making such an attempt he was, unbeknown to himself, following in the footsteps of the very young Ludwig Boltzmann.[31] In 1866, at the age of 22, Boltzmann published his first scientific paper titled: 'On the Mechanical Meaning of the Second Law of Thermodynamics'. Boltzmann contrasted the understanding of the first law in terms of energy, with that of the second law, with its 'peculiarly exceptional position' and 'the roundabout and uncertain methods' by which physicists sought to establish it.

His clear and certainly ambitious aim was: 'to give a purely analytical, completely general proof of the second law of thermodynamics, as well as to discover the theorem in mechanics that corresponds to it'. However, in this paper of 1866, although he produced several interesting results, he fell far short of his target. His most important theorem, which he claimed as giving a mechanical account of the second law, was proved only for cyclic phenomena, very far from the irreversible systems in which we are most interested in thermodynamics.

Interestingly, almost 20 years later, Helmholtz[32] was to produce his own mechanical expression of the second law, again for 'monocyclic systems'. By this stage, persuaded by Maxwell, Boltzmann had long since given up the idea that there could be a mechanical *explanation* of thermodynamics, but Helmholtz presented his idea as merely a mechanical *analogy*, and Boltzmann found the approach fascinating, and indeed carried it further.

Clausius made two attempts to provide a mechanical foundation for the second law, the first in connection with the virial theorem in 1870, and the second the following year, when, ignorant of Boltzmann's work of 1866, he produced many of the same results, though expressed in Clausius' own notation, which was different from that of Boltzmann. Clausius accepted Boltzmann's general priority, though there were some remaining differences between the two men.[31]

Maxwell regarded these attempts as totally ill-conceived. In 1873, he wrote[33] to Peter Guthrie Tait, his old schoolmate and now Professor of Natural Philosophy at Edinburgh University. Tait (Fig. 9.1(c)) had been preferred to Maxwell himself for that post because of his perceived superior teaching. The letter included the words:

I have also a great respect for the elder of those celebrated acrobats, Virial and Ergal, the Bounding Brothers of Bonn ... But it is rare sport to see those learned Germans contending for the priority of the discovery that the 2nd law of $\theta\Delta$cs [thermodynamics] is the Hamiltonsche Princip, when all the time they *assume* that the temperature of a body is but another name for the vis viva [effectively kinetic energy] of one of its molecules, a thing which was suggested by the labours of Gay Lussac, Dulong, etc, but first deduced from dynamical statistical considerations by $\frac{dp}{dt}$ [Maxwell's way of describing himself, from the equation $\frac{dp}{dt}$ = jcm (James Clerk Maxwell)]. The Hamiltonsche Princip, the while, soars along in a region unvexed by statistical considerations, while the German Icari flap their waxen wings in nephelococcygia amid those cloudy forms which the ignorance and finitude of human science have invested with the incommunicable attributes of the invisible Queen of Heaven.

Expressed more simply, Maxwell was saying that the second law of thermodynamics is not a mechanical law but a statistical one. When small enough collections of molecules are considered, the second law is constantly, in effect, being violated[34] in spontaneous fluctuations by which heat is passed from slower molecules to faster ones by random motion and collisions of the molecules. However, we do not describe this as a genuine violation of the second law, because, as originally set out by Clausius and Thomson, this relates to observable heat flow not the random fluctuations of molecules.[35] When we move to larger groups of molecules, such spontaneous processes become enormously less probable, but they are still conceivable, and certainly not ruled out by mechanical considerations. It is statistics not mechanics that dictates that we will not see the second law violated.

'No pure dynamical statement would submit to such an indignity', Maxwell wrote to Tait in 1876. Maxwell was pointing out that the processes described by the second law as impossible are perfectly consistent with the laws of dynamics. He is saying that no mechanical law could be violated just because the number of particles being studied is small.[36]

Perhaps Maxwell's clearest summary of his argument occurred in a letter to John Strutt, later Lord Rayleigh,[37] in 1870. He wrote: 'Moral: The 2nd law of thermodynamics has the same degree of truth as the statement that if you throw a tumblerful of water into the sea, you cannot get the same tumblerful out again'. It is clear that there is certainly no mechanical principle that prohibits getting the same tumblerful out, even that in principle it is not impossible that it should occur; all that one may say is that to do so is statistically so unlikely that even in an enormous number of lifetimes of the universe, it would not happen.

Maxwell's demon was very much a part of his argument for the statistical nature of the second law, and it will be discussed in the next section. However here we discuss a few of the reactions to the argument of this section.

Clausius, as we have seen, never really came to terms with the statistical approach to the second law. On the other hand, Boltzmann did so very speedily, As Klein[31] says, in 1871, while Clausius was still debating the rather small differences between his version of the mechanical approach and that of Boltzmann, Boltzmann himself offered an entirely new approach to the equilibrium case of the law, for which the entropy is constant, and in the following year he did the same for the non-equilibrium case, where the entropy increases. This new approach was a statistical one, the forerunner of Boltzmann's extremely important and influential work in future years. Indeed, many of the essential features of Boltzmann's work were even to survive the transition from classical to quantum physics.

However it soon became clear that his claim to have demonstrated that entropy must *always* either stay the same or increase could not be absolutely valid. In the mid 1870s, Joseph Loschmidt drew his attention to an argument[36] that is sometimes called Loschmidt's paradox. This considers a situation where, at a particular time, the motion of all the particles in an isolated system, the entropy of which is increasing, or indeed in the whole universe, is reversed. The subsequent behaviour of the system must be to run backwards over the previous development, and entropy must now decrease.

Two points may be noted. First, Loschmidt's argument is actually a particular way of pointing out that any theory, such as Boltzmann's, that claims to develop an irreversible theory, a theory with an arrow of time, from Newtonian mechanics, must be faulty, since Newton's laws themselves are time-reversible; they say, for example, that if a film is shot of a proper physical event, and this film is shown backwards, the result will be another proper physical event. The second point is that William Thomson[38] had discussed Loschmidt's point privately many years before, at least as early as 1867, and he had published it in 1874.

Boltzmann's response[39] was twofold, both points relating to the initial conditions of the evolving system. First, he suggested that there are many more initial states that evolve with increasing entropy than with decreasing entropy. But he also suggested that, due to a fluctuation, we might just happen to be living in a period where the entropy has been unusually low and is now increasing. It is now thought that such a large fluctuation would be far too unusual to happen randomly, but we may believe that the Big Bang formed the universe in just such a state of extremely low entropy.

Loschmidt himself did not accept Boltzmann's argument.[36] Since his own argument disproved the unconditional validity of the second law, as stated by Clausius and Thomson, he felt this statement of the law must be untrue. He considered, much like Clausius and Boltzmann around 1870, that the proper basis for this law must be in traditional dynamical principles. In 1878, he stated his view as follows:

> Since the second law of the mechanical theory of heat, just like the first, should be a principle of analytical mechanics, then it must be valid, not only under the conditions which occur in nature, but also with complete generality, for systems with any molecular form and under any assumed force, both intermolecular and external.[39]

Loschmidt was far from despondent that, in his view, the original statements of the second law should be jettisoned. Rather, he was delighted that his argument removed the heat death:

> Therefore the terrifying *nimbus* of the second law, by which it was made to appear as a principle annihilating the total life of the universe, would also be destroyed; and mankind could take comfort in the disclosure that humanity was not solely dependent upon coal or the Sun in order to transform heat into work, but would have an inexhaustible supply of transformable heat at hand in all ages.

However, though Loschmidt's 'reversibility paradox' has been taken seriously, his conclusions have been disregarded, and the views of Maxwell and the later Boltzmann have held sway. It is relevant, though, to mention one more ardent supporter of a mechanical rather than a statistical approach to the second law; this is Ernst Mach,[40] the arch-positivist. With his well-known distaste for the atomic idea, it is not surprising that he disliked any approach based on molecules.

The Coming of the Demon

The first mention of Maxwell's demon occurred in a letter[41] to Tait written on 11 December 1867. Tait had himself written to Maxwell in connection with the book Tait was writing, to be called *Sketch of Thermodynamics*,[42] dealing as much with the history of the subject as the subject itself. The first two chapters had originally been written for private class list under the title *Historical Sketch of the Dynamical Theory of Heat*, and published in the *North British Review*, while the third is a compact account of the mathematical elements of the subject. Tait was planning to give the vast majority of the credit for the creation of the subject to the British Joule and Thomson, at the expense of the non-British Mayer and Clausius, and perhaps hoped to get a little support for this from the British Maxwell.

In his reply, Maxwell immediately ruled out such an idea. He said that he did not know in any controversial manner the history of thermodynamics, and so he could not make any assertions about priority without referring to the actual works, which he clearly did not intend to do. Rather he suggested that any contributions he might make would be of altering an occasional point of view for clearness or variety, or picking holes here or there to ensure strength and stability. The first example he gave was a suggestion of paying particular attention to the absolute scale of temperature. The second is as follows:

To pick a hole – say in the 2nd law of ΘΔcs., that if two things are in contact the hotter cannot take heat from the colder without external agency.

Now let *A* and *B* be two vessels divided by a diaphragm and let them contain elastic molecules in a state of agitation which strike each other and the sides.

Let the number of particles be equal in *A* and *B* but let those in *A* have the greatest energy of motion. Thus even if all the molecules in *A* have equal velocities, if oblique collisions occur between them their velocities will become unequal, and I have shown that there will be velocities of all magnitudes in *A* and the same in *B*, only the sum of the squares of the velocities is greater in *A* than in *B*.

When a molecule is reflected from the fixed diaphragm *CD* no work is lost or gained.

If the molecule instead of being reflected were allowed to go through a hole in *CD* no work would be lost or gained, only its energy would be transferred from the one vessel to the other.

Now conceive a finite being [Fig. 9.2] who knows the paths and velocities of all molecules by simple inspection but who can do no work except open and close a hole in the diaphragm by means of a slide without mass.

Fig. 9.2 A poetic vision of the demon from Lerner, A.Y., *Fundamentals of Cybernetics* (Plenum, New York, 1975).

Let him first observe the molecules in *A* and when he sees one coming the square of whose velocity is less than the mean sq. vel. of the molecules in *B* let him open the hole and let it go into *B*. Next let him watch for a molecule of *B*, the square of whose velocity is greater than the mean sq. vel. in *A*, and when in comes to the hole let him draw the slide and let it go into *A*, keeping the slide shut for all other molecules.

Then the number of molecules in *A* and *B* are the same as at first, but the energy in *A* is increased and that in *B* diminished, that is, the hot system has got hotter and the cold colder and yet no work has been done, only the intelligence of a very observant and neat-fingered being has been employed.

Or in short if heat is the motion of finite portions of matter and if we can apply tools to such portions of matter so as to deal with them separately, then we can take advantage of the different motion of different proportions to restore a uniformly hot system to unequal temperatures or to motions of large masses.

Only we can't, not being clever enough.

The letter has a pencilled annotation reading: 'Very good. Another way is to reverse the motion of every particle of the Universe and to preside over the unstable motion thus produced.' Cargill Gilston Knott, Tait's biographer of 1911, reports that this in Thomson's writing, though Brush[43] thinks it is that of Tait. Thomson's biographers, Crosbie Smith and Norton Wise,[44] and most other commentators certainly assume it is that of Thomson.

Maxwell wrote along similar lines to Strutt,[37] and it is highly likely that there was a considerable amount of discussion of the issue among the leading British scientists in the period up to its publication, in 1871 in Maxwell's *Theory of Heat.*[45] *Theory of Heat* was one of series of books intended for schoolchildren or the 'interested layperson', and is sub-titled: 'Adapted for the use of artisans and students in public and science schools'. It presented an excellent account of the subjects of heat and thermodynamics and, on page 69, introduced the extremely useful set of equations now known as Maxwell's Relations, but it is fair to say that Maxwell was not good at deciding what would be comprehensible to any but the brightest and most knowledgeable students, and *Theory of Heat* could definitely not be described as introductory material!

The first 21 chapters deal with the properties of heat and macroscopic thermodynamics. Only in the last chapter, titled 'Molecular Theory of the Constitution of Bodies', does Maxwell study microscopic matters. He introduces the idea of the molecule and argues that, in the gaseous state, molecules must be in rapid motion. He proceeds to a brief history of kinetic theory and a thorough, though not very mathematical, account of the theory and its successes. He then has a short section titled 'Limitation of the Second Law of Thermodynamics', and remarks that the second law is 'undoubtedly true as long as we can deal with bodies only in mass, and have no power of perceiving or handling the separate molecules of which they are made up'.

There follows an account of the unnamed 'finite being' with 'sharpened faculties' very much along the lines of the latter to Tait; the violation of the second law is 'without expenditure of work'. Maxwell then remarks that:

This is only one of the instances in which conclusions which we have drawn from our experience of bodies consisting of an immense number of molecules may be found not to be applicable to the more delicate observations and experiments which we may suppose made by one who can perceive and handle the individual molecules which we deal with only in large masses.

Three years later, Thomson[38] presented a fairly short paper in which he amplified Maxwell's discussion somewhat. He also used the term 'demon', though he remarks that the definition of the term had been given by Maxwell as 'an intelligent being endowed with free will, and fine enough tactile and perceptive organisation to give him the faculty of observing and influencing individual molecules of matter'. In this paper, Thomson uses an army of cricket-bat wielding demons stationed across a plane within the gas, and able to inhibit any move to thermodynamic equilibrium. In order to show how small the possibility of a genuine violation of the second law in a macroscopic system is, Thomson also calculated the possibility of examining a system of 2×10^{12} molecules of oxygen and 8×10^{12} molecules of nitrogen, and finding all the oxygen molecules in a section $\frac{1}{5}$ the size of the entire vessel. The probability was indeed extremely small, 1 in 10^n, with n itself of the order of 10^{12}. In this paper, Thomson also gives a brief account of what we have called Loschmidt's paradox.

The last item we shall mention in this section is another communication[46] from Maxwell to Tait, which is undated but was presumably written after the publication of this paper of Thomson. It reads:

Concerning demons

1. Who gave them this name? Thomson
2. What were they by nature? Very small BUT lively beings incapable of doing work but able to open and shut valves which move without friction or inertia.
3. What was their chief end? To show that the 2nd Law of Thermodynamics has only a statistical certainty.
4. Is the production of an inequality of temperature their only occupation? No, for less intelligent demons can produce a difference in pressure as well as temperature by merely allowing all particles going in one direction while stopping all those going the other way. This reduces the demon to a valve. As such value him. Call him no more as demon but a valve like that of the hydraulic ram, suppose.

It is interesting to realize that the style of this document is exactly that of a catechism, and indeed the Westminster Shorter Catechism, which was completed in 1647 by the Westminster Assembly, and which still serves a part of the doctrinal standards of many Presbyterian Churches, begins: 'What is the chief end of man? Man's chief end is to glorify God and to enjoy him forever . . .'. Again Maxwell is succeeding in stirring together scientific ideas, religious references, and a little humour.

The Demon and the Statistical Nature of the Second Law

As Maxwell wrote, the main point of the demon argument is to show that the second law is statistical, and this is the line that we shall follow in this section of the chapter. In other words, the focus is on what the demon tells us about thermodynamics. Later we shall return to think about the demon itself.

It is interesting that discussion of the demon was often linked with that of Loschmidt's paradox, particularly by Thomson,[38] and especially if the emendation to Tait's letter was indeed by his hand. At this level then, the message seems to be the fairly general one that the statistical approach to thermodynamics could, if the system were tweaked somewhat by a demon or by a reversal of velocity, give rise to phenomena totally discordant with the classical thermodynamic approach.

Yet Maxwell certainly meant something more than this. He stressed that the demon did no work, and this was so that he could argue that he was not essentially changing the dynamics of the system, so from the point of view of Clausius and the early Boltzmann, there should have been no change in behaviour, and yet, of course, he was able to demonstrate quite the reverse. Thus the second law could not be mechanical in nature.

Did he intend to destroy the second law? It is possible to argue this way. Tolstoy[47] says that: 'appealing to statistical validity was not the way out of the dilemma; as Maxwell suggested in another letter it actually cast doubt on the very validity of the second law'. And even Maxwell's own phrase – 'to pick a hole' – might be interpreted as meaning to find a genuine fault in the law. Yet it is probable that he meant the phrase in a less dramatic way – more like to give a cause for further thought.

His own view was stated in his review[48] of Tait's book:

> The second law must either be founded on our actual experience in dealing with real bodies of sensible magnitude, or else deduced from the molecular theory of these bodies, on the hypothesis that the behaviour of bodies consisting of millions of molecules may be deduced from the theory of the encounters of pairs of molecules, by supposing the relative frequency of different kinds of encounters to be distributed according to the laws of probability.

With his demon, Maxwell's aim was not to destroy the second law, but to show its limitations, 'to ensure strength and stability', as he remarked in his letter to Tait.

It is true that when Tait suggested to Clausius that the demon argument was fatal to the latter's general argument on thermodynamics, Clausius merely replied[31] that; 'I believe that I can summarize my reply to Mr Tait's objection in the brief remark that my law is concerned, not with what heat can do with the help of demons, but rather with what it can do by itself.'

However, in 1893, Henri Poincaré[49] wrote that: 'To see heat pass from a cold body to a warm one, it will not be necessary to have the acute vision, the intelligence, and the dexterity of Maxwell's demon; it will suffice to have a little patience.' Similarly, in his well-known text of 1904, James Jeans[50] wrote that: 'Maxwell's sorting demon could effect in a very short time what would probably take a very long time to come about if left to the play of chance. There would, however, be nothing contrary to natural laws in one case any more than the other.'

To sum up, Maxwell's argument was that, since the second law was statistical, there was a possibility that results quite contrary to those stated originally by Clausius and Thomson might ensue. However, for macroscopic systems, the probability of such results was so unimaginably small that the original statements were without any doubt empirically true. The demon served to demonstrate the statistical element much more clearly.

Is the Size of the Demon Important?

In his excellent biography of Maxwell, Basil Mahon[51] comments that Maxwell actually gives two explanations of why the demon may succeed in its task. The first is, as we have stressed, the fact that the second law is a statistical one. The second, Mahon suggests, 'although apparently light-hearted, turned out to be even more penetrating'. Mahon refers to Maxwell's comment, in the letter to Tait, that the demon must be 'neat-fingered'; we cannot achieve the feats of the demon, 'not being clever enough'. Interestingly, even in his 'Concerning demons' homily to Tait, Maxwell writes of demonstrating the statistical nature of the second law as the 'chief end', suggesting that that there was at least one other.

It is natural to feel that the laws of classical physics, such as Newton's Laws, must be independent of the nature or even the existence of any observer, and to feel that the coming to the fore of the observer concept was an innovation of quantum theory, and a most unwelcome one at that.

Most interpretations of quantum theory[52] would say that, in quantum theory, the observer is an *active participant* in the evolution of the system, over and above the actual physical interaction between measured and measuring sub-systems; indeed it is often said that the measurement actually *creates* the value of the quantity it measures. In thermodynamics, there is no question of the observer as active participant or creator in this sense. However it may be said that the *description* given by the observer will depend on his size and 'neat-fingeredness'.

It is clear that the very statement of Thomson's version of the second law assumes a macroscopic observer, indeed a technologist, who has set up equipment to produce useful work from fuel. Clausius' statement is less obviously intrinsically macroscopic, since it does not necessarily relate to technical machinery, but it does imply the impossibility of an *observed* flow of heat from colder to hotter. (Of course one can make Clausius' statement more analogous to that of Thomson by speaking explicitly of a *refrigerator*, and stressing that, to make it work, we do need to plug it in and provide work.)

Both these statements rely totally on the distinction between work and heat. Here, work is defined as energy that can be made use of by human beings, while heat cannot be so used. It is immediately clear that the distinction is purely in terms of a macroscopic being.

Maxwell[31] wrote in terms of observation rather than control:

According to the molecular theory, the only difference between these two kinds of communication of energy is that the motions and displacements which are concerned in the communication of heat are those of molecules, and are so numerous, so small individually, and so irregular in their distribution, that they quite escape all our methods of observation, whereas when the motions and displacements are those of visible bodies consisting of large numbers of molecules moving altogether, the communication of energy is called work.

Indeed he also wrote:[53]

[T]he notion of dissipated energy could not occur to a being who could not turn any of the energies of nature to his own account, or to one who could trace the motion of every molecule and seize it at the right moment. It is only to a being in the intermediate stage, who can lay hold of some forms of energy while others elude his grasp, that energy appears to be passing inevitably from the available to the dissipated state.

Another well-known description of the difference between work and heat is that for the former, the energy is, in some sense, ordered, while in the latter, it is disordered or random. It is perhaps not quite so obvious that a description in terms of order and disorder is not generally applicable, but is useful only for macroscopic observers such as humans, but Maxwell points out that this is indeed the case:

The idea of dissipation of energy depends on the extent of our knowledge ... A memorandum book does not, provided it is neatly written, appear confused to an illiterate person, or to the owner who understands it thoroughly, but to any other person it appears to be inextricably confused.[53]

And again:

[T]he idea of dissipation of energy depends on the extent of our knowledge. Available energy is energy which we can direct into any desired channel. Dissipated energy is energy which we cannot lay hold of and direct at pleasure, such as the confused agitation of molecules which we call heat. Now, confusion, like the correlative term order, is not a property of material things in themselves, but only in relation to the mind that perceives them.

So Maxwell argues that the concept of order depends very much on the properties of the observer. Thus, given the well-known relationship, broadly due to Boltzmann, between entropy and disorder, it may be seen that the statement of the second law in terms of the non-decrease of entropy, like the statements of Thomson and Clausius, is only appropriate to a range of observers. We may say that[54] the principle of irreversibility cannot then be regarded as an absolute law of nature but only as a consequence of human inability to deal with individual molecules; that the second law describes not the system itself, but the observer's knowledge of the system.[55]

When we use the term 'properties' of the observer, we naturally include size – perception of order must require the ability to take in information at the appropriate scale. But it is clear that Maxwell is also moving in the direction of the importance of the mind of the observer, in particular its ability to take in and use information.[56] In the later parts of this chapter, we shall explore the first ideas of information theory and their relation to Maxwell's demon.

To conclude this section of the chapter, we point out that the difficulty of distinguishing between work and heat at the molecular level inevitably caused great problems for Clausius and Boltzmann in their attempts to obtain the second law from the mechanics of the molecules themselves. Klein[31] describes their valiant attempts, but Maxwell regarded these attempts as unsatisfactory, and considered this failure to be a necessary and fundamental reminder that their schemes were flawed in principle.

The Wider Context

Maxwell's demon, like Schrödinger's cat and possibly Bertlmann's socks,[57] has become one of those elements of physics that, partly because of its intrinsic and dramatic interest, partly perhaps because of a catchy name, has attracted attention and discussion well beyond its specific scientific context. Much has certainly been written on the demon that goes totally beyond anything to do with Maxwell's actual ideas.

In this section we discuss only a few arguments that, while outside Maxwell's published discussions, follow the ideas of the brothers William and James Thomson, and Maxwell himself into the regions of theology, and in particular the place of man in the physical universe. We start with two quotes, the first one being from William Thomson[58] in his 1879 discussion of Maxwell's demon:

The conception of the 'sorting demon' is purely mechanical, and is of great value in purely physical science. It was not invented to help us deal with questions regarding the influence of life and of mind on the motions of matter, questions essentially beyond the range of mere dynamics.

The second from Clarke[55] expresses rather a different point of view:

[T]he resilience of Maxwell's Demon rests not on its technical possibility or impossibility, but on its power to operate in the interstices between physical theories and cultural desires. Maxwell's 'very observant and neat-fingered being' immediately departed its scene of origin and began its cultural extension as an allegory of thermodynamics, when Tait forwarded the manuscript of this letter to their mutual colleague, William Thomson.

The Thomson brothers and Maxwell were part of an informal group of physicists mostly located in the Scotland in particular, but also the northeast of Ireland;[59] other members included Fleeming Jenkin and Balfour Stewart, and the group had two particular concerns over the application of the ideas of physics to more fundamental matters. The first concerned the heat death. Though the group contained some of those mostly responsible for the emergence of thermodynamics, there was some reluctance to accept that it necessarily led to this depressing conclusion.

Indeed, Clarke[55] suggests that in the late-Victorian period:

> [T]hermodynamic entropy became cultural allegory by revealing a plausible version of last things. In some circles the second law of thermodynamics was interpreted as God's withdrawal from the material universe ... [T]he projected outcome of universal entropy is a scientized form of the melancholy ruin.

There was also a plausible crossover for the term 'dissipation' between the physical and the moral spheres. Katherine Hayles[60] has suggested that the universal tendency for entropic heat loss could be seen as analogous to a similar tendency for the worsening of deplorable personal habits. And if a dissolute individual might be 'redeemed', then perhaps the universe might be saved from eventual disaster in the form of heat death by a development in human knowledge.

In a recent book, Helge Kragh[61] has written also of analogous debates about the beginning of the universe. Since the entropy of the universe is decreasing, it could be argued that the universe must have had a finite lifetime, and so indeed a beginning in the form of divine creation. There were intense debates on the matter in the late nineteenth century, in which religiously minded physicists and also theologians were opposed, in particular, by Friedrich Engels, the communist theoretician.

The second concern for the Scottish natural philosophers and Christians was the problem of free will in a universe which science appeared to be showing more and more clearly was governed by determinism. Any room for a special role for living creatures, or in particular human beings, was rapidly diminishing. Even Helmholtz,[62] the German physiologist and physicist, with whom William Thomson and his friends remained on good terms, argued from the conservation of energy to the removal of any 'vital principle' and so to an absence of any clear scientific distinction between inorganic and organic matter.

The members of the so-called X club, such as Thomas Huxley, Joseph Hooker, John Tyndall and Herbert Spencer,[63] were more explicit, and were exemplified in the famous or notorious Presidential address of Tyndall,[64] the arch-opponent of the Scottish group, especially William Thomson and Tait, to the British Association for the Advancement of Science in Belfast in 1874. Tyndall followed Huxley in putting forward a theory of 'scientific naturalism', which claimed the victory of a materialist mechanical worldview. Scientific naturalism[65] was based on the developments of science in the nineteenth century – evolution by natural selection, geological advances, a belief in the uniformity of nature, and most of all the conservation of energy, which, according to Huxley and Tyndall, brought vital as well as physical phenomena under the power of determinism. They claimed to detach the scientific methods and ethos from any theological arguments.

In his 1874 address, Tyndall said that, believing in the continuity of nature, he could not stop abruptly when his microscope ceases to be useful, but must cross the boundary to deeper considerations, to express the supremacy of science, and to take from religion any right to infringe upon the domain of science. While, just as Huxley claimed to be an agnostic rather than an atheist, Tyndall suggested the truth might lie in pantheism, he was most often categorized, by supporters as well as opponents, as a materialist.

At the same Belfast meeting, Huxley argued that animals, including humans, should be regarded as automata,[54] machines governed solely by natural laws. He said that there was no proof that changes in the state of consciousness of an animal can cause any change in its motion. In other words, consciousness ceases to indicate the existence and functioning of a soul. Where *was* the soul? Where was free will?

Let us now see how the Scottish group responded to these two challenges, and how they attempted to make Maxwell's demon a part of their armoury. Smith and Wise[59] discuss an important interaction between the Thomson brothers in the years before the invention of the demon, which attempted to work out the limitations, if any of the second law.

In his 1851 paper on thermodynamics, William Thomson had been extremely careful at least to allow for the possibility of important limitations. His statement of the second law began: 'It is impossible, by means of inanimate material agency . . .'. Thus living beings were excluded, and in a paper of the following year he also excluded vegetative or chemical action. The most significant question was whether human or animal free will might be able, not to create energy, thus violating the first law, but to direct it, conceivably violating the second law.

Around 1862, the Thomson brothers considered these matters further. James, in particular, was keen that the possibility that mind – or vitality or life – could reverse the second law should not, at the very least, be dismissed out of hand. Human mind can certainly affect the course of events during which the second law is obeyed, so is it not possible that it might do more and actually *reverse* the process?

It is clear that these thoughts have some relevance, not just to how or why Maxwell's demon was invented, but to how it might be discussed or used following its invention. As we have seen, in 1874 Thomson[38] described the demon as 'an intelligent being endowed with free will and fine tactile and perceptive organization'. In a later important paper of 1879, Thomson[66] described the demon as intelligent, a living animal, 'differ[ing] from real living animals only in extreme smallness and agility'.

As Smith and Wise remark, the demon has the same faculties as Thomson ascribes to man, and it natural to suppose that he believes free will is an essential aspect of the demon, and thus of violating the second law. It is true that, as quoted at the beginning of this section, Thomson denied that the idea of the demons can help us deal with questions as to the influence of life and mind on matter. Nevertheless, Smith and Wise[67] insist that the historical origin of the demon lay in Thomson's and Maxwell' emphasis on free will and its ability to direct energy.

Maxwell, though, was uneasy about too many faculties being ascribed to the demon. In his note to Tait, he talked of 'less intelligent' demons, and in a letter to Strutt he even suggested that 'intelligence might . . . be dispensed with and the thing made self-acting'. As he says 'Call him a valve'.

Looking forward, it is interesting to note that a valve will not work! In the famous *Feynman Lectures on Physics*, Richard Feynman[68] shows that any simple mechanism designed as Maxwell's demon, such as a ratchet and pawl, must ultimately heat up. Eventually, he says, the mechanism itself will shake so much from Brownian motion that it certainly cannot tell whether molecules are coming or going.

So something is needed over and above Maxwell's 'valve'. As we shall see, what is required is not intelligence or free will – a mechanical device may perform the task. What is actually required is the ability to take in information, to act on it, and then, equally important, to discard it.

Let us, though, return to Maxwell and his response to Tyndall and his associates. Central here is what Maxwell calls the railway 'pointsman'[53] – who today would be more often be called a 'signalman'. Maxwell would not dispute with Tyndall that the laws of physics, in particular the law of conservation of energy, held sway in the physical sphere. He would not even dispute with Helmholtz that there was good reason to suppose that the human mind was subject to the same laws as the physical world – 'that animals, including humans, were constrained by the same energy laws that explained steam engines and water wheels'.[53] He could even appreciate the argument of Spencer that: '[A]ll actions whatever must be determined by those psychical connections which experience has generated . . . in his constitution,' which was apparently another blow for the idea of free will.

So Maxwell, as a Christian and a scientist, was by no means inclined to trim the science to oblige the theology. Neither, though, was he inclined to demand that basic theological concepts dance to a would-be scientific tune. In his writings, there was as much criticism of those Christians who, for example, suggested that the soul takes part in the creation or annihilation of energy, as of Herbert Spencer, who claimed to deduce consciousness as an extrapolation of fundamental physics.

Maxwell certainly believed in the soul, but he recognized that it could not possibly be the moving force or store of energy for the body; apart from any incongruity, on such a picture the soul would inevitably run down, rather than remaining as a guide to feeling and living through this world, and, Maxwell believed, the next.

Yet, if the soul could not take part in exchanges of energy, how could it play such a role? Maxwell's tentative answer[53] was given in a letter to his close friend and later biographer, Lewis Campbell.

> There is action and reaction between body and soul, but it is not of a kind in which energy passes from one to the other, - as when a man pulls a trigger it is the gunpowder which projects the bullet, or when a pointsman shunts a train it is the rails that bear the thrust.

Clearly the pointsman did not physically push the train; his expenditure of energy was negligible. Equally clearly his freedom of manouvre was limited; the train moved along rails provided by the company. And yet his role was crucial; without him the train just would not get to its intended destination.

Maxwell's solution to the problem of free will was based on the idea of the pointsman. He wished to find a process that begins with consciousness, may affect the physical outcome of events, but does not require provision of energy. He came up with tentative solutions in Balfour Stewart's concept of unstable systems, in which an infinitesimal amount of energy may tip in one of a number of directions; and alternatively in the more mathematical discovery of singular states, points in the solution of differential equations for the motion of particles where it is impossible to predict which route the particle will take.[53]

It is not clear that these solutions are any more (or less) successful than those of today's quantum physicists with a religious bent, who may see the possibility of divine intervention in the indeterminism of quantum theory. More important and interesting is the general pointsman idea – a fully deterministic universe, but with room for guidance by human will. Maxwell considered his approach essential both for theologians, who should not feel overwhelmed by the tide of scientific confidence, and for scientists, who should not be so misled by the successes of energy physics and determinism that they feel they have squeezed out any reference to the human being as a human being, with soul and free will.

We now move to a very different sphere of Maxwell's thoughts – the demon; and we may note, perhaps with surprise, that in the letter to Strutt[37] concerning the demon that we have already referred to, Maxwell uses exactly the same metaphor:

[A]ppoint a doorkeeper, very intelligent and exceedingly quick, with microscopic eyes but an essentially finite being . . . In this way the temperature of *B* may be raised and that of *A* lowered without any expenditure of energy, but only by the intelligent action of a mere guiding agent (like a pointsman acting switches who should send the express along one line and the goods along another).

Clearly Maxwell's view is that just as the pointsman exchanges no energy with the railway system, but nevertheless plays a crucial guiding role in determining the passage of the trains; just as human will has no direct physical interaction with the body, but may play a central part in the actions of the human; similarly the demon does not interact directly with the molecules of the system, and so there are no exchanges of energy between demon and system, but the demon affects very significantly the evolution of the system.

There is an interesting parallel between the task of the pointsman in assimilating and acting on information, and that of the demon being exactly the same. There is maybe a hint here of Maxwell's understanding of the primacy of information in physics, and, to a very limited extent, anticipating the development of information theory. Incidentally, Stanley[53] adds in the suggestion that the human will may also process information and act on it.

It might be thought appropriate at this point to distinguish between James Thomson's idea of a being, later to be immortalized as the demon, that could actually have free will, and Maxwell's pointsman, which was merely an analogy or metaphor for free will. Yet perhaps this is unnecessary, and we might agree with Stanley[53] that:

The demon itself did not have to be a supernatural being and it was intended to solve a technical problem in thermodynamics. But its function was based on a metaphor that was also part of a chain of reasoning based on manifestly religious premises. Thus the pointsman was both a scientific and religious entity, and in both cases it was intended to raise difficult questions about the knowledge available from the current state of science.

Faced with the physical, emotional and religious problems of thermodynamics, Victorian scientists came up with a bewildering range of 'pictures' of what was going on. They may be described as analogies, allegories, metaphors, models or even genuine beliefs. Categorization was rather unimportant; what was important was the transfer and discussion of ideas and views. Clarke notes that:

[T]he discourse of Victorian thermodynamics was immediately subjected to earnest moral amplifications, and its cultural dissemination was mediated at all points by acts of allegorical extension. Following quickly upon the formulations of Carnot, Helmholtz, Thomson, Clausius and Maxwell, thermodynamic theories gave rise to a complex, often contradictory range of social ideologies and creative responses. The scientists developing the physical theories also initiated the allegorization of thermodynamics by packaging the mechanical implications of thermodynamic ideas in a series of strong rhetorical creations.

We may go even further and follow Edward Daub[36] in thinking that:

Maxwell's demon was not mortal because he was made in the image of God. And like God, he could see without seeing and hear without hearing. In short, he could acquire information without any expenditure of energy.

Yet Daub considers that Maxwell would have severed the demon's ties to theology had he known of the work of Leo Szilard in the 1920s.

Maxwell's Demon and the Coming of Information Theory

One genuine puzzle concerning the demon, left from the nineteenth century, was to explain how it related to the second law. It has been seen that it shed a great deal of light on the law, and indeed that, from the point of view of the microscopic demon, the law is irrelevant, as the distinction between heat and work does not apply. However, from the point of view of the macroscopic human observer of the scene, it is certainly relevant, and we should be able to understand the relationship between the demon and the law.

The credit for taking a major step towards understanding the relationship goes to the young Leo Szilard[69] (Fig. 9.3a) in his University of Berlin doctoral thesis of 1925 and a related paper of 1929. This work is also credited with the demonstrating the first appreciation of information theory.

Szilard[70] was a scientist of immense ability. In his mid twenties, and while his friend, fellow Hungarian émigré and later Nobel Prize winner, Eugene Wigner, was nervous of approaching the élite members of the Berlin physics community such as Einstein, Szilard felt himself able to meet them as equals. His later scientific career, though, was not perhaps as illustrious as it might have been. After the early work on Maxwell's demon, his most important work was the invention of the chain reaction in the physics of nuclear fission, but he was to spend most of his life working for nuclear disarmament in the Cold War.

(a) (b)

Fig. 9.3 Szilard and Brillouin linked the question of Maxwell's demon to information theory. (a) Leo Szilard in 1953. Courtesy of Kark Maramorosch, from L. Hargittai, *The Martians of Science* (Oxford: Oxford University Press, 2006). (b) Leon Brillouin. American Institute of Physics Emilio Segré Visual Archives, Leon Brillouin Collection.

In his analysis of the demon, Szilard considers a gas of a single molecule. In the usual demon scenario, it is observed or measured to be on one side or other of the partition. Depending on which side it is located, the partition is moved to the other end of the volume so that the molecule has the freedom of the whole volume. In this procedure, others things being equal, it can be shown that there is a decrease of entropy and so the second law is violated.

But Szilard stresses that, fortunately (at least if we wish to see the second law remain intact), other things are by no means equal. He notes that the measurement is made by an 'intelligent being', and that the measurement makes the system behave in a manner distinctly different from how it would behave if it were left to itself. He argues that there is a 'measurement faculty' that avoids an overall decrease in entropy, because the measurement itself is accompanied by an increase in entropy which exactly cancels out the decrease in entropy in the process following the measurement.

Rolf Landauer,[71] one of the great names in information theory in recent years, has said that Szilard's work should indeed be regarded as the start of information theory. The initial measurement provides information on the whereabouts of the molecule, which is stored in the memory of the experimenter, and used in the decision of which way to move the partition.

Szilard reports that the increase in entropy caused by the measurement is $k \log_e 2$, where k is the ubiquitous Boltzmann's constant, which appears in all the formulae relating to the entropy and energy of particles such as molecules. The log is to the base e, the exponential number, and $\log_e 2$ is equal to 0.693. If the measurement is carried out at temperature T, then, according to Szilard, the energy expended in performing a measurement is equal to $kT \log_e 2$.

In retrospect, it is clear Szilard had made tremendous advances in understanding both the demon itself and the primacy of information – Leff and Rex[72] comment that:

He identified the three central issues related to Maxwell's demon as we understand them today – measurement, information, and memory – and he established the underpinnings of information theory and its connection with physics.

However, his discussion was not totally clear, and although his identification of the quantity $kT \log_e 2$. was extremely important, he had actually located it in the wrong place, as we shall see in the following section. It is probably not coincidence that his work was performed shortly after the arrival, in 1925–26, of modern quantum theory, in which, as has been said, measurement appeared to play a central role. While previously measurement could be thought of as a possibly passive recording of a quantity already in existence, the quantum position, most clearly enunciated by John von Neumann,[73] was that, in most cases, the value of the quantity being measured did not even exist before the measurement, and that the measurement caused a 'collapse of wave-function' to a function corresponding to the value obtained in the measurement.

In this climate it seemed very natural to assume that any measurement was a 'special' event that might indeed require an expenditure of energy. In the case of the demon, it could be suggested that the demon had to pick out a particular molecule against background noise that would certainly increase with temperature.

Again, it was not surprising that von Neumann himself came up with exactly the same value, $kT \log_e 2$, for the energy expended in computing. In a lecture[74] at the University of Illinois, given in 1949 but not published until 1966, he announced that this was the amount of energy required to manipulate one bit of information, in other words to make a decision between two paths in a computation. If energy were expended in every computer operation, clearly computers could not be *physically* reversible. Also any straightforward computational scheme cannot be *logically* reversible because, for example, $3 + 8 = 11$, but given the 11, we obviously cannot recover the 3 and the 8.

The connection between physical and logical reversibility in computation is interesting and instructive, though not totally straightforward.[75] Processes that are logically irreversible may actually be physically reversible for certain initial distributions of the relevant parameters.

As computers developed, von Neumann' suggestion was to become important. For many years, the amount of energy used by actual computers was very many orders of magnitude greater than this supposed theoretical limit; most of the energy was required to keep different parts of the computer at different voltages. However, following the empirical 'Moore's law',[52] as the density of transistors in the computer became higher and higher, so even this minimum energy did become a potential concern.

More important, though, was the fact that quantum computation, which was discussed from the 1980s, would require reversible techniques, because quantum theory is itself reversible. Von Neumann's suggestion implied energy loss at each stage of the process, and would seem to rule out the possibility of constructing a quantum computer. For several decades, the solutions of Szilard and von Neumann were accepted by most physicists and computer scientists; the problem of Maxwell's demon was considered to be finally resolved, and quantum computation thought to be probably impossible.[52]

In 1949, Claude Shannon[76] built on Szilard's suggestions to develop the subject of information theory. He introduced a quantity called *information entropy*, which bore a very strong resemblance to thermodynamic entropy, though the connection between the two concepts was warmly debated. Some were unhappy to regard even as analogous the quantities of thermodynamic entropy, which they described as a fixed physical quantity, and information, which apparently related to the possibly subjective knowledge of an individual. However, even earlier in this article, it was suggested that thermodynamic entropy was less straightforward than sometimes imagined!

The well-known physicist, Leon Brillouin[77] (Fig. 9.3b), attempted to clarify matters. In an important book of 1956 he stated that gaining information about a physical quantity decreases its entropy. He defined *negentropy* as the negative of entropy, and then included information in the second law, his version of which becomes: (negentropy + information) never increases; in a reversible change it is constant, while in an irreversible change it decreases.

Earlier, Brillouin[78] had claimed to confirm directly Szilard's idea that a measurement required an expenditure of $kT \log_e 2$ of energy. The demon, for example, will receive a photon from any particular molecule, and it must be able to distinguish it from the background radiation. Detailed calculations agreed with Szilard's earlier arguments. Although Planck's constant, h, the central quantity of quantum theory, did not appear in the all-important expression, it is clear that the argument was quantum-mechanical, depending on the particle-like nature of light.

And there the matter lay for some decades.

The Problem is not Measurement but Erasure!

Gradually the consensus was questioned, both for computation and then for Maxwell's demon. As early as the 1960s, Landauer[79] (Fig. 9.4a) decided to ignore the focus on 'measurement' as a primary term, and to analyse how computations were actually carried out. To his surprise, he came to realize that, provided all the operations were carried out extremely slowly, just as in textbook thermodynamics, there need be no expenditure of energy during the computing process.

Fig. 9.4 (a) Rolf Landauer. American Institute of Physics Emilio Segré Visual Archives, Physics Today Collection. (b) Oliver Penrose. Courtesy of the London Mathematical Society. (c) Charles Bennett. Copyright Zentralbibliothek für Physik, Renate Bertlmann.

The computation may be reversible. However, when information is erased or discarded, there must be a dissipation of energy, the amount being, as one might expect, $kT \log_e 2$, an amount now known as Landauer's constant, the whole idea being called Landauer's principle. So Szilard had been broadly right, but had assumed that the expenditure of energy took place at the incorrect point in his procedure. It takes place not at measurement but at the discarding of the information.

Simple physical models show why erasure requires dissipation of energy. We may imagine erasure taking the form of setting a register to zero, but the register may read any value before this procedure, and it is not possible to design a unique reversible physical procedure to achieve this; at best the reading may oscillate around zero. What must be done is to include a damping term that will result in the oscillation being damped and decaying towards zero, but that, of course, corresponds to exactly the expenditure of energy and so the irreversibility that Landauer requires.

Equally, when discussing logical irreversibility, we may note that, when calculating $3 + 8 = 11$, it is not the calculation of the 11 that causes the irreversibility, but the subsequent erasure of the 3 and the 8.

It might seem to be impossible to design a working computer that would not have to discard information, but Charles Bennett[80,81] (Fig. 9.4c), one of a group of physicists and computer scientists centred on Feynman in the 1970s, worked out how to do it. Any computation should be performed in three parts. In the first the desired answer is obtained, but no information is discarded so the process may be reversible. For example, if the sum $3 + 8 = 11$ is performed, one might retain the 11 but also the 3 in a double register. In the second stage the actual answer is copied onto a blank tape; this stage is also reversible. Finally, in the third stage, the first stage is essentially reversed. Thus all the excess information generated by the first stage is automatically removed, but this is achieved by the computer itself, not by erasure, so this stage too is reversible. Overall the only change achieved by the entire program is that the required answer is on the formerly blank tape.

The arguments of Landauer and Bennett were so much contrary to received opinion that there was great reluctance to accept that reversible computation was possible even in theory. Marvin Minsky,[82] a leading computer scientist, has reported that he was asked to referee one of the crucial papers. He read it every day for a month, and then advised the journal that the result seemed impossible, but since he could not find the mistake they should publish the paper.

Feynman, too, was highly suspicious of the result, but came up with several independent proofs, which encouraged him to consider how a reversible classical computer might be designed. It is clear that all the gates must themselves be reversible, which implies that they must have as many outputs as inputs. The theory of effective gates was constructed by Feynman and his associates, Ed Fredkin and Tom Toffoli, and Fredkin and Toffoli then proceeded to design a so-called billiard-ball computer.[83]

In this arrangement, billiard balls were to be rolled along a smooth table, making fixed angles with a principal axis. Timings were arranged so that billiard balls collided appropriately and were reflected by suitably arranged mirrors. The whole scheme was arranged to achieve reversible computation. Of course, in practice, friction could not be avoided, and timings and directions would have small errors, so the scheme was not intended to be a practical one. What it did, though, was to make it perfectly clear that there could be no theoretical objection to a reversible classical computer.

Equally, there could be no intrinsic reason why quantum computers could not be designed and built, and, of course, an immense amount of work in this direction is being carried out today, though the difficulties are formidable.[57,84,85]

Let us now turn to the analogous resolution of the problem of the Maxwell demon, which was also obtained by Bennett[86] and Oliver Penrose[87] (Fig. 9.4b); Leff and Rex[88] call it the Landauer–Penrose–Bennett solution, and it follows on directly from the ideas on quantum computation.

This solution recognizes that, although, contrary to Szilard, there is no increase in entropy when the demon performs a measurement, the decrease in entropy caused as a result of the operation of the demon is balanced by the increase in entropy when the demon discards the information obtained from the measurement. Thus there is no overall decrease of entropy, and so no violation of the second law.

Of course, one might ask why the demon has to discard the information – could it not retain it and avoid the increase in entropy on erasure? Apart from the awkwardness of the memory of the demon ever increasing, we should remember that thermodynamics requires the use of cyclic processes. We must leave the demon in the same state after dealing with a molecule as it was before the encounter. Information must be discarded. The second law cannot be beaten!

Conclusions

So we are now fairly confident (though presumably no more confident than Brillouin was in the 1950s!) that we have fully solved all the problems that Maxwell set for us 140 years ago.

Over those years, the demon has provided, painfully in some cases, considerable understanding of fundamental areas of physics, and the relationship between physical systems, information and memory. Though clearly Maxwell could hardly have anticipated the various directions in which ideas would emerge, he nevertheless deserves enormous credit for realizing the true nature of thermodynamics, and for appreciating the difficulties that remained for scientists of his generation and later ones to grapple with.

As in so many areas, it seems that Maxwell had the greatest gift among nineteenth-century physicists, not just to create outstanding physical theories, but also to appreciate the fundamental meaning, significance and limitations of the structure of contemporary physics, and thus to anticipate the areas where classical physics would need to be transformed in the twentieth century.

CHAPTER 10

Maxwell's Contributions to Electricity and Magnetism

DANIEL M. SIEGEL

James Clerk Maxwell's contributions to the science of electricity and magnetism have had a broad and profound influence in modern science and technology. The eponymous *Maxwell's equations* for the electromagnetic field have provided a lasting foundation for electromagnetic theory; aspects of Maxwell's field theory – invariance and signal-transmission properties – played crucial roles in the development of relativity theory; Maxwell's approach in completing his field equations, through the introduction of the *displacement current*, has provided a basic paradigm for the practice of theoretical physics; and his idea of electromagnetic waves has formed the basis for a variety of technologies that define the modern world, from communications networks to lasers, from imaging technology to particle physics apparatus.

Maxwell's intensive work on electromagnetic theory began in 1855, the year in which he turned 24 and became a Fellow of Trinity College, Cambridge. His first publication on electromagnetic theory, 'On Faraday's Lines of Force', was presented to the Cambridge Philosophical Society in December of 1855 and February of 1856, and his work on electricity and magnetism then continued for 25 years, issuing in the monumental *Treatise on Electricity and Magnetism* of 1873; work on a second edition was halted by Maxwell's death in 1879, and that edition appeared posthumously in 1881.[1]

Maxwell's Goals in Electromagnetic Theory

The themes that were to dominate Maxwell's research in electromagnetic theory for 25 years were defined at the outset, in 1855, in the introduction to his first publication on electricity and magnetism, 'On Faraday's Lines of Force'. Maxwell had three goals: first, to develop a theory that would encompass all the known facts of electricity and magnetism; second, to accomplish this unification by using Michael Faraday's approach to electricity and magnetism – the *lines of force* or *field* approach; and third, to have the new theory go beyond what was known from experiment, filling in the gaps, so as to achieve theoretical completeness. This final goal, in Maxwell's own words, was to have theory 'afford the means of calculating the effects in . . . cases where the known formulae' – from experiment – did not yet reach. This was, at the time, a novel perspective on the responsibilities of the theoretical physicist.[2]

These were ambitious goals that Maxwell, at age 24 and just embarking on his scientific career, had set for himself. In his initial publication of 1855 he was not yet ready to take on the full task; this was to be just a start – a limbering-up exercise. But, by the time Maxwell published his *Treatise on Electricity and Magnetism* 18 years later, he had accomplished all three goals, with major implications for the development of both science and technology.

The first goal, to give a unified account of the phenomena of electricity and magnetism, had become pressing as a result of accumulating nineteenth-century discoveries. Before 1800, the study of electricity and magnetism had centred on electrostatics and magnetostatics, dealing with the attractions and repulsions of electrically charged objects, as well as the attractions and repulsions of magnetic poles. Beginning in 1800, however, a dizzying array of new phenomena were discovered. It started with Alessandro Volta's work in Italy on the first electric battery, formed from multiple cells of dissimilar metals connected by aqueous chemical media. This produced, for the first time, continuous electric currents, opening a new field of studies.

The study of current electricity soon led to another fundamental discovery – in 1820, by Hans Christian Oersted in Denmark – of the magnetic effects of electric currents: electric currents were found to exert forces on magnetic poles in their vicinities, causing the magnetic poles to tend to go in circles around the current-carrying wires.

Oersted's discovery was immediately taken up, later in 1820, in two European capitals: André Marie Ampère, in Paris, undertook further experimental studies as well as sophisticated mathematical analyses of the forces involved; while Michael Faraday, in London, undertook a more qualitative conceptual analysis, focusing on experimental and technological aspects, and resulting in the first electric motor.[3]

Faraday went on to explore the possibility that if electric current could cause magnetic effects, then the reverse process should also happen, with magnetism producing electric current. His search for such an effect was not immediately successful, but eventually, in 1831, he was rewarded with the discovery of electromagnetic induction, in which magnetic changes – such as produced by a moving magnet or a changing electric current – induce an electric current in a wire. This was the first electric generator.

Faraday continued on his path of discovery, investigating the connections between electricity and chemistry in voltaic and electrolytic cells, and discovering new magnetic effects in paramagnetic and diamagnetic materials. Faraday's investigations were repeated, elaborated and followed up experimentally and theoretically all over the world, and by the time that Maxwell came to confront the large and still growing list of electromagnetic phenomena, the scientific community had developed a severe case of intellectual indigestion, trying to accommodate and assimilate all of this new material.[4]

Maxwell's training, first at the University of Edinburgh and then at Cambridge University, had prepared him admirably to take on this daunting task. At Cambridge, he had learned the mathematical techniques he would need to bring all of this experimental material under theoretical control; the Cambridge approach, however, tended to concentrate on the sciences that were traditionally more mathematical – such as mechanics, optics and astronomy – and did not encourage study of the newer subjects, such as electricity and magnetism and heat, which were seen as perhaps not yet ripe for mathematization. Maxwell's experiences at Edinburgh, however, had provided a different orientation: from Professor James D. Forbes at Edinburgh, Maxwell had picked up a more encyclopedic approach to science, in which one wanted to give attention to all of science, and neighbouring technologies as well.

Putting the two traditions together, Maxwell formulated the project of bringing the mathematical methods of Cambridge to bear on some of the encyclopedic subject matter of Edinburgh – namely, electricity and magnetism, where the unifying properties of mathematical treatment were so sorely needed. By showing how to treat all known electromagnetic effects in terms of a well-defined set of mathematical equations, Maxwell was to achieve a notable success in the direction of unification.[5]

Maxwell's second goal was to accomplish the unification of electromagnetic phenomena through the approach that had been developed by Faraday in the years from the 1820s to the 1850s. In his studies of electricity and magnetism, Faraday had addressed a question that had been bothering physical scientists for over a hundred years. Isaac Newton, back in the seventeenth century, had studied the gravitational forces active in the solar system, and had treated gravitational force as a force acting at a distance: the Earth, for example, exerts a gravitational force on the moon, 250,000 miles away, and this was treated as a directly-acting force, without any discussion of how the force got from here to there. Newton justified this on grounds of an empiricist scientific methodology: all we know is that the force does get from here to there, and that its intensity is governed by a mathematical law (the inverse-square law); that is all we can say. This approach, *action at a distance*, was then applied also to electrical and magnetic forces in the eighteenth and nineteenth centuries. When the time came, in the mid-nineteenth century, to think about ways to unify electricity and magnetism, the action-at-a-distance approach seemed to many the most natural way to do this. Wilhelm Weber, in Germany, attempted this, accommodating the great variety of electromagnetic phenomena by postulating complex force laws for these distance forces.[6]

Faraday's example, however, suggested a different approach. Beginning with his follow-up of Oersted's work in 1820, Faraday had taken on the question of how electrical and magnetic forces are propagated through space. Beginning with considerations of how iron filings arrange themselves in the neighbourhood of magnets, Faraday had developed a notion of *lines of force*, whose orientation in space indicated the direction of the given electrical or magnetic force, with the intensity of the force indicated by the number of such lines of force. In this way, the lines of force indicated conditions in the space intervening between the interacting bodies, by means of which the force was transmitted from one to the other. Later, Faraday began to refer to the set of lines of force as a *field* of force (etymologically, from *field of action*), and that terminology has stuck: since then, one refers to *electric and magnetic fields* as the basis of this approach to electromagnetic theory.[7]

Maxwell, following Faraday, set out to use the field idea in unifying electromagnetic theory. Maxwell's equations are stated in terms of field quantities – the electric field, which we shall denote by a vector $\mathbf{E}$, and the magnetic field, denoted by the vector $\mathbf{H}$.[8] The field idea has been central not only to electromagnetic theory, but to all of physics ever since. In particular, as Einstein himself pointed out most trenchantly, and as historians and philosophers of science whole-heartedly agree, the idea of the propagation of forces at the velocity of light through the electromagnetic field, as described by Maxwell's equations – rather than instantaneously at a distance – formed the necessary basis for relativity theory.[9]

Maxwell's third goal was to use theoretical analysis to go beyond what had been established by electrical and magnetic experiments. Put another way, this was a goal of completeness: even though electromagnetic experiments left some gaps in the account of electromagnetic phenomena, the theory Maxwell was looking for would not be limited by this circumstance, but would fill in the gaps. This was a new conception of the duties of the theoretical physicist, and one that not all of Maxwell's contemporaries were willing to accept. But Maxwell took this duty seriously, and it was in the context of this commitment that Maxwell undertook – on theoretical grounds alone, with no experimental input whatever – to modify one of the basic equations, through the addition of a new term in the equation; this new term – called the *displacement current*, because it had the effect of an electric current and involved the displacement of certain particles – had the form

$$-\frac{1}{4\pi c^2}\frac{d\mathbf{E}}{dt}$$

where c is a constant involved in relations between electrical units, and $d\mathbf{E}/dt$ expresses the rate of change with time t of the electric field **E**.[10]

The introduction of the displacement current did two things. First, it made Maxwell's equations into a complete and coherent set of equations, which has functioned as the basis of electromagnetic theory ever since. Second, it led to the idea of electromagnetic waves, which are waves of alternation of electric fields **E** and magnetic fields **H**. The first use of the idea of electromagnetic waves, by Maxwell himself, was to identify light as an electromagnetic wave, and that has been our understanding of the nature of light since that time. Beyond this, Heinrich Hertz, in 1888, was able to generate and detect electromagnetic waves with laboratory electrical apparatus, creating new vistas in both science and technology.[11]

The introduction of the displacement current by Maxwell is properly seen as a paradigm case of theoretical innovation in science: an innovation with zero experimental basis, which then had profound effects in both science and technology. This strikingly successful use of theory to advance physics has stood as a model and inspiration for theoretical physics ever since. For Maxwell himself, the introduction of the displacement current was central to the achievement of his three goals for electromagnetic theory – unifying electromagnetic theory in terms of a coherent foundation; doing this in terms of forces propagated in time through the electromagnetic field; and going beyond experiment, filling in gaps and predicting new phenomena.[12]

Maxwell and Mechanical Modelling in Electromagnetic Theory

But what was the secret of Maxwell's success – how did he manage to accomplish all three of his ambitious goals, with profound consequences for modern science and technology? In order to answer this question, we will have to look at his work in some detail. The basics of Maxwell's electromagnetic theory, including the displacement current and the electromagnetic theory of light, were worked out in a ten-year period, from 1855 to 1865, in a series of three substantial papers: 'On Faraday's Lines of Force' (FLF) (1855–56); 'On Physical Lines of Force' (PLF) (1861–62); and 'A Dynamical Theory of the Electromagnetic Field' (DT) (1864–65).[13]

In order to fully understand Maxwell's work in these three papers, we will have to be careful not to judge his nineteenth-century approaches and procedures by our own, twenty-first-century, standards: this would cause us to undervalue and neglect the very modalities that were most helpful to Maxwell in making progress toward his impressive achievements. In our own time, we look at electric and magnetic fields – our **E** and **H** vectors – as primitive quantities, not to be reduced to something else. They are defined in terms of their effects on electric charges and magnetic poles, and their properties are defined by Maxwell's equations. In part, this very abstract view of electromagnetic fields, involving no visualizable picture of what is going on out there in the field, is Maxwell's legacy: Beginning with the third of the crucial electromagnetic papers, 'A Dynamical Theory of the Electromagnetic Field' – and even more emphatically in later years – Maxwell adopted this abstract view. The equations were the essence of Maxwell's legacy, as most trenchantly expressed by Heinrich Hertz, who first produced and detected the electromagnetic waves that Maxwell's theory predicted: 'To the question, "What is Maxwell's theory?" I know of no shorter or more definite answer than the following: – Maxwell's theory is Maxwell's system of equations.'[14]

In Maxwell's earlier work on electromagnetic theory, however, as in the papers, 'On Faraday's Lines of Force' (FLF) and 'On Physical Lines of Force' (PLF), he was thinking not in the abstract terms of present-day field theory, but rather in terms of concrete and pictorial images, involving mechanical models, in which material systems – particles, fluids, elastic solids – interacted and changed with time in accordance with the laws of Newtonian mechanics. In FLF, the electric and magnetic lines of force were pictured as the flow lines of a fluid, something like water seeping through an underground rock stratum. In PLF,the mechanical modelling was yet more elaborate: in what he referred to, in the subtitles to the successive installments of the paper, as 'The Theory of Molecular Vortices', he first pictured small regions of whirling motions, or vortices, in a fluid medium that was conjectured to fill space. He then altered and refined this picture, finally imagining the mechanical medium underlying the electromagnetic field – the 'magneto-electric medium' – as a cellular medium, with rotating parcels of elastic material within the cells (which he continued to call 'vortices'), and monolayers of small, spherical particles at the surfaces of the cells – as in Maxwell's own diagram of the magneto-electric medium, shown in Fig. 10.1.[15]

In Maxwell's account of the behaviour of this magneto-electric medium, he acknowledged that the monolayers of small particles, and their function, had been suggested by industrial machinery:

> In mechanism, when two wheels are intended to revolve in the same direction, a wheel is placed between them ... and this wheel is called an 'idle wheel.' The hypothesis about the vortices which I have to suggest is that a layer of particles, acting as idle wheels, is interposed between each vortex and the next.

In modern technology, these particles would be identified as ball bearings; in any case, as one Continental commentator lamented, one thought that, in reading Maxwell and like-minded British physicists, one would be 'entering the tranquil and neatly ordered abode of reason, but, [instead] we find ourselves in a factory'.[16]

Many twentieth- and twenty-first-century readers of Maxwell – both scientists and historians, looking back from their modern vantage-point – have been dismayed, even embarrassed, to find the great Maxwell, their culture hero, using such 'strange', 'bizarre' and altogether 'outlandish' mechanical models. Even though he described these models in detail, say these later commentators, he must not really have believed in them: believing in them would be ridiculous, and Maxwell is not to be considered ridiculous! Relatedly, this line of retrospective thinking continues, these fanciful mechanical models couldn't really have helped Maxwell in his innovations in electromagnetic theory.[17]

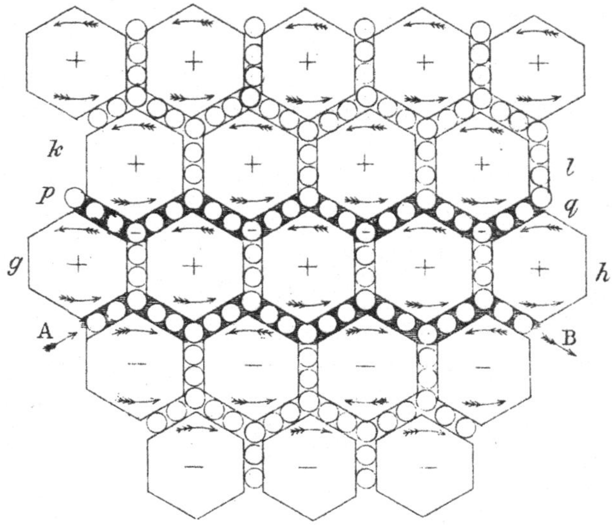

Fig. 10.1 Maxwell's diagram of the structure of the magneto-electric medium (from Maxwell, 'Physical Lines,' Plate VIII, Figure 2, opposite p. 488, explained on pp. 477, 489.) The medium is cellular, with the cells pictured, in cross-section, as a plane-filling array of hexagons. Correspondingly, in three dimensions, one envisions a space-filling array of polyhedra – perhaps the irregular dodecahedra specified by the hexagonal close packing of spheres; in any case, Maxwell approximated the shapes of the polyhedral cells to spheres for purposes of calculation. In the final form of the molecular-vortex model, the material within a cell was to be envisioned as a single chunk or blob of elastic material. Rotations of these chunks of elastic material, indicated by the curved arrows in the diagram, constituted the 'vortices' at this point. The 'cell-walls', according to Maxwell, are 'composed of [monolayers of spherical] particles which are very small compared with the cells'.

These kinds of retrospective judgements have led most modern commentators on Maxwell's electromagnetic theory to give only very superficial attention to the mechanical models that Maxwell described in FLF and PLF, and they have judged, on this cursory basis, that these mechanical models – these preposterous creations of Maxwell's – are not relevant to an understanding of his creative process in electromagnetic theory. There is, however, an alternative tradition in Maxwell scholarship and commentary. In the work of some British commentators and followers of Maxwell,[18] and especially in the work of Ludwig Boltzmann, a German follower of Maxwell, one finds serious attempts to fully understand Maxwell's models, and to see how they might have been helpful in facilitating his remarkable achievements.

Ludwig Boltzmann is best known for his work in statistical mechanics, where he was following up Maxwell's work in that area; the resulting statistical treatment of gas dynamics is known eponymously as the *Maxwell–Boltzmann statistics*. In electromagnetic theory, Boltzmann did not add to the research-front achievements of Maxwell, but he did publish editions of Maxwell's papers in electromagnetic theory, with commentary, and lectures on Maxwell's theory. The analysis that follows is in the tradition of Boltzmann's approach, but is based directly on the Maxwell texts, rather than the filtering of these through Boltzmann.[19] As Boltzmann pointed out, and as is clear enough from careful reading of Maxwell, Maxwell's attitudes toward the mechanical pictures in FLF and PLF were quite different. In 1855 in FLF, Maxwell was just starting out in his work on electromagnetic theory, and he felt that it was not yet the time for him to attempt the kind of serious, comprehensive theory to which he ultimately looked forward. So the mechanical representations that he presented in FLF were presented just as aids to the imagination, intended to accomplish a 'simplification and reduction of the results of previous investigations to a form in which the mind can grasp them'. Further, 'By the method which I adopt [in FLF], I hope to render it evident that I am not attempting to establish any physical theory of a science in which I have hardly made a single experiment'.

Maxwell was a young man of 24, without standing and reputation in science, and he was attempting to take further steps in a science that had been dominated by Michael Faraday's successful endeavours for over 30 years. Furthermore, Maxwell himself had not done any experiments in electricity and magnetism, and the role of the theoretical physicist – one who thought about physics but did not do experiments – had not yet been established; indeed Maxwell's own work was to be central in first establishing that role. In this situation, Maxwell made only very modest claims for the mechanical representations that he put forward in FLF: the images of fluids flowing through porous materials, with streamline flow lines analogous to the lines of electric or magnetic force, were to be taken as nothing more than analogies, with no claim whatever to physical reality, and no claim to explanatory power.[20]

There was precedent for this kind of use of mechanical representation in an analogical sense. William Thomson, seven years Maxwell's senior, had, in the decade before Maxwell's work on FLF, proposed various mechanical analogies for the purpose of gaining a clearer understanding of Faraday's approach, and this work provided important background for Maxwell's own work in FLF. Thomson, however, had eventually moved on from the agnosticism of a purely analogical approach, and had begun to think about what the implications of these mechanical analogies might be concerning the actual basis of the electromagnetic field. Thomson's thoughts about the possibility of moving forward in this way were also to be highly suggestive for Maxwell.[21]

In FLF, Maxwell had begun with an analogical approach, as had Thomson, but, from the very beginning, Maxwell was looking forward to something more. Eventually, Maxwell wrote in the introduction to FLF, he wanted to move on to a 'mature theory, in which physical facts will be physically explained'. And what, precisely, did Maxwell mean by using the phrase 'mature theory' to characterize his ultimate aim? Maxwell's rhetorical stance at this point, in FLF, was a modest one – not making strong claims concerning his ultimate goals. But Boltzmann, in his translation of FLF for a German edition, interpolated a phrase to make Maxwell's intent clear: The ultimate theory to which Maxwell looked forward was to be 'über das Wesen der Dinge'–'a mature theory ... *concerning the nature of things*'. Boltzmann here was using his knowledge of what Maxwell would subsequently do and say, in order to clarify the language in FLF – in the process altering the rhetorical force of Maxwell's statement, making it bolder than Maxwell had intended. But, as we shall see in our discussion of Maxwell's second paper on electromagnetic theory, PLF, Boltzmann was exactly correct in parsing the full meaning of Maxwell's somewhat guarded statement in FLF: in his second effort on electromagnetic theory, Maxwell was prepared to talk about what was 'real[ly] going on in the [electro]magnetic field'.[22]

The signal to Maxwell that the time had come, in mechanical representations of electromagnetic phenomena, to go beyond mere analogy and begin talking about the true nature of things, came again from William Thomson. In a paper published one year after Maxwell's FLF, and in subsequent publications, Thomson gave attention to Faraday's discovery, in 1845, that a beam of light going through a piece of glass situated in a magnetic field would experience a rotation of its plane of polarization. Analysis of the technical details of this phenomenon, with respect to the handedness of these rotations – left-handed or right-handed, depending on the particularities of the experimental situation – convinced Thomson that the actual mechanical basis for this effect was some kind of rotary, vortex motion going on in the magnetic field. The vortices would have their axes of rotation aligned along the magnetic field lines, and Thomson's assertive conclusion was that 'a certain alignment of axes of revolution in this [vortical] motion IS *magnetism*. Faraday's magneto-optic experiment makes this not a hypothesis, but a demonstrated conclusion.' Thomson's typographical cues, in the capitals and italics of 'IS *magnetism*', make it crystal clear that he is talking about the reality of the mechanical basis of the magnetic field, and not some mere analogy.[23]

Maxwell was powerfully impressed by this: 'Professor Thomson has pointed out that the cause of the magnetic action on light must be a *real* rotation going on in the magnetic field', Maxwell wrote approvingly; the time had come to go beyond the analogical approach and begin constructing, on the basis of Thomson's picture of vortices oriented along magnetic field lines, something like the 'mature theory' to which Maxwell had looked forward.[24]

The Molecular-Vortex Model

The 'mature theory' to which Maxwell had looked forward finally made its appearance in his paper 'On Physical Lines of Force', published in a series of four installments, respectively in March and April of 1861, and January and February of 1862. It is clear that when Parts I and II appeared early in 1861, Maxwell did not anticipate the innovations that were to appear in Parts III and IV, early in 1862. The successive parts of the paper thus actually document the development of Maxwell's thought on the subject, rather than merely displaying a rhetorical strategy. As a result, we will be able to see how Maxwell developed and extended the theory, first to cover all known electromagnetic phenomena, and then to predict in cases where experiment did not yet reach.

In the subtitles to the successive parts of the paper, Maxwell referred to the mechanical model that he was developing and extending as 'The Theory of Molecular Vortices'. This terminology was most apt at the outset, in Part I of the paper, where he was envisioning the mechanical medium underlying the electromagnetic field – what he called the 'magneto-electric medium' – as a fluid medium, and the 'molecular vortices' were small whirlpools or eddies in the fluid. By Part III of the paper, Maxwell had changed to a view of the magneto-electric medium as a cellular medium, in which monolayers of small particles defined the cells, while the interior of each cell was a parcel of rotating elastic material, as we have seen in Fig. 10.1. He retained the terminology of 'molecular vortices', however, as now referring to the whirling, vortex-like motions of the rotating blobs.[25]

Maxwell intended that the theory of molecular vortices would get at the reality of the electromagnetic field, rather then merely serving as a useful analogy, and there were parts of the theory that he regarded as fully meeting that reality standard. The fundamental idea that Maxwell had taken up from Faraday's experiments and Thomson's theoretical analysis – that the actual, physical basis for the magnetic field was rotational motion in the magneto-electric medium – was something that Maxwell firmly believed all the way through, and this belief stayed with him to the end of his life. He actually carried out experiments to try and detect these rotations mechanically. He had an instrument constructed, with a gimbal-mounted electromagnet that was free to rotate about both horizontal and vertical axes, and looked for effects of the rotations of the molecular vortices on the mechanical behaviour of this electromagnet. (He was unable to detect any effect, which indicated to him that the vortices must be small, rather than calling into question their existence.) As for electric fields, Maxwell was to be unwavering in his belief that elastic deformations in the magneto-electric medium constituted their true basis, beginning with his introduction of this idea in Part III of PLF, and continuing through the *Treatise on Electricity and Magnetism*.[26]

The idea of the monolayers of small particles at the surfaces of the cells, on the other hand, was to be an embarrassment to Maxwell all along – something that he himself could not really believe in. We have seen that the idea of these small particles, functioning as idle wheels or ball bearings, was suggested by industrial machinery; as a result, the magneto-electric medium was starting to look like the factory setting that Maxwell's critics were to ridicule.

Maxwell was well aware of this problem, and it is worth quoting him at some length on this matter, from the closing section of Part II of the paper, where he addressed, specifically, the problem of the rolling contact of the idle-wheel particles with the surfaces of the vortices, but also, by implication, the more general problem of the artificiality of these particles as an 'electrical hypothesis':

> The conception of a particle having its motion connected with that of a vortex by perfect rolling contact may appear somewhat awkward. I do not bring it forward as a mode of connexion existing in nature, or even as that which I would willingly assent to as an electrical hypothesis. It is, however, a mode of connexion which is mechanically conceivable, and easily investigated, and it serves to bring out the actual mechanical connexions between the known electro-magnetic phenomena; so that I venture to say that anyone who understands the provisional and temporary character of this hypothesis, will find himself rather helped than hindered by it in his search after the true interpretation of the phenomena.[27]

The small idle-wheel particles were thus place-holders in the theory: there had to be some mechanism for coupling the rotatory motions of the parcels of elastic material contained within the cells of the magneto-electric medium, and the small particles illustrated how this connecting mechanism had to function; but the small particles themselves were clearly unrealistic, a 'provisional and temporary' expedient. As a good friend from Cambridge days wrote to Maxwell, it would be hard to 'get people to [believe in] a little file of particles squeezed ... between rows of wheels'. As even this good friend had to tell Maxwell, if one were to judge his theory by this part of it, it would come across as ridiculous. Maxwell's interpreters and commentators through the generations who have concentrated their attention on this part of the molecular-vortex model, to the exclusion of the context, have understandably concluded that the model is not worthy of the careful attention of the historian or philosopher of science.[28]

But the model was, in fact, very successful as an engine of discovery. What is important in understanding Maxwell's success in using this model, even though it did include some admittedly awkward and provisional place-holders, is that he was doing his best to make the model as realistic as possible, in order to reflect the 'actual ... connexions' between the various electromagnetic phenomena as faithfully as possible. The proof of the pudding was in the eating: the proof of the faithfulness of the model, and the justification of Maxwell's Herculean efforts in trying to get at the reality of the electromagnetic field, was that the model turned out to be good enough – though admittedly far from perfect: it was good enough to motivate and guide Maxwell's introduction of the displacement current, and it was good enough to produce the first form of the electromagnetic theory of light.

What the model needed to do, in order to be good enough, was to properly embody the content of the equations of electricity and magnetism. In this sense, it is Maxwell's equations that remain at the centre of the story. Maxwell used the equations as a guide to the development of the molecular-vortex model, and the result was that the model was able to give a complete and coherent reflection of the content of those equations, so as to make clear what their implications might be.

Three of the set of four equations that we now regard as the canonical set of Maxwell's equations were used explicitly in developing the molecular-vortex model in 'Physical Lines'. These equations may be written, in vector notation, as follows:

$$\mathbf{j} = \frac{1}{4\pi}\text{curl}\,\mathbf{H} \tag{10.1}$$

$$-\text{curl}\,\mathbf{E}_1 = \mu\frac{d\mathbf{H}}{dt} \tag{10.2}$$

$$\rho = \frac{1}{4\pi c^2}\text{div}\,\mathbf{E}_2 \tag{10.3}$$

The fourth of the canonical Maxwell's Equations was tacit in PLF, while more explicitly treated later, in Maxwell's *Treatise*, where it was expressed as

$$\text{div}\,\mathbf{B} = 0 \tag{10.4}$$

The first of these equations, Equation (10.1), was Maxwell's expression, in terms of the field-theory approach that he had adopted from Faraday's work, of Oersted's observation that electric currents are accompanied by magnetic effects in a circular pattern about the conducting wire. Here, **j** represents the electric current density, **H** represents the magnetic field strength, and curl represents a mathematical operation describing how the field quantity – here the magnetic field **H** – is distributed in space. It was Maxwell who gave this mathematical operation the name curl; in the present case the curl operator tells you how the magnetic field **H** curls around the electric current **j** in circles, as Oersted had discovered experimentally.[29]

Equation (10.2) expresses the generator effect that Faraday had discovered in 1831, in which a changing magnetic situation causes an electric current to flow in a wire. Here, **H** is again the magnetic field strength; $d\mathbf{H}/dt$ indicates the rate of change with time of **H**; μ indicates the effects of any magnetic materials in the neighbourhood; and $\mathbf{E}_1$ is the electric field that is the result of the changing magnetic field and produces the current in the wire.[30]

Equation (10.3) expresses, in the language of field theory, the most basic electrical effect, the inverse-square force between electrical charges, positive or negative. Here, ρ represents electric charge; $\mathbf{E}_2$ is the electrostatic field that arises from this charge and exerts a force on another charge; c is the factor we met before relating to electrical units; and div is a mathematical operation that here measures the outward flow, or *divergence*, of the electric current density **j**.[31]

Equation (10.4), finally, expresses the circumstance that all magnetic fields – here represented by a somewhat different measure of magnetic field strength, **B** – are caused by electric currents, as in Equation (10.1), and not by magnetic charges or monopoles. Mathematically, this is expressed in terms of the operation div, which here measures the divergence of the magnetic field lines **B** from any point, and is set to 0 to express the absence of magnetic charges.[32]

With the formulation of these equations, Maxwell had made great progress toward the first two of his goals for electromagnetic theory – namely, to give a unified account of the plethora of discoveries in electricity and magnetism, and to accomplish this in the context of Faraday's field-theory approach. Equation (10.1), however, which is the form in which Maxwell presented this equation in the initial parts of PLF, is somewhat problematic: It applies only to closed electric currents – as in wires that form closed loops – and therefore cannot deal with situations in which electric charge is accumulating. For this reason, Equation (10.1) is not fully compatible with Equation (10.3). So this set of equations, with which Maxwell began in PLF, is not yet a full and consistent basis for electromagnetic theory.

What made this set of equations into a full, consistent, and permanent foundation for electromagnetic theory was Maxwell's addition of an extra term in Equation (10.1) – namely, the displacement current – in Part III of PLF. The introduction of the displacement current in turn then led Maxwell to the first form of the electromagnetic theory of light, also presented in Part III of PLF. This crucial advance in electromagnetic theory was the crowning achievement issuing from Maxwell's development of the molecular-vortex model in PLF. He used Equations (10.1)–(10.3) in developing the model; and the model, as it developed, in turn showed the way toward the modification of Equation (10.1), with consequences of immense significance for the development of modern science and technology. Accordingly, we'll look at the development of the molecular-vortex model in some detail, in order to understand how this event came to pass.

Maxwell began, in Part I of PLF, with a discussion of magnetic forces. At this point he was regarding the magneto-electric medium as a fluid medium. The magnetic field, **H**, characterized a region of the magneto-electric medium in which there were vortex motions, with their rotation axes oriented along the direction of the magnetic field lines. Owing to the centrifugal forces of the vortex rotations, the vortices would tend to contract along their rotation axes and expand laterally. This would give rise to mechanical stresses within the fluid that would produce, according to Maxwell's calculations, all of the known magnetic forces: the inverse-square forces between magnetic poles; the forces on the diamagnetic and paramagnetic materials identified by Faraday; and the forces exerted among electric currents and magnetic poles, as treated by Oersted and Ampère. As a result of this successful accounting for all magnetic forces, Maxwell felt confident in his identification of the magnetic field **H** with the rotations of the vortices.[33]

Having established this mechanical interpretation of the magnetic field **H**, Maxwell brought other electromagnetic quantities into the theory according to their linkages with the magnetic field quantity, **H**, through the field equations. The next quantity to be included in the theory was the electric current density. The electric current density **j** is linked to the magnetic field **H** through the first of the four Maxwell's equations to be treated in PLF:

$$\mathbf{j} = \frac{1}{4\pi}\text{curl } \mathbf{H} \tag{10.1}$$

Maxwell now searched for a mechanical interpretation for the electric current density **j** that would be consistent with Equation (10.1). He accomplished this by introducing the small particles interposed between adjacent vortex cells, as shown in Fig. 10.1, then identifying the motions of these particles – as they roll on the surfaces of the vortices and are 'squeezed along' between neighbouring vortex cells – with the electric current density **j**. Maxwell was able to calculate, from the mechanics of the connections between the rotations of the vortices and the motions of the small particles rolling on their surfaces, that Equation (10.1) would be satisfied – given this mechanical interpretation of the electric current density **j** – so that Equation (10.1) was now properly incorporated into the model.[34]

The next quantity to be included in the theory was the field $\mathbf{E}_1$, which is linked to the magnetic field **H** through the second of the four Maxwell's equations to be treated in PLF:

$$-\text{curl } \mathbf{E}_1 = \mu\frac{d\mathbf{H}}{dt} \tag{10.2}$$

Here the task was to find a mechanical interpretation for the electric field $\mathbf{E}_1$ that would be consistent with Equation (10.2). This was accomplished by identifying the forces that the vortex cells exert on the small particles at their surfaces as the electric field $\mathbf{E}_1$. Maxwell was able to calculate, from the mechanics of the relationship between the forces exerted at the surfaces of the vortices and the changes in the rotational velocities of the vortices, that Equation (10.2) would be satisfied – given this mechanical interpretation of the electric field $\mathbf{E}_1$ – so that Equation (10.2) was now properly incorporated into the model.[35]

Maxwell's final task was to extend the theory to electrostatics, as encapsulated in the third of the four Maxwell's equations,

$$\rho = \frac{1}{4\pi c^2} \text{div } \mathbf{E}_2 \qquad (10.3)$$

relating ρ, the electric charge density, to the electric field $\mathbf{E}_2$. But in order to continue to use his standard procedure for the inclusion of new electromagnetic quantities in the theory, it was necessary for Maxwell to find electromagnetic equations linking the new variables ρ and $\mathbf{E}_2$ back to the magnetic field variable **H**, as had already been done for the quantities **j** and $\mathbf{E}_1$. To get a linkage to ρ, the electric charge density, Maxwell would have to use the equation of continuity,

$$\text{div } \mathbf{j} = -\frac{d\rho}{dt} \qquad (10.5)$$

which expresses conservation of electric charge, and links ρ to the electric current density **j** – which in turn links back to the magnetic field **H** through Equation (10.1).[36]

The problem for Maxwell was that Equation (10.1) for the electric current had been built into the theory, and it applies only to closed circuits, where the electric current density **j** has no points of divergence or convergence, so that no electric charge is accumulating, and the left- and right-hand sides of Equation (10.5) are identically zero at all times. So Maxwell's mechanical model, as it had been developed to this point, with Equation (10.1) built in, could not supply any mechanical representation of Equation (10.5), except to say that the equation reduces to $0 = 0$ in this case, which does not furnish any useful information; Equation (10.3) then also reduces to $0 = 0$, as no electric charge ρ can accumulate.[37]

Maxwell had first written down Equation (10.1) back in FLF, where he had acknowledged that its application was 'limited to closed currents'. His apology at that point for such a limitation was that 'we know little of the magnetic effects of any currents which are not closed'. That is, the brief, transient magnetic effects of unclosed currents had not yet been well measured, so lack of theoretical treatment of this case by Equation (10.1) was not a major problem.[38]

The validity of this excuse, however, for the limitations of Equation (10.1), had expired when Maxwell had undertaken, in PLF, to present a serious theory, rather than the mere analogies of FLF. This was to be the 'mature theory', to which Maxwell had been looking forward, and he had committed himself to the requirement that such a theory would have to go beyond experiment, filling in the gaps where experiment did not yet reach. Equation (10.1) would thus have to be modified in order to satisfy the completeness criterion that represented one of Maxwell's three basic goals in electromagnetic theory.

Beyond this, Equation (10.1), because of its limitation to closed circuits, was now stymieing Maxwell's attempt to extend the theory of molecular vortices to electrostatics; this was the proximate motivation for modifying the equation. But how was the equation to be modified, if there was no experimental data to offer guidance? In contradistinction to the previous extensions of the theory, where known experimental results were used to indicate what the mechanical model had to accomplish, it now was the mechanical model that had to lead: what did the molecular-vortex model have to suggest for the modification of Equation (10.1)?

It was the introduction of elasticity into the model, in Part III of PLF, that allowed for the modification of Equation (10.1) and indicated how to do it. Equation (10.1) was calculated on the basis of the assumption that the rotational (angular) velocity throughout a rotating parcel of vortex matter was constant – essentially, that it rotated as a rigid body. Maxwell now, in Part III of PLF, introduced the notion that the parcels of ethereal matter – the interiors of the cells in Fig. 10.1 – were elastic blobs. If so, then elastic distortions of the surfaces of the blobs would be another contributor to the motions of the small idle-wheel particles at their surfaces, and that would supply an extra term on the right-hand side of Equation (10.1). This additional contribution to the electric current due to elastic displacements was the 'displacement current'. These elastic displacements would be associated with forces at the surfaces of the elastic parcels, similar to those envisioned in the considerations leading to Equation (10.2), except that these would relate to the electric field in electrostatic situations, $\mathbf{E}_2$. This would add an extra term, the displacement current, to the equation, so that Equation (10.1) became Equation (10.1'),

$$\mathbf{j} = \frac{1}{4\pi}\text{curl}\,\mathbf{H} - \frac{1}{4\pi c^2}\frac{d\mathbf{E}_2}{dt} \qquad (10.1')$$

This, finally, was the promised extension of the theory to embrace the case of the open circuit, with theory – as embodied in the molecular-vortex model – leading experiment in the introduction of the displacement current, rather than following it.[39]

Maxwell's immediate use of the new Equation (10.1') was to facilitate the extension of the molecular-vortex model to electrostatics. Mathematically, he substituted the new Equation (10.1') into Equation (10.5), the equation of continuity, to get Equation (10.3), the basic equation of electrostatics. Conceptually, accumulations of electric charge ρ were now seen as centres of elastic strain in the magneto-electric medium.[40]

Equations (10.1'), (10.2) and (10.3), three of the four Maxwell's equations, were now explicitly treated in the model. Including the fourth, implicit, equation expressing the absence of magnetic monopoles (later explicit as Equation (10.4)), Maxwell now had a consistent set of four equations that have furnished the foundation of electromagnetic theory ever since. Maxwell had begun with the task of mathematizing Faraday's field approach to electromagnetic theory, on the basis of known experimental results; he had gone beyond this task in modifying one of the resulting equations, Equation (10.1), for the sake of theoretical completeness; and the result was a complete, coherent and enduring foundation for electromagnetic theory.

Light and Electromagnetic Waves

A new theory of light was implied also in the full and final form of Maxwell's molecular-vortex model. This was not yet a truly electromagnetic theory of light – that was to be another two years in coming. Historian Joan Bromberg has felicitously denoted this first form, in 1862, of Maxwell's new theory of light, an 'electromechanical' theory of light, going through the mechanics of the molecular-vortex model, rather than proceeding directly from the electromagnetic equations.[41]

In the mechanics of the molecular-vortex model, electrical and magnetic forces were correlated, respectively, with the elasticity and density of the magneto-electric medium: the elasticity determined the forces exerted by the vortex blobs on the small particles at their surfaces – the mechanical basis of the electric field $\mathbf{E}_2$; the density determined the centrifugal forces associated with the whirling of the vortices – the mechanical basis of the magnetic field **H**. The ratio of the strengths of electrical and magnetic forces had been measured quite carefully, in connection with the standardization of electrical units, and, in Maxwell's theory, this specified the ratio of elasticity and density of the magneto-electric medium. But the square root of this ratio, according to standard mechanical theory, would give the velocity of elastic waves in the medium.[42]

Computing the velocity of elastic waves in the magneto-electric medium in this manner, using up-to-date data on the ratio of electrical units from the work of Wilhelm Weber and Rudolph Kohlrausch in Germany, Maxwell found that V_W, the wave velocity, would be[43]

$$V_W = 193{,}088 \text{ miles per second}$$

This was a familiar-looking sort of number. Looking up the measurement of the speed of light, V_L, that was authoritative at the time – by French experimenter Hippolyte Fizeau, in 1849 – Maxwell found[44]

$$V_L = 195{,}647 \text{ miles per second}$$

Some other relevant estimates of the speed of light were 192,000 miles per second and 193,118 miles per second.

The agreement of Maxwell's calculated velocity of elastic waves in his magneto-electric medium, V_W, with the measured velocity of light, V_L, was very close – at the 1% level – and the other two measurements of the velocity of light bracketed Maxwell's value for V_W. Maxwell's statement of his conclusion was an occasion for italics:

> The velocity [V_W] of . . . undulations [waves] in our hypothetical medium . . . agrees so exactly with the velocity [V_L] of light calculated from the optical experiments of M. Fizeau, that we can scarcely avoid the inference that *light consists in the . . . undulations of the same medium which is the cause of electric and magnetic phenomena.*[45]

This was not yet the electromagnetic theory of light. In Maxwell's theory at this point, electromagnetic fields were the manifestations of one set of mechanical processes in the magneto-electric medium, namely, rotations and elastic deformations of the material within each cell of the medium. Light waves were the manifestations of a different set of mechanical processes in the medium: large-scale undulations – back and forth motions – of whole regions of the medium, encompassing many cells. The connection between electromagnetic fields and light was through the mechanical constants of the medium – the elasticity and the density. So, at this point, the magneto-electric medium was the necessary go-between, linking electromagnetism and light.

But Maxwell realized that many of his readers in the scientific community would not be very impressed with his hypothetical magneto-electric medium – with its elastic blobs and its monolayers of ball bearings. For this reason, Maxwell was highly motivated to try to liberate his theory of light from the particulars of the molecular-vortex model, making it follow directly from the field equations.[46]

Mathematically, that was easy enough to do, as a direct result of the introduction of the displacement current. In empty space, where there is no electric conduction current density **j**, Equation (10.1'), with displacement current, can be rewritten as

$$\text{curl}\,\mathbf{H} = \frac{1}{c^2}\frac{d\mathbf{E}}{dt} \qquad (10.6)$$

and in empty space Equation (10.2) becomes

$$-\text{curl } \mathbf{E} = \frac{d\mathbf{H}}{dt} \tag{10.7}$$

(Subscripts 1 and 2, distinguishing the different mechanical aspects of the electric field **E**, are absent here, reflecting the purely electromagnetic treatment at this point.)

Mathematically, these two equations can be combined to give a wave equation, which describes waves of alternation of electric field **E** and magnetic field **H**, propagating at velocity *c*, the ratio of electrical units. In qualitative, intuitive terms, Equation (10.5) – because it has the displacement current in it – tells you that a changing electric field **E** will give rise to a magnetic field **H** in its neighbourhood; then, that magnetic field **H**, as it varies, will, according to Equation (10.7), give rise to another electric field **E**; and so on, in a series of leapfrogging **E** and **H** fields constituting a propagating electromagnetic wave. So, once Maxwell had introduced the displacement current, the mathematics of electromagnetic waves dropped out easily.[47]

In conceptual terms, however, Equation (10.6) would not have made sense for Maxwell. In the understanding of field theory that Maxwell had taken over from Faraday, fields are the sources – the causes – of electric charges and electric currents, in contradistinction to a modern view in which the opposite is true – namely, electric charges and currents are the sources, or causes, of fields. Consequently, in Maxwell's understanding at that point, Equation (10.6), in which displacement current is the source, or cause, of a magnetic field, would not have made sense. In order to be able to make use of Equations (10.6) and (10.7) together, and the picture of leap-frogging electric and magnetic fields that went with that mathematics, Maxwell had to make some basic changes in his conceptualization of charges, currents, and fields.[48]

This required some difficult wrestling with foundational issues, but Maxwell was highly motivated: he knew that the result he had obtained from the mechanical theory of the magneto-electric medium, concerning the propagation of waves in that medium at the velocity of light, was a discovery of the first order, and he was determined to liberate the essence of that result from the mechanics of the medium.

He accomplished that in 'A Dynamical Theory of the Electromagnetic Field', sent to the Royal Society of London in October of 1864, and published in their *Transactions* thereafter. Using, essentially, the mathematics of Equations (10.6) and (10.7), Maxwell calculated that electromagnetic waves – alternations of electric and magnetic fields – would propagate through space at a velocity given by *c*, the ratio of electrical units, indicating, for the velocity of these waves,

$$V_W = 310{,}740{,}000 \text{ metres per second}$$

On the other hand, 'The velocity of light', as established by Fizeau's experiments, was

$$V_L = 314{,}858{,}000 \text{ metres per second}$$

These were actually the same numbers that Maxwell had used in PLF, expressed in metric units; here – because the intermediary of the magneto-electric medium was no longer required – the conclusion was more direct and definitive: 'light is an electromagnetic disturbance propagated through the field according to electromagnetic laws'.[49]

The significance of Maxwell's concept of electromagnetic waves, as subsequent history has shown, goes far beyond its application to light. Depending on the wavelength of the electromagnetic waves, they comprehend a vast variety of phenomena. Understanding of the electromagnetic waves with wavelengths bracketing the visible spectrum – infrared and ultraviolet – came rapidly. Electromagnetic waves of longer wavelengths, which form the basis of all modern broadcast as well as cable and cellular communication technologies, were first produced in 1888, by Heinrich Hertz in Germany, leading directly to the communications technologies of the twentieth century and beyond. Intermediate between the communications wavelengths and infrared are the radar waves and microwaves, developed during World War II. Shorter wavelengths are manifest as X-rays and the gamma rays of nuclear physics. The whole spectrum of electromagnetic waves, from radio waves at the long wavelengths to the gamma rays at the shortest wavelengths, comprises a substantial portion of modern science and technology.[50]

Beyond this, electromagnetic waves are the paradigm case for all transmission of signals through the electromagnetic field at the speed of light, and this is the basis for the ubiquity of the speed of light in the equations of relativity theory; as Albert Einstein was always to emphasize in his discussions of the historical roots of relativity theory, Maxwell's field theory, and the propagation of signals at the speed of light as required by that theory, are the bedrock foundation of relativity theory.[51]

Conclusion

The importance of Maxwell's innovations in electromagnetic theory was immediately apparent throughout the international scientific community, but his theory was differently approached and assimilated in the Anglo-American and Continental contexts.

In Britain and the United States, the main vehicle for the dissemination of Maxwellian ideas was his *Treatise on Electricity and Magnetism*, first published in 1873. Cambridge University was the nexus for intensive study and further development of Maxwell's ideas, in a form that was quite faithful to all of the elements of Maxwell's approach, including the commitment to the electromagnetic field as the primary basis of electromagnetic phenomena, and the use of mechanical models to illustrate and understand the functioning of the electromagnetic field. Starting at Cambridge, and then expanding through the English-speaking world, a group of followers of Maxwell, the 'Maxwellians', sustained and developed the Maxwellian tradition.[52] In this book, the Maxwellians are discussed in Chapter 11 by Chen-Pang Yeang.

On the Continent, the crucial figure in the assimilation of Maxwell's ideas was Hermann von Helmholtz. It was in 1870 that Helmholtz published his seminal paper, in which he translated Maxwell's system of equations and general approach to electromagnetic theory into the action-at-a-distance language that was dominant in Germany, France and the other Continental scientific communities. Helmholtz's fusion of the field-theory and action-at-a-distance traditions was a bit awkward, but it paved the way for broad acceptance and further development of Maxwell's work throughout Europe in the later nineteenth century.

Subsequently, Maxwell's accomplishment in electromagnetic theory has become paradigmatic for modern physics in a variety of ways. Maxwell's equations form the permanent basis of electromagnetic theory, both the classical form – the basis of all electrical technologies – and the quantum form, paradigmatic for all of fundamental physical theory. Maxwell's use of purely theoretical considerations to innovate in physics – especially considerations of completeness and coherence of theory – are paradigmatic for the field of theoretical physics in general.

Of interest to historians and philosophers of science, perhaps more than to scientists themselves, is the issue of Maxwell's use of mechanical modelling to make progress in science. Viewed with hindsight, from a twenty-first-century vantage point, we would have to regard Maxwell's molecular-vortex model as merely an analogy: we don't today believe that there is any real mechanical basis for the electromagnetic field. Viewed in that way, Maxwell's very successful use of the model as an engine of discovery shows the value of analogical thinking in physics, and in science more generally. The particulars of how Maxwell constructed and developed the theory, using the linkages of variables specified by Maxwell's equations, provide a case study of the kinds of strategies for the construction of models envisioned, for example, by physicist–philosopher Norman Robert Campbell.[53]

Maxwell's strong belief that he was discovering the true and real mechanical basis for the electromagnetic field may seem, from our modern vantage point, a quaint, nineteenth-century conceit. But one must keep in mind that this notion of getting at the reality of the electromagnetic field was what motivated Maxwell's dedicated and persistent efforts to develop a mechanical model that would reflect the actual behaviour of the electromagnetic field. This model was sufficiently faithful to the actual behaviour of the field as to furnish a basis for the first, 'electromechanical' form of the electromagnetic theory of light; and this soon led to Maxwell's formulation of a true electromagnetic theory of light, deriving electromagnetic waves directly from the field equations. We cannot regret the fact that Maxwell actually believed in his molecular-vortex model: he was a nineteenth-century scientist, thoroughly imbued with the mechanical worldview characteristic of that century; on the basis of that mechanical orientation, he invested a great deal of effort in developing the molecular-vortex model; and the success of that effort propelled science and technology into the twentieth and twenty-first centuries.

CHAPTER 11

The Maxwellians: The Reception and Further Development of Maxwell's Electromagnetic Theory

CHEN-PANG YEANG

Introduction

James Clerk Maxwell has the reputation of being the founder of modern electromagnetic theory. Today, physicists and engineers take Maxwell's equations to be the foundation of classical electromagnetism. It is generally believed that such equations – comprising Coloumb's Law, Gauss's Law, Ampère's Law and Faraday's Law – were synthesized at Maxwell's hand as he turned these scientists' empirical rules into vector-formed differential equations, which have been synonymous with the axioms of electricity and magnetism since the late nineteenth century.

Maxwell's theory of electricity and magnetism was indeed influential in the late nineteenth century. However, contrary to the popular view, this theory had fundamental differences from the doctrines of modern classical electromagnetism. The nineteenth-century theory was based on certain ontological presumptions that are quite alien to us: the assertion that electric and magnetic effects are results of mechanical actions in the complex 'luminiferous aether' (ether) and the marginal roles of microscopic charged particles, and so on. In addition, Maxwell's theory operated with several mathematical apparatuses – including Hamilton's principle and Lagrange's equations – that modern students of electrical science and engineering rarely use. Moreover, Maxwell did not single-handedly accomplish this theory. The approach he launched in the 1860s and early 1870s did not reach full maturity until two decades later, owing to the efforts of quite a number of other physicists. As a salient example, Maxwell did not present the modern vector form of the so-called Maxwell's equations; this was rather the work of a British scientist, Oliver Heaviside, in 1884, five years after Maxwell's death.[1]

This chapter reviews the reception and further development of Maxwell's electromagnetic theory after his last major publication, *A Treatise on Electricity and Magnetism*, in 1873. Our focus is thus not on Maxwell himself, but on a generation of researchers after him. Mostly Britons, these individuals converted Maxwell's work from a promising but esoteric and somehow puzzling theory into a full-fledged research programme. They filled significant conceptual gaps in Maxwell's theory, employed the theory to explain exotic magneto-optical phenomena, and extended Maxwell's mathematical approach in the treatment of physical problems. German physicist, Heinrich Hertz's, laboratory demonstration of electromagnetic waves in 1886–87 was a great triumph of the Maxwellian programme, as the experiment confirmed Maxwell's most important prediction in a purely electromagnetic, non-optical realm. The years from 1888 to 1894 were the Maxwellian physicists' heyday. During this period, they enjoyed unprecedented prestige, worked closely as a research community, started to investigate the behaviour of the electromagnetic field associated with moving bodies that set the stage for relativity, and explored the application of Hertzian waves to wireless communications.

From the mid 1890s on, however, the Maxwellian programme received an increasing challenge from the corpuscular microphysics that was popular in Continental Europe and was setting foot in the UK. In contrast to the Maxwellian theory, in which electric and magnetic fields were primary while electric charge and current were only epiphenomenal, corpuscular microphysics asserted that all the electric and magnetic effects were results of interactions between microscopic charged particles known as 'ions' or 'electrons'. Against the Maxwellian programme's longstanding difficulty to account for electric conductivity, the corpuscular theory was successful in understanding a series of magneto-optical experimental results in 1894–1900, and with Cambridge physicist John Joseph (J. J.) Thomson's cathode-ray experimental demonstration of the electron in 1897, the corpuscular theory eventually gained the upper hand. By the beginning of the twentieth century, while Maxwell's equations remained central, it was clear that his physical interpretation lacked explanatory power and required supplementation by the electronic microphysics, which became the canonical theory of electromagnetism

The Maxwellians

Who were the followers and further developers of Maxwell's electromagnetic theory? Historian Bruce Hunt called these people 'Maxwellians'. According to historian Jed Buchwald, there were about three dozen such Maxwellians active in the last quarter of the nineteenth century.[2] The majority of this community was British. Many of them graduated from Cambridge University. They were either the first group of students at Cambridge's Cavendish Laboratory, or those who had studied Maxwellian electrodynamics under the university's idiosyncratic pedagogy of mathematical physics in preparation for the Tripos exam. This is hardly surprising, given Maxwell's and the Maxwellian theory's close relation to Cambridge. Maxwell himself was a product of the Tripos system (he was the Second Wrangler in 1854, i.e. he was ranked second in the Tripos exam) and was the first Director of the Cavendish Laboratory. His electrodynamics was rooted in a British tradition of field theory, started by Michael Faraday, that was being established in Cambridge by the mid nineteenth century. In the last quarter of the nineteenth century, Maxwell's

Treatise provided both topics for Tripos exam questions and materials for training experimenters at Cavendish.[3]

Yet, the Cambridge men did not monopolize the interpretation and development of Maxwell's electromagnetic theory. About a third of the Maxwellians received training from other places, including Trinity College in Dublin, University College in London, University of Edinburgh, Oxford, Johns Hopkins, Rensselaer Polytechnic Institute and self-education.[4] According to Hunt, none of the three leaders of the Maxwellians was a Cambridge man. George FitzGerald was an alumni and faculty member of Trinity College in Dublin. Oliver Lodge was a graduate of University College in London and a professor at University College in Liverpool. Oliver Heaviside learned electrical science by himself and did not hold any academic position. (His only working experience was as a telegraph operator.)[5] The fact that these individuals did not have significant contact with Maxwell or the central institution of the Maxwellian electrodynamics (Cambridge University) did not hinder their grasp of Maxwell's theory.

Learning and Teaching Maxwell's Theory

Maxwell's *Treatise* was the most important textual resource for physicists to learn his electromagnetic theory. To all Maxwellians, studying the book closely was a *sine qua non* for mastering Maxwell's approach and models. But the *Treatise* was not an ideal textbook for this purpose. First, the book comprised long summaries of mathematical techniques for solving electrostatic and magnetostatic problems, extensive discussions on electrical experiments and measuring apparatuses, and Maxwell's new theory of the electromagnetic field. In terms of volume, the mathematics of electrostatics and magnetostatics, experiments, and instruments overshadowed the novel field theory of ether displacement. Therefore, to many inexperienced readers at the time, the work read like a textbook on mathematical physics or metrology. Moreover, systematic exposition of physical concepts through careful mathematical analysis was not Maxwell's strength; the reasoning in the *Treatise* was difficult to follow from time to time. Finally, there were a significant number of errors in the book. These shortcomings created notable barriers for physicists to learn Maxwell's new theory from the *Treatise*. Not surprisingly, a contemporary Cambridge student called the text 'a very hard nut to crack'.[6]

For some Maxwellians, devoted self-study was the only way to crack this nut. Although learning Maxwell's theory exclusively from the text without having contact with him was difficult, it was not impossible to do so. Heaviside outside academia and Henry Rowland and Edwin Hall in the USA (see below) are cases in point. A few Maxwellians, including FitzGerald, Lodge and Heaviside, befriended and corresponded closely with one another. Through private communication, they exchanged research ideas, helped each other to clarify concepts, and challenged their understanding, all of which enhanced their learning of Maxwell's theory. In addition to this informal social network, FitzGerald also provided his students with more formal training on Maxwellian electromagnetism at Trinity College in Dublin.[7]

Nevertheless, a much more institutional and systematic form of learning Maxwell's theory was fostered at Cambridge University. As stated, Cambridge produced almost two thirds of the Maxwellian researchers in the late nineteenth century. Maxwell himself was an obvious facilitator for this pedagogical regime, for he was appointed as the university's first Professor of Experimental Physics and the first Director of the Cavendish Laboratory. But according to historian Andrew Warwick, Maxwell's role in teaching his novel field theory was quite limited. He did give public lectures on electricity and magnetism in the early 1870s, but those lectures were not well attended and their contents were not sufficiently organized to highlight the original aspects of Maxwell's theory. Rather, three pedagogical undertakings at Cambridge, which were independent of or only marginally related to Maxwell, had more significant influences on students' reception and application of the new theory. First, the Tripos examiners in the 1870s and 1980s, including Maxwell himself, Lord Rayleigh, and other prominent physicists, gradually made electricity and magnetism questions derived from the *Treatise* a central part of the exam. To prepare for the Tripos, therefore, students had to master the novel discussions on electric displacement in Maxwell's theory. Second, the private 'coaches' (tutors), who offered supplementary teaching to ambitious students aiming for the top rank of the Tripos exam, developed syllabuses and pedagogical materials that were notably more systematic introductions to Maxwell's theory than the *Treatise*. Third, the intercollegiate lectures on electricity and magnetism at Trinity College by W. D. Niven further introduced and exposed Maxwell's theory (and with more clarity than the *Treatise*) to students who did not hire coaches. The Tripos exam, private tutoring and Niven's lectures at Trinity helped make Cambridge a training centre for Maxwellians.[8]

Central Elements of Maxwell's Theory

To understand the basis on which the Maxwellians built up their electromagnetic framework, it is useful to highlight briefly a few central elements of Maxwell's theory by the early 1870s. (For more details about Maxwell's electromagnetism, see Chapter 10 of this volume, Daniel Siegel's *Electricity and Magnetism: The Development of a Theory.*)

A salient characteristic of Maxwell's electromagnetic theory was his use of mechanical models. Like many British physicists since the first half of the nineteenth century, Maxwell treated electric, magnetic and optical phenomena as mechanical effects of an imponderable, ubiquitous medium known as the ether. In 'On Physical Lines of Force' published in 1861, he proposed his famous 'vortex and idle wheels' model to illustrate such mechanical effects (Chapter 10, Fig. 10.1). In this model, the magnetic field $\mathbf{H}$ was represented by tiny vortices in the ether, with the axis and amount of the vortex rotation at a specific location identical to the direction and strength of the local magnetic field. Between the vortices lay numerous tiny particles like 'idle wheels', which served to link magnetism and electricity. When the magnetic field was uniform in space, the amounts and directions of rotations in all vortex cells were constant, and thus the idle wheels in between remained stationary. When the magnetic field was spatially heterogeneous, however, the differences in the amounts and directions of vortex rotations drove the idle wheels into circular motions. Such motions of idle wheels could be conceived as electric current, for the motion $\mathbf{J}$ under Maxwell's model was a kind of spatial derivative of the magnetic field $\mathbf{H}$ (i.e. vortex rotation) that followed Ampère's Law $\nabla \times \mathbf{H} = \mathbf{J}$ (in today's vector form).[9]

Yet, the idle wheels did not only undergo circular motions due to the heterogeneity of the magnetic field. While the presence of vortices implied that the ether resembled a fluid, in other aspects of Maxwell's model the ether had the properties of an elastic solid. In particular, the jelly-like ethereal substrate that embedded the idle wheels could bring the tiny particles into lateral displacements when it was compressed or elongated by a force. According to Maxwell, such a force (or tension) represented electric field $\mathbf{E}$ (or electromotive force). Like the case in other elastic materials, the lateral displacement $\mathbf{D}$ responding to the force incurred on the idle wheels was proportional to the force, $\mathbf{D} = \varepsilon\mathbf{E}$, with the dielectric constant ε equal to the inverse of the ether's elastic constant. Moreover, the time variation of the electric displacement $\partial\mathbf{D}/\partial t$ on the idle wheels corresponded to another kind of electric current known as 'displacement current'. Finally, when a conductor was present, the freedom of electric displacement in the ether was constrained at the boundary between the conductor and ether, which would result in a tension (electric field) in space along particular directions and the accumulation of the tiny idle wheels on the boundary – the latter represented electric charge.[10]

Four years later, Maxwell attacked the same problem from a different perspective in *A Dynamic Theory of the Electromagnetic Field*. In his 1865 article, he replaced contrived mechanical modelling with a particular type of mathematical formulation. Here, he no longer mentioned the idle wheels and quietly abandoned the attempt to work out the mechanical details of ethereal vortices. Rather, he only claimed generically that the electric effects were due to the lateral displacement of the continuous ethereal medium and the corresponding tension, while the magnetic effects were due to the rotational motion of the ether. Without knowing the exact mechanism, therefore, Maxwell identified the electric field energy (as an integral of energy density over a given volume in space)

$$U = \iiint d^3x \frac{1}{2}\mathbf{D}\cdot\mathbf{E}$$

with the ethereal system's potential energy stored in the electric displacement. He also identified the magnetic field energy density

$$T = \iiint d^3x \frac{1}{2}\mathbf{B}\cdot\mathbf{H}\ (\mathbf{B} = \mu\mathbf{H} \text{ is the magnetic flux density})$$

with the ethereal system's kinetic energy manifested in vortex rotation. Then he followed a common practice in mechanics known to his contemporaries as 'dynamic': he obtained the system's Lagrangian $L = T - U$, took its time integral from an initial to a final instant to get the system's action

$$\int_{t_1}^{t_2} L dt,$$

and applied Hamilton's principle to assert that among all the possible routes in the state space between $t = t_1$ and $t = t_2$, only the one that minimized the action obeyed the laws of mechanics. Using the calculus of variation, such minimization of the functional resulted in a set of differential equations known as Lagrange's equations, which could serve as the system's equations of motion. In the case of Maxwell's electromagnetism with

$$T = \iiint d^3x \frac{1}{2}\mathbf{B}\cdot\mathbf{H}$$

and

$$U = \iiint d^3x \frac{1}{2}\mathbf{D}\cdot\mathbf{E},$$

Lagrange's equations of the ethereal system were nothing but Maxwell's equations for the electric and magnetic fields.[11]

Maxwell's 1865 theory, which he developed more comprehensively in the *Treatise*, had several important advantages: it promised to unite various empirical laws in electricity, magnetism, and optics; it subjugated the science of electricity, magnetism and optics under mechanics (which was considered the most rigorous and basic physical science); it predicted a new effect – electric displacement – that could be tested empirically; and the presence of electric displacement led Maxwell's equations to a wave equation, suggesting to Maxwell that optical waves could be a kind of electromagnetic oscillation.

Yet, Maxwell's framework had a problem in coping with the most elementary electrical phenomenon: electric current in a conductor. We will defer the examination of this problem to the next section and simply sketch Maxwell's idiosyncratic treatment of conduction current here. In Göttingen physicist, Wilhelm Weber's, corpuscular theory, that was popular in Continental Europe around the same time, conduction current was nothing but a flow of electrified microscopic particles in the conductor. In Maxwell's theory, however, all the electromagnetic phenomena were manifestations of field effects, and conduction current was no exception. Under the Maxwellian framework, a conductor actually constrained or even expelled field actions. (For example, an electric field could hardly penetrate a metal cage.) And this was so because the electric field inside the conductor was quickly dissipated. Thus, what a conductor did to an electric field was to deplete its energy through thermal dissipation via Joule's Law $P_{diss} = \mathbf{C} \cdot \mathbf{E}$ (where **C** is conduction current density and **E** is electric field). In other words, the electric displacement in the conductor was 'relaxed' and the potential energy stored in the displacement disappeared. When there was no continual supply of energy from an external source, the relaxation of the electric displacement and the disappearance of the potential energy in the conductor were permanent. (The transient discharge of a Leyden jar was exactly this effect.) When the conductor was connected to an energy source that provided external electromotive force (like a battery), the relaxed displacement was restored and the dissipated potential energy resupplied. And this process of dissipation–restoration occurred at an enormously fast rate. The net effect of this rapidly oscillating process was that the conduction current **C** followed Ohm's Law $\mathbf{C} = \sigma\mathbf{E}$ and could be measured by an ammeter.[12]

Maxwellians' Major Technical Work

From the 1870s to the 1890s, the Maxwellian physicists in Great Britain and the United States further developed their electromagnetic theory. In this section, we review several major topics that comprised their focus of research: the issue of the electromagnetic field's energy flow, the generation and detection of electromagnetic waves other than light, the Hall effect and other magneto-optic phenomena, and the problem with conduction current.

Energy Flow and the Poynting Vector

The direction of energy flow in an electric circuit was an important question to the Maxwellian electromagnetic theory, as its answer to the question differed remarkably from that of the Weberian corpuscular theory. According to the corpuscular theory, the electric current in a circuit was a flow of electric particles. Like a hydrodynamic system, therefore, the direction of transporting electric energy associated with this circuit should be identical to the direction of the material flow, which was the direction of the electric current. Yet, this conclusion was far from inevitable in the Maxwellian approach, according to which the current was just epiphenomenal and the genuine energy was stored in the electric and magnetic fields not only inside the circuit's wire but also in the space *outside* the wire. Thus the question became the following: given an electric and a magnetic field in space, what was the corresponding energy flow?

The Maxwellians worked on this question during the late 1870s and early 1880s, but to no avail. FitzGerald, for example, searched for a proper characterization of the electromagnetic field's energy flow when he tried to derive the radiation intensity from an oscillating circuit; but his result was inconclusive.[13] It is generally believed that English physicist, John Henry Poynting, gave the first definite answer to this question. Poynting was a student at Cambridge University from 1872 to 1876, became a disciple of Maxwell at the Cavendish Laboratory in the late 1870s, and assumed a professorship at Mason College in Birmingham in 1880. Although Poynting was a canonical 'Maxwellian', Warwick points out that his discovery of the electromagnetic energy flow density vector (later known as the Poynting vector) did not result from systematic research on specific electromagnetic problems. Rather, the finding originated from his investigation on the propagation of sound waves when he was preparing for teaching.[14] In 1884, he published the discovery in 'On the Transfer of Energy in the Electromagnetic Field', in *Philosophical Transactions of the Royal Society of London*.[15]

Poynting's starting point was three of the major field equations in Maxwell's theory. In today's vector form, they were:

$$\nabla \times \mathbf{E} = -\frac{\partial \mathbf{B}}{\partial t} \quad \text{Faraday's Law}$$

$$\nabla \times \mathbf{H} = \mathbf{C} + \frac{\partial \mathbf{D}}{\partial t} \quad \text{Ampère's Law}$$

$$W = \iiint \frac{1}{2}(\mathbf{E}\cdot\mathbf{D} + \mathbf{H}\cdot\mathbf{B})\, d^3x \quad \text{Field energy expression}$$

Under the assumption that $\mathbf{D} = \varepsilon\mathbf{E}$, $\mathbf{B} = \mu\mathbf{H}$, and the permittivity ε and permeability μ were constant, Poynting substituted Faraday's Law and Ampère's Law into the time derivative of the field energy expression. After algebraic manipulations and using the vector identity

$$\mathbf{E}\cdot\nabla\times\mathbf{H} - \mathbf{H}\cdot\nabla\times\mathbf{E} = -\nabla\cdot(\mathbf{E}\times\mathbf{H}),$$

he obtained an equation of energy conservation:

$$\frac{\partial W}{\partial t} + \iiint d^3x(\mathbf{E}\cdot\mathbf{C}) = -\iiint d^3x\nabla\cdot(\mathbf{E}\times\mathbf{H}).$$

From Gauss's theorem in vector calculus, the volume integral at the right-hand side of the above equation equalled a two-dimensional integral over a closed surface wrapping the volume, and the equation of energy conservation became

$$\frac{\partial W}{\partial t} + \iiint d^3x(\mathbf{E}\cdot\mathbf{C}) = -\oint\!\!\oint_A d\mathbf{s}\cdot(\mathbf{E}\times\mathbf{H}).$$

To Poynting, the interpretation of this equation led to the expression of field energy flow: the first term $\partial W/\partial t$ on the left-hand side denoted the rate of field energy change in a given volume, the second term $\iiint d^3x(\mathbf{E}\cdot\mathbf{C})$ on the left-hand side represented the rate of thermal dissipation due to conductivity. To maintain energy conservation, the term $-\oint\!\!\oint_A d\mathbf{s}\cdot(\mathbf{E}\times\mathbf{H})$ on the right-hand side must be the flow of energy out of the volume. Thus the vector $\mathbf{S} \equiv \mathbf{E} \times \mathbf{H}$, later known as the 'Poynting vector', must be the surface density of energy flow in the electromagnetic field (i.e. the integral of the vector over a surface was the amount of energy across the surface per unit time).[16]

The expression of the Poynting vector $\mathbf{S} = \mathbf{E} \times \mathbf{H}$ indicated that the direction of the energy flow of an electromagnetic field was neither along the electric field nor along the magnetic field, but perpendicular to both; and the magnitude of the flow's surface density was proportional to the electric and magnetic field intensities as well as the sine of the angle between the two fields. This conclusion had a crucial implication to the Maxwellians: it was *proven* that the energy flow of an electric circuit was not along the direction of the electric current, but was determined by the electric and magnetic fields in the space outside the circuit's conductors. The significance of this implication can be demonstrated from an example. Poynting considered a simple circuit comprising a parallel-plate condenser (capacitor) that could be connected by a conducting wire (Fig. 11.1). Suppose the condenser is charged electrically in the first place and then the two plates are connected via the wire. Upon connection, the circuit undergoes a discharge and a transient conduction current is generated along the wire, until the energy is consumed by heat dissipation due to the wire's resistance. Since the wire is actually along a line of electric field that produces the conduction current during the discharge, the energy, according to the expression of the Poynting vector, should not flow *along* the wire, but instead, should flow *normal to* the wire. After analysis, Poynting concluded that the energy of such a system flowed out from the space between the condenser plates, following the equipotential lines outside, arriving perpendicularly at the wire, and is eventually dissipated into heat.[17]

Poynting went further to develop a physical interpretation of the surface density of electromagnetic energy flow. According to him, the Poynting vector $\mathbf{S} = \mathbf{E} \times \mathbf{H}$ was not a mere mathematical expression obtained from manipulating field equations. It actually represented the lateral motions of the 'tubes' of electric displacement **D** and/or magnetic induction **B**, and hence denoted the flow of real, physical entities.[18] Physical interpretation aside, the most important lesson Poynting's theory taught the contemporary British physicists was a reaffirmation of Maxwell's dictum that in an electrical system the apparent flow of charge was epiphenomenal, while the genuine energy intake came from the electromagnetic field in space. Thus, the actual energy in a telegraphic or an electric power circuit did not flow with the conduction current *along* the cable, but flowed *into* the cable through the electric and magnetic fields surrounding the cable induced by the battery cells at the transmitter or the power generator. Soon after its publication in 1884, Poynting's theory was accepted and used by the majority of Maxwellians.

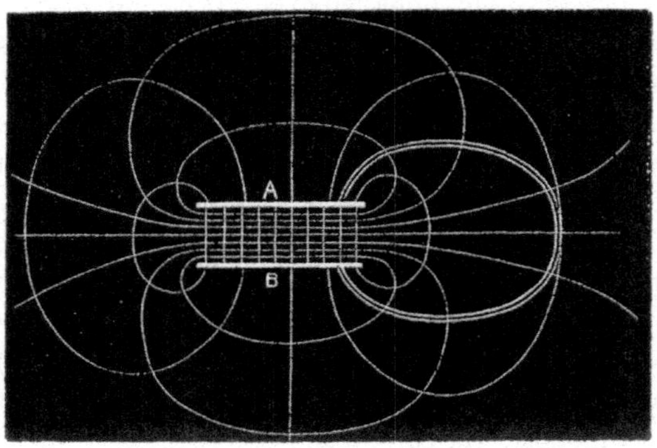

Fig. 11.1 Poynting's demonstration of field energy flow surrounding a discharged capacitor.

Source: Poynting, 'On the Connexion' (1885), p. 286, Figure 2

It is worth noting that Poynting was not the only individual to obtain the results on energy flux. Around the same time in 1884, Oliver Heaviside also derived, independent of Poynting, the formula $\mathbf{S} = \mathbf{E} \times \mathbf{H}$. Although Heaviside might come up with the idea a little earlier than Poynting (according to Heaviside himself), the ex-telegraphic operator published his work later than the Cambridge-trained physicist and did so in the considerably less visible trade journal *Electrician* rather than in *Philosophical Transactions*. Not surprisingly, the contemporaries gave the credit of discovery to Poynting, not Heaviside.[19]

Generation and Detection of Electromagnetic Waves

Maxwell's most famous theoretical prediction was the existence of electromagnetic waves and their identity with light. It is generally believed that this prediction not only unified two major areas of nineteenth-century physics – optics and electromagnetism – but also opened up the avenue for the invention of various electromagnetic-wave technologies – from radio to radar – in the twentieth century.

Interestingly, Maxwell himself did not delve much into this important prediction he made. In the *Treatise*, he simply asserted that (i) *a* (not *the*) solution to the electric and magnetic field equations had the form of propagating waves; (ii) the speed of such waves in free space equalled the square root of the free space's permeability over permittivity $\sqrt{\mu/\varepsilon}$; (iii) the numerical order of this theoretical velocity (obtained from the contemporary data on the measurement of μ and ε) was comparable to that from the contemporary measurement of the speed of light; and (iv) thus light might just be the predicted electromagnetic waves. Maxwell did not indicate how to generate such electromagnetic waves with electrical means that did not involve optics, or more fundamentally, whether it was possible to do so. Between the publication of the *Treatise* in 1873 and his death in 1879, he basically remained silent on the topic.[20]

Maxwell's inability to generate electromagnetic waves with an electric circuit seems to be related to his belief in the nature of light. Light (including infrared and ultraviolet) was the only form of electromagnetic wave Maxwell knew at the time. To him, therefore, the question was really how to produce light with electrical means, or whether it was possible to do so. Maxwell's thought was that even though the propagation of light followed electromagnetic laws, the generation of light was not necessarily electromagnetic in character. Although Maxwell's equations implied that an oscillating electric current was a source of electromagnetic waves, he nonetheless seemed to view the production of light as a result of vibrations of molecules in matter and the associated motions in ether, which were not necessarily connected to electricity and magnetism. Because this microscopic mechanical process involving ether and matter was inaccessible to observers, Maxwell did not pay much attention to the generation of light.[21]

The first Maxwellian to conceive the feasibility of generating electromagnetic waves via electromagnetic means was Oliver Lodge. In the late 1870s, Lodge was considering a slightly different ether model in which the electrical displacement was the strain rather than the physical body movement of the ether and light resulted from the periodic change of such strain. Since Lodge believed that the nature of displacement and vibration was identical in both electromagnetism and optics, it was quite possible to generate 'electromagnetic light' with a rapidly oscillating electric current. Lodge suggested a few methods in a paper given in 1879 at a British Association Meeting, including the passage of a current through a fast spinning disc with a spring contact or microphone (so that the disc rotation could modulate the current and turn it into oscillations with high frequencies), the passage of a current through a vacuum tube or an electrolytic solution (utilizing the intermittent, discontinuous character of current in these devices), and a chain of coil frequency multipliers. None of these schemes turned out to work. But he did propose a method which became more important for future research: the production of electromagnetic waves by the electrical discharge of a condenser. (It was known at this time that the discharge of a condenser was not a gradual release of current, but was strongly oscillatory.[22])

Lodge's idea stimulated his friend FitzGerald's interest on the topic. In 1879, FitzGerald gave a paper titled 'On the Possibility of Originating Wave Disturbances in the Ether by Means of Electric Forces' at the Royal Dublin Society. Despite the presence of the word *possibility* in the title, FitzGerald's paper actually argued for the impossibility of generating electromagnetic waves with an oscillating electric current. FitzGerald began with an observation that, according to Maxwell, a propagating electromagnetic wave in free space was generated under the initial condition in which the vector potential of the electromagnetic field **A** and its time derivative were non-zero in a source region and zero in the rest of the space. Yet, FitzGerald asserted that from Gauss's theory, 'the potential of any system of attracting bodies [like the vector potential] cannot have a zero value throughout one part of space and another value in any other communicating part'. Thus, Maxwell's condition for generating electromagnetic waves with electrical means could not be met.[23]

Then, FitzGerald argued for the impossibility from another perspective. He considered a generic system comprising a number of conductors and non-conductors, and argued that according to Maxwell himself, the Maxwellian theory of action through mediation of ether and the Continental theory of direct action-at-a-distance should give the same results in terms of the system's electric and magnetic behaviour. But when an oscillating current was present in one of the conductors in this system, from the theory of action-at-a-distance all the energy and electric effects should be confined in the conductors, which meant that the energy did not propagate in the non-conducting space. Given that the Maxwellian and Continental approaches were identical in effect, Maxwell's theory should not allow the propagation of electromagnetic waves in the non-conducting space, either.[24]

FitzGerald soon realized that his reasoning had a critical flaw: both Gauss's theorem and the identity between the Continental and Maxwellian systems applied only to the electrostatic case. When the electric displacement and displacement current were introduced into Maxwell's theory, Gauss's theorem no longer held and the equivalence to the Weberian or Neumannian framework broke down. The reasons for FitzGerald's objections were faulty.[25]

Identifying these faults did not immediately change FitzGerald's view, however. In a follow-up paper in 1880, he acknowledged his previous mistakes, but still argued against the possibility. This time, he focused on Maxwell's wave equation for electric potential V in the neighbourhood of a conductor excited by an oscillating current: $\nabla^2 V - \mu\varepsilon\partial^2 V/\partial t^2 = 0$. FitzGerald expressed the solutions for this equation in terms of a Fourier series expansion $V = \sum_n [A_n \sin(nt) + B_n \cos(nt)]$ and tried to match the coefficients $\{A_n, B_n\}$ with the boundary condition determined by the shape of the conductor. The general form of this solution, FitzGerald insisted, was not a travelling wave, but rather a group of standing waves like acoustic oscillations inside a pipe. The energy of these waves did not propagate.[26]

FitzGerald soon found the problem with this solution as he turned to Lord Rayleigh's *Theory of Sound* for help. Rayleigh's monumental work comprised a comprehensive mathematical treatment of acoustic waves, which served as a useful analogy to the study of electromagnetic waves. FitzGerald acknowledged that Rayleigh's solution to a similar wave equation took the form of a series expansion of travelling waves $V = \sum_n [A_n \sin(nt - mr) + B_n \cos(nt - mr)]$ instead of standing waves (where $m = n/c$ and c is the wave velocity). In the new form, the electromagnetic energy did propagate outwards. Thus, FitzGerald acknowledged in writing in 1882 that he was wrong, and he now believed that it was possible to generate electromagnetic waves with an oscillating electric current.[27]

With a converted view, FitzGerald now examined the characteristics of electromagnetic waves due to an oscillating current. In 1883, he tried to find the radiating power of such waves from a loop circuit (in today's term, a 'loop antenna') carrying an alternating current. Using L. V. Lorenz's method of 'retarded potential' that supposed the solutions to the electromagnetic wave equation (i.e., the potentials) to be in the form of $f(t \mp r/c)$, FitzGerald obtained an expression in which the radiating power was proportional to the fourth order of the current's oscillating frequency.[28]

This result had an important implication. The fourth-order dependence indicated that the radiating power increased rapidly with frequency. So it was possible to generate sufficiently strong waves with wavelengths much longer than those of visible light (wavelength is inverse proportional to frequency), meaning that the production of electromagnetic waves with an oscillating current was feasible in laboratory experiments. Good news as it was, FitzGerald's prediction nonetheless did not specify *how to detect* such invisible electromagnetic light. For Lodge, in 1879, this was not really a problem, since he aimed to produce visible light from an oscillating current. But as it became clear from energy calculations that electrical apparatuses were most likely to generate electromagnetic waves much longer than light (later known as 'radio waves'), detection became a more pressing issue among interested researchers in the rest of the 1880s. Researchers also considered using the bolometer, a kind of thermocouple, to detect electromagnetic waves. FitzGerald's own suggestion was a sensitive galvanometer or an electrodynamometer. But none of the Maxwellians built a workable wave detector before 1888.[29]

In fact, even though the Maxwellians acknowledged the possibility of generating electromagnetic waves with electrical apparatus by 1883, whether this prediction could be confirmed experimentally was far from clear to them. J. J. Thomson's finding is a case in point. As a graduate of Cambridge and a lecturer of the university's Trinity College, Thomson heard FitzGerald's presentations at the British Association Meetings in the early 1880s. The Irish physicist's work prompted the Cambridge don to examine the generation, propagation, and detection of electromagnetic waves. Thomson's focus was on the most promising wave generator at the time: the discharging condenser. He considered the electromagnetic radiation from an electrically charged hollow spherical conductor when it was discharged from inside or outside. The inside case was not of interest, as all the energy was confined within the cavity in the form of standing waves. The outside case raised more hope. But Thomson discovered that the discharge occurred within a very short duration. Even though the current underwent oscillations fast enough to generate invisible 'electromagnetic light', the duration was so short that such radiation was only sustained for a very brief period of time. In other words, the discharge from a conductor would only radiate a short pulse, not a steady train of waves. And the former was inevitably much harder to detect than the latter.[30]

The difficulty of finding a workable electromagnetic wave detector and the theoretical prediction on the transient, elusive character of radiating electromagnetic waves from a discharged condenser obstructed the Maxwellians from further progress on this topic although it was so central to Maxwell's theory. It turns out that the real breakthrough was made by a German physicist trained in a Continental, not the Maxwellian, paradigm. In 1887–88, Heinrich Hertz, a student

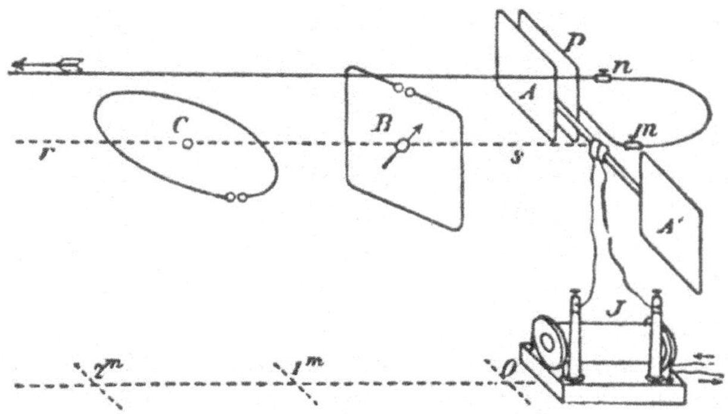

Fig. 11.2 Hertz's setup for the interference experiment.

Source: Hertz, *Electric Waves* (1900), p. 108, Figure 25

of Hermann Helmholtz in Berlin and then a professor at Karlsruhe Technical College, reported the success of generating and detecting invisible electromagnetic waves with electrical apparatus. Hertz's experiment is beyond the scope of this chapter. It suffices to point out that his generator comprised two metal arms separated by a narrow gap and charged by coils and batteries, while his detector was a loop with an adjustable gap called a 'resonator'. When the generator was sufficiently charged, the two arms would undergo intermittent discharge indicated by the presence of a spark across the gap; and when discharge occurred, a second spark was produced across the resonator's gap even when it was metres away from the spark-gap generator. Hertz demonstrated the wavelike character of this effect when he placed the resonator between the generator and a reflective board and observed that the intensity of the secondary spark varied with the resonator's location in a way similar to the interference pattern of standing waves between the generator and the board (Fig. 11.2).[31]

The difference between Hertz's device configuration and the Maxwellians' reveals another reason regarding why the latter failed to generate and detect non-optical electromagnetic waves before the German's. The electromagnetic oscillators that Lodge, FitzGerald and Thomson considered all had close-circuit configurations, such as a loop or a discharged sphere. These configurations were good at retaining and confining electromagnetic energy, but were not efficient in radiating such energy out. In other words, they were excellent resonators, but poor transmitters. In contrast, Hertz's spark-gap set with two stretched arms had an open-circuit configuration. Its performance in retaining resonance was meagre, but its radiation efficiency was excellent. According to Buchwald, it was the failure to recognize the fundamental distinction between a resonator and a generator that misguided the Maxwellians in the experimental work on electromagnetic waves.[32]

Hall Effect and Magneto-Optics

Although the study of electrical and optical effects in stationary magnetism is less known today than the Poynting vector and radio waves, nineteenth-century Maxwellians devoted no less effort on it. The topics under study included the presence of an extra electromotive force normal to a conduction current when the latter was exposed in a static magnetic field (the Hall effect), the changing polarization of light reflected from a magnet (the Kerr effect), and the rotation of light polarization when a static magnetic field was applied along the light's travelling direction (the Faraday effect). Exotic as they were, these effects turned out to offer a fertile ground for the Maxwellians to explore the interactions between an external magnetic field and the electromagnetic field associated with light or an electric circuit. Moreover, the way they studied these effects was characteristic of the Maxwellians and fundamentally different from today's macroscopic approach based on the Lorentz force on electrons and ions. To give a sense of the Maxwellians' approach without having to delve into technical details, we concentrate on the research on the Hall effect in this section.

The work on the Hall effect was one of a very few cases in which American physicists contributed to the advancement of the Maxwellian programme. A native of Maine, Edwin Hall was a high school teacher and a Ph.D. student at Johns Hopkins University in Baltimore during the second half of the 1870s. His supervisor, Henry Rowland, was a leading experimental physicist in the USA, a former student of Hermann Helmholtz, and the first professor of physics at Hopkins (the first American institution in higher education modelled after the German research university). Both Hall and Rowland were interested in Maxwell's theory of electromagnetism, but they were not in direct contact with Maxwell, the Cambridge circle, or other British Maxwellians. Like many physicists at the time, they learned Maxwell's theory by reading his *Treatise*. When Hall tried hard to grapple with the *Treatise*, he came across a puzzling statement: Maxwell claimed that the presence of a magnet or a time-invariant electric current did not affect another electric current in its neighbourhood. The conductor carrying the second current received a mechanical force due to the external magnetism, as Ørsted and Ampère had established decades ago. But if that conductor were held tight so that the Ampèrian electrodynamic force did not move it, then the intensity and direction of the current would not change. Hall believed from intuition that this was wrong, and he could prove it experimentally.[33]

Hall first set up a simple electric circuit and an electromagnet, with both held closely at fixed positions, and connected the circuit to a Wheatstone bridge. If the electromagnet incurred a direct action on the current, then after turning on the electromagnet the current measured by the bridge should change. After ruling out possible sources of errors in his experimental results, Hall nonetheless could not claim any definite change.[34]

Real progress was made soon, however. In May 1879, Rowland advised Hall to work on a different setup in which the two ends of a piece of gold leaf were connected to batteries, a static magnetic field was applied normal to the leaf, and another pair of the leaf's ends perpendicular to the battery charging ends were connected with a galvanometer for measuring possible transverse current (Fig. 11.3). This arrangement produced the effect Hall was looking for: when a conduction

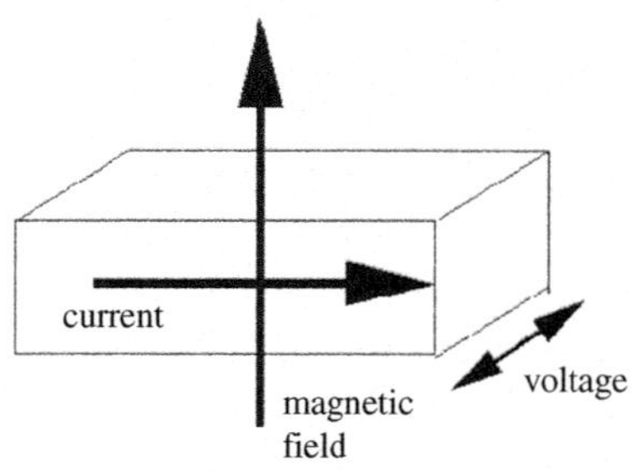

Fig. 11.3 The Hall effect.

current was established along the x direction and a uniform and time invariant magnetic field was applied along the y direction, an additional conduction current along the z direction could be detected (x, y, z were mutually orthogonal).[35]

Hall's discovery prompted Rowland to develop a theoretical interpretation. The Maxwellian character of Rowland's interpretation becomes clear when it is compared with the present theory of the Hall effect. The present theory is based on the presence and behaviour of microscopic charge-carrying particles in the conductor, such as electrons in a regular metal or 'holes' in a semiconducting material. The conduction current along x is a flow of the charge carriers. Suppose the electric charge of a carrier in this flow is q and its velocity is $\mathbf{v}$, and the external magnetic field along y is $\mathbf{B}_T$. Then the moving carrier in the magnetic field receives an additional mechanical force, the 'Lorentz force', which can be expressed as $\mathbf{f} = q\mathbf{v} \times \mathbf{B}_T$. The Lorentz force moves the carriers along a direction perpendicular to both the external magnetic field and the main current and hence establishes the Hall current.

In contrast, Rowland's theory did not involve charged particles and the Lorentz force at all. Like other Maxwellians, he believed that conduction current was not a flow of particles, but a state of ether displacement and relaxation, which could only be changed by an electromotive force. So the Hall current was actually an outcome of a transverse electromotive force. Moreover, since the electromotive force applied to *both* conduction current in a conductor and displacement current in an insulator, the Hall effect should be present in displacement current as well. In other words, if we could produce a displacement current (changing electric displacement over time) in a dielectric and place it in an external magnetic field (e.g. a discharged condenser between the two arms of a horseshoe magnet), then we should be able to observe an additional transverse electric field in the dielectric, even though it was not conducting. Thus, the Hall effect represented the generation of a new electric field from the interaction between the electric displacement and an external magnetic field. That meant the Hall effect was a fundamental field effect, and Maxwell's equations had to be modified.[36]

Following this reasoning, Rowland came up with a revision of Maxwell's equations. With the Hall effect, the electric field of an electric system under an external magnetic field had an additional term $h(\mathbf{B}_T \times \partial\mathbf{D}/\partial t)$ (h was a constant coefficient). Plugging the new electric field into Ampère's Law and substituted the vector potential $\mathbf{A}$ into the expression, Rowland obtained a modified equation for $\mathbf{A}$:

$$\mu\varepsilon\frac{\partial^2\mathbf{A}}{\partial t^2} = \nabla^2\mathbf{A} - \varepsilon\mu h(\mathbf{B}_T \cdot \nabla)\left(\nabla \times \frac{\partial\mathbf{A}}{\partial t}\right)$$

Note that the equation without the Hall term – the second term on the right-hand side – is Maxwell's wave equation for the vector potential.[37]

Hall's experiment and Rowland's theory raised British Maxwellians' interest. They found that the Hall effect might reveal the presence of the similar kind of electromagnetic field energy responsible for two major phenomena in magneto-optics. In 1845, Michael Faraday directed a linearly polarized light beam through a birefringent crystal (like an Iceland spar, which could split the beam into two with distinct polarizations) and employed a constant magnetic field along the direction of the light. He discovered that the polarization of the light beam leaving the crystal rotated in direction with respect to the beam entering the crystal, and the amount of this rotation was proportional to the intensity of the external magnetic field. In 1877, Reverend John Kerr in Glasgow, Scotland, found another interesting magneto-optic effect when he directed a linearly polarized light beam onto the surface of a magnetic bar. The reflected light from the magnet turned out to have a tiny but detectable polarization in addition to its original polarization; the additional component made the reflective light elliptically polarized; and the direction of the component reversed when the magnet was flipped. The British Maxwellians believed that the Americans' work offered a clue to unify the Hall effect with the Faraday rotation and the Kerr effect under the same theoretical framework.

The first British Maxwellian to grapple with this task was Richard Glazebrook. Glazebrook was a core member of the Maxwellians circle at Cambridge University. A graduate and fellow of Trinity College, Fifth Wrangler in 1876 and student of Maxwell at Cavendish Laboratory, Glazebrook was a demonstrator for the laboratory and lecturer for the university in the early 1880s. In 1881, he reworked the theory of the Hall effect. Rowland's derivation, albeit interesting, was only phenomenological to the British Maxwellians, for it was not based on ether mechanics, which was what a rigorous Maxwellian was supposed to undertake. Glazebrook did just that. Following Maxwell's derivation of the Faraday rotation, Glazebrook started with the dynamic equations for ether displacement when certain permanent vortices (corresponding to the external magnetic field) were present. From these equations, he obtained the expression of some additional energy. Treating it as the field energy associated with the new effect, Glazebrook plugged the expression into the system's Lagrangian, and by employing the least-action principle, he was able to get a field equation (like Rowland's) that could account for the new effect, be it Faraday rotation or Hall current.[38]

Glazebrook's work laid solid ground for the further integration of the Hall effect and magneto-optics into Maxwellian electromagnetism. In 1888, Alfred Barnard Basset, a graduate of Trinity College in Cambridge and Thirteenth Wrangler in 1877, presented a theory in an attempt to explain the Kerr effect. The line of reasoning and mathematical procedure was similar to Glazebrook's for the Hall effect. But Basset encountered several technical problems which he could not resolve. In the late 1880s and early 1890s, J. J. Thomson worked on generalizing Glazebrook's theory and removing Basset's problems. In his *Notes on Recent Researches in Electricity and Magnetism* in 1893, he eventually provided a theory based on ether mechanics and Lagrangian dynamics that unified the Hall effect, Faraday rotation and the Kerr effect as different consequences of the same field energy terms.[39]

Mechanical Analogy and the Problem with Conduction Current

While the treatments of the Hall, Faraday and Kerr effects marked a success of the Maxwellian theory, the explanation of conduction current nonetheless continued to pose a severe conceptual problem to the approach. In brief, it was impossible to have a consistent correspondence between the ether mechanical variables and electromagnetic field variables of a physical system unless one ignored conductivity. Throughout the 1880s and 1890s, the Maxwellians were haunted by this problem and grappled with it by glossing over conduction current in terms of a macroscopic, average effect of electric displacement without having to specify its details. Far from satisfactory, this approach settled the problem only temporarily. As new experimental results were reported and interest in looking further into the details of conductivity increased, the problem with conduction current eventually led to the electron theory and the end of Maxwell's programme.

Glazebrook's and Thomson's works showed the fruitfulness of mechanical analogy in the Maxwellian framework. As indicated in the previous section, such an approach was based on the assumption that electric displacement **D**, magnetic field intensity **H**, permittivity ε, and permeability μ were manifestations of the ether's mechanical (more specifically, hydrodynamic) characteristics. A Maxwellian's task was thus to identify these electromagnetic variables with proper mechanical variables. Once the identification was done, he could equate various energy terms of the ethereal fluid to the corresponding electromagnetic field energy terms, obtain an expression of the system's action and employ the least-action principle on the expression to get Lagrange's equations. If the analogy was consistent, then these equations should be the right electromagnetic field equations that gave the correct physical results.

Maxwell developed the first mechanical analogy of this kind. In the 1850s, he followed William Thomson's suggestion to correspond the magnetic field intensity **H** with the ether's vorticity defined by $(1/2)\nabla \times (\partial \mathbf{u}/\partial t)$, where **u** was the ether's displacement vector field. In 1881, Glazebrook took up and further extended this model in his work on the Hall effect. According to Glazebrook, one could derive the electromagnetic field equations for the Hall effect from the ethereal fluid's equation of motion when **H** equalled $(1/2)\nabla \times (\partial \mathbf{u}/\partial t)$, **D** equalled $(-1/2)\nabla^2\mathbf{u}$, μ equalled the medium's density, and ε equalled the medium's elastic coefficient.[40]

The Maxwell–Glazebrook model was not the only vortex ether model Maxwellians conceived. In 1876–80, FitzGerald entertained another one to derive the laws of optical reflection and refraction from Maxwellian electromagnetics. FitzGerald's idea of deducing the laws of reflection and refraction from the dynamics of ethereal medium came from Irish physicist, James MacCullagh, at Trinity College in Dublin. FitzGerald corresponded **H** to $\partial \mathbf{u}/\partial t$ and **D** to $\nabla \times \mathbf{u}$. From this correspondence, he interpreted the potential and kinetic energy of the ether in terms of field energy terms, and obtained the system's Lagrangian and action. In order to follow the principle of least action, FitzGerald found, the field quantities on both sides of the boundary (of reflection or refraction) must satisfy a set of what are known today as 'boundary conditions'. From these boundary conditions, FitzGerald easily derived the laws of reflection and refraction.[41]

Useful as it was, the mechanical analogy of the electromagnetic field nonetheless had a significant shortcoming. Both the Maxwell–Glazebrook model and FitzGerald's model worked when there was no conduction current in the medium. If the medium was conductive and there was a conduction current that followed Ohm's Law $\mathbf{C} = \sigma \mathbf{E}$, then the analogy broke down; more specifically, the mapping of the electromagnetic field variables to ethereal mechanical variables in both models became inconsistent with either Maxwell's equations, or the hydrodynamic equations, or both. In other words, conduction current became a hindrance to the Maxwellians' aim to build electromagnetism on the foundation of mechanical models and dynamic equations.

In the 1880s and early 1890s, the Maxwellians' general strategy to cope with this problem was to treat conductivity as a macroscopic and average effect of some intermittent and microscopic processes whose details were not known and did not matter. In this regard, for example, conduction current was nothing but a measure of the energy loss in the frequent breakdown and restoration of electric displacement in a conductor. In so doing, one could bypass the issue of the mechanical basis of conductivity and model the dissipative effect of conductivity in an ad hoc, phenomenological way, similar to how nineteenth-century physicists dealt with friction. Both Poynting and J. J. Thomson undertook this approach in their mechanical modelling of electromagnetic fields.[42]

Yet, this strategy only covered up the problem rather than solving it. By the early 1890s, the Maxwellians had an increasing sense of crisis over the dilemma of conductivity. Oliver Heaviside made the situation even more critical when he demonstrated, in 1893, that one could not set up a consistent and physically reasonable correspondence between electromagnetic field variables and ethereal mechanical variables, unless there was no conduction current and conductive dissipation in the system. He showed that in all viable ways that he could think of that identified magnetic field with vorticity, the attempt to make such a correspondence led to the mismatch of at least one pair of variables if the conduction current was present. After all, conductivity seemed impossible to incorporate into the dynamics of electromagnetism. The problem with conductivity thus became a typical Kuhnian crisis to the Maxwellian paradigm by the 1890s.

Rise of the Electron Theory and Decline of the Maxwellian Programme

By the early 1890s, the Maxwellian electromagnetism as a research programme had reached its climax. Its theoretical development in energy flow, vortex ether modelling and dynamic Lagrangian formulation attained maturity. It recapitulated the laws of reflection and refraction in optics via electromagnetism. It gained significant success in explaining the Hall effect and other magneto-optic phenomena. And Hertz's spark-gap experiment in the late 1880s confirmed Maxwell's prediction on electromagnetic waves. While the Continental disciples of Weber, Neumann and Helmholtz still continued to hold the corpuscular theory of electricity, the majority of physicists in the UK and USA were converted to field theorists at this time.

In the following decade, however, the influence of the Maxwellian programme declined rapidly. At the beginning of the twentieth century, most scientists either dismissed or marginalized ether vortices, and replaced the Maxwellian field theory with certain microphysics in which the electric charge of a body was the sum total charge of its constituent particles and electric current was a stream of charged particle flow. In this new microphysics, charge and current regained the status of fundamental physical entities, and atomic interactions between field and matter replaced novel field energy terms as the main sources of exotic optic, electric, and magnetic effects. In this section, we give a very brief review of the electron theory that came to replace the Maxwellian programme in the 1890s.

This microphysical revolution actually began from within. The first scientist to propose a theory of electrons came from the Cambridge school of electromagnetism. Joseph Larmor was a graduate of St. John's College, Senior Wrangler, and lecturer of mathematical physics at Cambridge. Larmor had shown interest in the dynamic structure of Maxwell's electromagnetic theory. In 1893, he launched a research programme aiming to build the mechanical foundation of Maxwellian on the principle of least action. MacCullagh's rotational ether constituted Larmor's starting point. Following FitzGerald's adoption of MacCullagh's model, Larmor conceived the ether as a fluid medium with vortices, equated magnetic intensity **H** to ethereal body velocity $\partial \mathbf{u}/\partial t$ and electric displacement **D** to ethereal rotation $\nabla \times \mathbf{u}$, and identified electric field energy with ether's elastic potential energy and magnetic field energy with ether's kinetic energy. By employing the principle of least action on these localized energy terms, Larmor, like FitzGerald, obtained the boundary conditions necessary for the laws of optical reflection and refraction.[43]

But like Maxwell, Poynting and J. J. Thomson, Larmor encountered the problem of conductivity in the dynamic formulation of electromagnetic fields. The Maxwellian mechanical analogy treated a conducting material as a perfect conductor whose dielectric constant ε was infinity (or analogically, whose ethereal elastic constant was zero). The formulation failed to cover the cases involving conductors with *finite* conductivity and dielectric constant. As mentioned, Maxwell, Poynting and Thomson explained the dissipative effect of finite conductivity by assuming that the established electric displacement in such a conductor was suddenly relaxed or the tubes of electric displacement moved away and suddenly disappeared. Larmor's early model was under the same spirit of 'rupture mechanism', as he hypothesized that the ether elasticity in the conductor suddenly went to zero and restored quickly. But inspired by the contemporary work on electrolysis, in 1894 he found that this relaxation and restoration of ether elasticity was equivalent to the convection of a discrete electric charge, and the latter seemed more plausible and convenient. In the following months, Larmor took FitzGerald's advice to discard gradually the components of the continuous ether model from his dynamic theory. By the end of the year, Larmor came up with a final version of the theory in which all materials were constituted of positively or negatively charged particles – which he termed 'electrons' after G. J. Stoney – and electric current resulted from convection of the discrete electrons. At the end of Larmor's Maxwellian exploration of the dynamics of electromagnetism there emerged a fundamentally anti-Maxwellian theory: microphysics.[44]

Note that, despite its fundamental difference from the Maxwellian theory, Larmor's model of electrons nonetheless retained some Maxwellian characters. To him (and to the Maxwellians attempting to cope with his novel theory in the late 1890s), ether still existed and constituted the basis of field effects. The electrons were singular points into which electric lines of force (ether's elastic tension) converged or from which they diverged. In other words, electrons were embedded in ether; their presence did not get rid of ether. Along this line of reasoning, moreover, Larmor and the Cambridge physicists influenced by him constructed an electron theory of matter in which all materials were electromagnetic in origin since their constitutive elements – electrons – were parts of the ether structure.[45]

As Larmor was developing his electronic theory in the UK, similar efforts (albeit without any recourse to the ether) were being made in Continental Europe. In Germany, Holland and France, physicists were much less committed to the Maxwellian framework, but were rather operating under the tradition of Weber's corpuscular theory and the Neumann–Helmholtz potential theory that involved action at a distance. Throughout the 1880s and 1890s, European researchers, such as Hendrik Antoon Lorentz at Leyden University, proposed several corpuscular models to explain various magneto-optic effects experimenters had reported. But the best known result of this Continental research was inspired by a novel experimental finding. In 1896, a post-doctoral researcher, Pieter Zeeman, at Leyden discovered that the spectral lines of white-heated sodium became wider when the spectroscopic set was placed close to a strong magnet. Lorentz soon heard of this interesting experimental outcome and gave an explanation in terms of his magneto-optical theory. According to Lorentz, what Zeeman observed was actually the split of the sodium spectral lines due to an external magnetic field. And this occurred because the emitting light from sodium was actually electromagnetic radiation from the oscillating charged particles (what Lorentz called 'ions') confined within the metal's atoms. The applied magnetic field engendered a new force – the 'Lorentz force' $\mathbf{f} = q\mathbf{v} \times \mathbf{B}_T$ – on the ions. This force changed the course of the ions' movement (more specifically, it turned a linearly oscillating ion to one with a spiral precession), and the way such a change happened doubled the ions' oscillating frequencies. Since the spectral lines of sodium's electromagnetic radiation corresponded to the metal's oscillating frequencies, the split of spectral lines was consistent with Lorentz's prediction on doubling oscillating frequencies. Using his theory, Lorentz obtained the charge-to-mass ratio of the ions from Zeeman's experimental data.[46]

With Larmor's theory of electrons and Lorentz's theory of ions, physicists in the late 1890s turned away from the Maxwellian approach toward microphysics. Now they could use the charged particles to account for electric conductivity, the Hall effect, the Kerr effect, the newly found Zeeman effect, as well as the other electromagnetic and optical phenomena Maxwell's theory could explain. The final blow to the Maxwellian theory came with an even more startling experimental discovery. In 1897, J. J. Thomson, a once adamant Cambridge Maxwellian but now a converted microphysicist and Director of the Cavendish Laboratory, conducted several trials on cathode rays. Discovered in the 1860s, the cathode ray was believed to be an invisible beam of energy emitting from a high-voltage-charged cathode in an evacuated glass tube, and was detected through photography, thermocouple or galvanometer. Physicists had debated for long about whether the cathode ray was the outcome of ether vibration like light or flow of particles. Thomson's trials in 1897 aimed to settle this debate: he applied a coil-generated static magnetic field on a beam of cathode rays and found that the rays were deflected by the field. When he replaced the magnetic field with a static electric field produced between two charged parallel plates, the rays were deflected, too. The deflectability of cathode rays by both electric and magnetic fields strongly suggested to Thomson that the rays were a stream of charged particles, not ether vibration (since it was possible to conceive that light could be deflected by a magnetostatic field, but impossible to conceive that it could be deflected by an electrostatic field). From the data of his deflection trials, moreover, Thomson was able to calculate the particles' charge-to-mass ratio, which was numerically close to the charge-to-mass ratio of Lorentz's ions. Within a short period of time, physicists reached the consensus that Thomson's corpuscles in cathode ray were identical to Lorentz's ions and Larmor's electrons, and Thomson's experiment in 1897 was canonized as the moment of 'discovering' the electron.[47]

By the turn of the twentieth century, the Maxwellian's theory of electromagnetism had all but disappeared from scientists' intellectual scope. What replaced it was Larmor's and Lorentz's microphysics in which matter was made of electrons and electromagnetic effects were results of microscopic interactions between electromagnetic fields and matter. It is this microphysics, not Maxwell's and his followers' original theory, that constitutes at least part of the basis of today's 'classical' electromagnetism. But as soon as this microphysics prevailed among the scientific community, it received radical challenges. Albert Einstein's relativity in 1905 replaced Lorentz's and Larmor's electrodynamic theory of length contraction on a moving object with a more fundamental, kinetic theory rooted in a redefinition of space and time. On the other hand, Max Planck's, Einstein's, Niels Bohr's and Arnold Sommerfeld's quantum theory constrained the motions of electrons in ways that were unintelligible to the classical microphysics. After the relativitistic and quantum revolutions in the first three decades of the twentieth century, Maxwell's original theory became even more alien. From that point on, he has been remembered mainly as the formulator of the field equations, and though the equations remain unchanged, modern field theory is very different from how Maxwell originally envisioned it.

CHAPTER 12

The Fluid Dynamics of James Clerk Maxwell

KEITH MOFFATT

Introduction

We tend to think of Maxwell's genius entirely in relation to his creation of electromagnetic theory (as summarized by Maxwell's equations) on the one hand, and to his seminal development of the kinetic theory of gases (leading to the Maxwell–Boltzmann distribution) on the other. To do so is perhaps to neglect his extraordinary insights in other directions, notably in the dynamics of fluids, a field in which his publications were few but nevertheless ground-breaking in their originality. Some of these insights were not published at the time, but are now available through the wonderful three-volume publication by Peter Harman of *The Scientific Letters and Papers of James Clerk Maxwell*,[1] on which I draw heavily in the following discussion. Maxwell's insights refer to centrifugal instability, to fluid-dynamical stability both in general and in particular application to the problem of Saturn's rings, to Kelvin's knotted vortices (in relation to which Maxwell played the part of amused commentator), and to the problem of Lagrangian particle displacement in an unsteady Eulerian flow, for which he provided perhaps the first explicit example. Fluid-dynamical thinking was in fact intrinsic to Maxwell's development of electromagnetic theory in the 1850s, and provided the initial framework upon which he constructed this theory, as evidenced by his first major paper in the subject 'On Faradays' Lines of Force'. His endeavours to express observed phenomena in mechanical terms lay at the heart of his thinking, and may help us to understand the development of a mind of extraordinary boldness and originality.

Centrifugal Instability

In Volume II of Harman's *Scientific Papers and Letters of James Clerk Maxwell,*[1] one may find the following *Draft question on the stability of vortex motion** as set by Maxwell who had been appointed a Moderator (i.e. External Examiner in modern parlance) for the Cambridge Mathematical Tripos 1866:

A mass M of fluid is running around a circular groove or channel of radius a with velocity u. An equal mass is running around another channel of radius b with velocity v. The one channel is made to expand and the other to contract till their radii are exchanged. Show that the work expended in effecting the change is

$$-\frac{1}{2}\left(\frac{u^2}{b^2}-\frac{v^2}{a^2}\right)(a^2-b^2)M$$

Hence show that the motion of the fluid in a circular whirlpool will be stable or unstable according as the areas described by particles in equal times increase or diminish from centre to circumference.

The final sentence of the question (the sting in the tail!) is particularly noteworthy: readers familiar with the theory of centrifugal instability will recognize that it embodies what is generally known as *Rayleigh's criterion,*[3] namely that a flow of an ideal fluid with velocity field $\mathbf{u}=(0,v(r),0)$ in cylindrical polar coordinates (r,θ,z) is stable or unstable to axisymmetric disturbances according as the circulation $k(r)=2\pi rv(r)$ increases or decreases with distance r from the axis. Rayleigh discusses the problem of stability first through analogy with the simpler problem of the stability of hydrostatic equilibrium of an incompressible fluid of variable density in which the density is a function of height. Then, appealing implicitly to the circulation theorem of Kelvin,[4] he goes on to say:

We may also found our argument upon a direct consideration of the kinetic energy T of the motion. For T is proportional to $\int v^2r\,dr$, or $\int k^2\,dr^2/r^2$. Suppose now that two rings of fluid, one with $k=k_1$ and $r=r_1$ and the other with $k=k_2$ and $r=r_2$ where $r_2>r_1$ and of equal areas dr_1^2 or dr_2^2 are interchanged. The corresponding increment in T is represented by

$$(dr_1^2=dr_2^2)\left(\frac{k_2^2}{r_1^2}+\frac{k_1^2}{r_2^2}-\frac{k_1^2}{r_1^2}-\frac{k_2^2}{r_2^2}\right)=dr^2\left(k_2^2-k_1^2\right)\left(r_1^{-2}-r_2^{-2}\right),$$

and is positive if $k_2^2>k_1^2$; so that a circulation always increasing outwards makes T a minimum and thus ensures stability.

Here Lord Rayleigh (Nobel Laureate 1904) in effect provides a correct solution the problem that had been set by Maxwell for undergraduates taking the Tripos examination half a century earlier; he would presumably have been given an α-mark for his answer, had he provided it then. Actually, Lord Rayleigh (then John William Strutt) had been a candidate for the Mathematical Tripos just one year earlier (1865) and was Senior Wrangler (i.e. top of the list) in that examination.

When Maxwell set his examination question in 1865, he of course did not have access to Kelvin's circulation theorem, proved four years later. He presumably thought in terms of conservation of the angular momentum of the fluid in his 'circular groove', and this is equivalent to conservation of circulation for the axisymmetric displacements that he envisaged. His phrase 'the area described by particles ...' clearly means the area swept out by a radial line ($z=$ constant, $\theta=$ constant) from the axis of symmetry to the particle in question, that is $rv(r)\delta t$ in time δt

(reminiscent of Kepler's phrase for planetary motion 'equal areas in equal times'), so again proportional to circulation. Furthermore, Maxwell posed this question in terms of the 'work done' in effecting the displacement, but since this work done must convert to a corresponding change in the kinetic energy of the fluid (assumed incompressible), Maxwell's criterion for stability is in effect identical with that of Rayleigh.

Viscosity has a stabilizing influence for this problem, as recognized by Rayleigh, so that for a viscous fluid, the condition that circulation decreases outwards is necessary but not sufficient for stability to axisymmetric disturbances. The effect of viscosity was brilliantly analysed and demonstrated experimentally by G. I. Taylor,[5] in a work that has provided the impetus for a huge body of research on the problem of centrifugal instability (see for example the works of Stuart,[6] Lin,[7] Drazin and Reid,[8] Koschmieder[9] and Iooss[10]). Maxwell's 1866 question for the Mathematical Tripos reveals his seminal insight into this important branch of hydrodynamic stability theory.

Stability of Steady Euler Flows

But now let us go back a bit earlier to 1854 when Maxwell, then aged 23, was himself subjected to the rigours of the Mathematical Tripos; in this, he came Second Wrangler, second to E. J. Routh who subsequently distinguished himself as the greatest Cambridge mathematics tutor of his day. Following the Tripos, Maxwell continued to work as a graduate student, aiming at a Fellowship of Trinity College, to which he was elected just one year later.

One of his preoccupations during this formative year of graduate study was the *Stability of Fluid Motion*, on which subject he revealed his ideas in a remarkable letter to William Thomson (later Lord Kelvin) dated 15 May 1855 (article No 66 in Volume I of Harman's collection[1]). Here Maxwell first considers the steady two-dimensional flow of an ideal incompressible fluid for which the velocity fluid $\mathbf{u}$ is given in terms of a streamfunction $\psi(x, y)$ by $\mathbf{u} = (-\partial\psi/\partial y, \partial\psi/\partial x)$, a notation introduced by Stokes, Lucasian Professor of Mathematics, at whose feet Maxwell sat during his undergraduate years. Maxwell first re-derives Stokes' condition[11] for steady flow $\nabla^2\psi = f(\psi)$ for some function $f(.)$. He then writes:

> To determine the stability or instability of this steady motion we must give it a perfectly general derangement & determine whether it will or will not tend to return to its original state.

It is difficult to follow Maxwell's subsequent reasoning, but by physical arguments that do not bear too close examination he in fact arrives at the correct conclusion, namely that

$$f'(\psi) > 0 \text{ is the condition of stability}$$

This is altogether astonishing, because the result as stated remained unpublished (and therefore unknown) for more than a century, and was not rediscovered and rigorously proved until the work of Arnold[12] (see also Moffatt[13]) through consideration of the second variation of kinetic energy K under 'isovortical perturbations', that is perturbations that transport the vortex lines of the basic flow as though 'frozen in the fluid', in conformity with the constraint of Euler's equations.[14] The first variation of kinetic energy $\delta^1 K$ vanishes for any steady Euler flow $\mathbf{u}^E$ with vorticity $\boldsymbol{\omega}^E = \nabla \times \mathbf{u}^E$ (this much was known to Kelvin), and the second variation under an arbitrary volume-preserving isovortical displacement $\boldsymbol{\eta}(\boldsymbol{x})$ is given by

$$\delta^2 K = \frac{1}{2}\int \left[\left(\boldsymbol{\eta} \times \boldsymbol{\omega}^E\right)_s^2 - \left(\boldsymbol{\eta} \times \boldsymbol{\omega}^E\right)_s . \nabla \times \left(\boldsymbol{\eta} \times \mathbf{u}^E\right)\right] \mathrm{dV},$$

where the suffix s denotes the 'solenoidal projection' of the vector field concerned, and the integration is over the fluid domain. In the case of two-dimensional perturbations of a two-dimensional flow with streamfunction ψ satisfying $\nabla^2\psi = f(\psi)$, this expression reduces to

$$\delta^2 K = \frac{1}{2}\int \left[(\boldsymbol{\eta}\times\boldsymbol{\omega}^E)_s^2 + (\boldsymbol{\eta}\times\mathbf{u}^E)^2\,(f'\psi)\right] dxdy,$$

so that a sufficient condition for $\delta^2 K$ to be positive for all admissible $\boldsymbol{\eta}$ (and so for the flow to be stable) is indeed that $f'(\psi) > 0$.

So here again, Maxwell demonstrates profound physical insight; although his argument is obscure to the point of incomprehensibility (perhaps inevitably because this was well before Helmholtz's recognition[14] of the key role played by vorticity), he nevertheless arrives at the correct conclusion, as ultimately confirmed more than 100 years later by Arnold, who was quite unaware of Maxwell's prior deliberations on this problem. During the same pivotal year, 1855, Maxwell wrote his first major paper on electromagnetism,[15] 'On Faraday's Lines of Force', in which he develops the analogy between the lines of force of a magnetic field and the streamlines of the flow of an incompressible fluid. He introduces this fluid with caution:

> The substance here treated of must not be assumed to possess any of the properties of ordinary fluids except those of freedom of movement and resistance to compression. It is not even a hypothetical fluid which is introduced to explain actual phenomena. It is merely a collection of imaginary properties which may be employed for establishing certain theorems in pure mathematics in a way more intelligible to many minds and more applicable to physical problems than that in which algebraic symbols alone are used. The use of the word 'Fluid' will not lead us into error, if we remember that it denotes a purely imaginary substance with the following property: *The portion of fluid which at any instant occupies a give volume, will at any succeeding instant occupy an equal volume.* This law expresses the incompressibility of the fluid and furnishes us with a convenient measure of its quantity, namely its volume.

In this way, Maxwell leads the reader to conceive of flow in the abstract, a concept that mathematicians a century later would come to describe in terms of 'volume-preserving diffeomorphisms'.[12] Many might still prefer Maxwell's more primitive terminology! The evolution of Maxwell's theory of electromagnetism over the years 1855 to 1862 was largely guided by his efforts to express the theory in fluid mechanical terms, although ultimately the great triumph of the theory was that he was able to discard the artificial and, as it turned out, irrelevant fluid-mechanical framework.

The Stability of Saturn's Rings

In October 1856, Maxwell took up his position as Professor of Natural Philosophy at Marischal College, Aberdeen, and immediately set to work on the problem of the motion and stability of Saturn's rings, the subject that had been posed for the 1857 Adams Prize at Cambridge University. He was the only candidate to make significant progress on the problem, and duly won the prize. He continued to work on the vexing problem of the stability of the rings, and the revised version of this work[16] was published two years later.

I wish to comment here on only a small portion of this work contained in section 23 of the paper,[**] in which Maxwell focuses on a model problem: the instability of a uniform layer of liquid of great extent compared to the thickness, subject only to the effect of self-gravitation. The best-known example of this type of instability is the 'Jeans[17] instability' of a large extent of compressible liquid subject only to self-gravitation; so here again, we have a situation in which Maxwell anticipates by several decades an instability mechanism of fundamental significance in cosmology, the difference being that, for Maxwell, it is the mobility of the free surfaces of the liquid, rather than its compressibility, that permits effective horizontal compression or dilation of the liquid layer. Self-gravitation promotes two effects: the propagation of (stable) gravity waves on both surfaces of the layer; and the horizontal 'bunching up' of the layer which can decrease gravitational potential energy and is therefore destabilizing. Maxwell recognized the competing influence of these effects and obtained the correct criterion for instability, a veritable *tour de force* at that time. If sinusoidal perturbations of the layer of wavelength $2\pi/m$ are considered, then it is found by linearized analysis that these perturbations (assumed symmetric about the central plane of the layer) grow like $\exp \sigma t$, where

$$\sigma^2 = 2\pi G\rho \tanh mb\,[1 + \exp(-2mb) - 2mb]$$

in which $2b$ is the thickness of the layer, ρ the density of the liquid, and G the universal gravitational constant. The function is positive for $2mb < 1.278$, indicating instability for this range of wave-numbers m. Maxwell gave the value 1.147 here, not very accurate, but his conclusion that the layer is gravitationally unstable to perturbations of wavelength $\lambda = 2\pi/m$ sufficiently large compared with the layer thickness $2b$ was certainly correct. The growth rate of the instability is in fact maximal when $mb = 0.294$, or $\lambda/2b = 10.69$.

When $mb > 1.278$, σ^2 is negative and the perturbation modes are oscillatory with frequency $\omega = i\sigma$; indeed for such large mb the above formula gives $\omega^2 \sim gm$, where $g(= 4\pi G\rho b)$ is the inward gravitational acceleration at the free surface of the liquid layer. This is just the familiar dispersion relation for gravity waves on deep water, a result already well-known to Stokes.[18]

(A full account of Maxwell's work on Saturn's rings is given in Chapter 6 of this book.)

Knotted Vortices

In 1867, Kelvin[19] published in the *Philosophical Magazine* his paper 'On Vortex Atoms', which must have caused quite a stir in the scientific community. It was in this paper that he mentions his visit to Peter Guthrie Tait's laboratory in Edinburgh, where he witnessed Tait's demonstration of the production of vortex rings by ejection of air from an orifice, the rings being visualized by smoke. Kelvin's subsequent realization that any knot in a vortex tube in an ideal fluid would be of permanent form, led him to hypothesize that the existence and character of the known elements of the periodic table might be explained at the atomic level in terms of 'vortex knots' of increasingly complex structure in the ether, an ideal fluid imagined to permeate all space; and that the frequency spectra associated with these elements might then be explained in terms of the frequencies of vibration of such vortex knots.

Maxwell refers to this paper with his usual pawky humour in his letter to Tait of 13 November 1867, which appears as article 275 of Volume II of Harman's collection:[1]

Thomson has set himself to spin the chains of destiny out of a fluid plenum ... and I saw you had put your calculus in it too. May you both prosper and disentangle your formulae in proportion as you entangle your worbles. But I fear that the simplest indivisible *whorl* is either two embracing worbles or a worble embracing itself . . .

And three weeks later (4 December 1867, article 276 in Harman's collection), is a further letter from Maxwell to Tait, mainly about the properties of certain plane curves; but he adds, as if by afterthought, *I have amused myself with knotted curves for a day or two.* He goes on to state the formula (due to Gauss, but which Maxwell had evidently here derived independently) for the linking number n of two closed curves C and C':

$$\int_C \int_{C'} \frac{(\mathbf{x} - \mathbf{x}') \, . \, d\mathbf{x} \times d\mathbf{x}'}{|\mathbf{x} - \mathbf{x}'|^3} = 4\pi n$$

He points out that although $n \neq 0$ implies linkage, the condition $n = 0$ does not necessarily imply that C and C' are unlinked, and he gives two examples of non-trivial links with $n = 0$ – examples that later became known as the Whitehead link and the Borromean rings. In the final paragraph of the letter, he gives a number of examples of knotted curves described in parametric form, and the manner in which the knot type can change as a parameter is varied. What an astonishing burst of creativity in that *day or two* of desultory amusement!

There is no doubt that Maxwell retained an interest in Tait's development of knot theory during the 1870s, but in a rather detached way 'from the sidelines'. It was doubtless the monumental study of Tait in 1877[20] as much as Kelvin's knotted vortices that stimulated Maxwell to compose his '*Paradoxical Ode*' *To* [the imaginary philosopher] *Hermann Stoffcraft PhD*, the first verse of which reads:

My soul's an amphicheiral knot
Upon a liquid vortex wrought
By intellect in the unseen residing
While thou dost like a convict sit
With marlinespike untwisting it
Only to find my knottiness abiding:
Since all the tools for my untying
In four-dimensional space are lying,
Where playful fancy intersperses
Whole avenues of universes:
Where Klein and Clifford fill the void
With one unbounded finite homaloid,
Whereby the Infinite is hopelessly destroyed

This poem, Maxwell's last, was written shortly before his death from cancer in 1879. It touches on current scientific and philosophical ideas in equal measure; an illuminating interpretation is provided by Silver.[21] The phrase 'whole avenues of universes' has a certain resonance with recent speculations (see, for example, Carr[22]) concerning the concept of the 'multiverse'; playful fancy indeed! (Stella Pratt-Smith also discusses this poem in her account of Maxwell's poetry in Chapter 13 of this book.)

Although Kelvin's vortex knots held great promise, he was unable to find any steady stable solutions of the Euler equations having a knotted character, and he was gradually forced to abandon the theory. If, instead of vortex tubes, he had chosen to focus on magnetic flux tubes in a perfectly conducting (i.e. ideal) fluid medium, the result might have been very different, for we know now that stable magnetostatic equilibria of arbitrarily complex topology do exist, albeit with imbedded tangential discontinuities.[23] All the ingredients for such a 'complementary' theory were already available in the 1870s as a result of Maxwell's development of electromagnetism; but the realization that the magnetic field is frozen in a perfectly conducting fluid (and so its topology conserved) had to wait the development of magnetohydrodynamics in the twentieth century, spearheaded by the work of Alfvén,[24] and the discovery of the conservation of magnetic helicity and its interpretation as a measure of knottedness of magnetic flux tubes.[25,26]

Lagrangian Particle Displacement

In an unstable fluid flow, the streamline pattern is continuously changing in time, and the (Lagrangian) trajectory of a fluid particle is quite distinct from the instantaneous streamline on which it finds itself at any instant. This was recognized by Maxwell,[27] who provided perhaps the first explicit calculation of the Lagrangian displacement of a fluid particle in an unsteady flow field. This was the two-dimensional flow generated by the motion of a circular cylinder whose axis is parallel to the z-axis with steady velocity along the x-axis, in the frame of reference in which the fluid velocity is zero at infinity. The cylinder passes any fixed point in this frame, carrying its streamline pattern with it; each fluid particle following a trajectory depending on its initial distance y from the x-axis, the net displacement being a positive function $d(y)$ in the x-direction. Maxwell succeeded in calculating the particle paths in terms of elliptic functions, and in plotting those out by hand, with the comment: '*The curves thus drawn appear to be as near the truth as I could get without a much greater amount of labour*'. The labour already invested was certainly sufficient to demonstrate quite beautifully the essential distinction between an unsteady streamline pattern and the resulting field of particle displacements.

A further example of Maxwell's originality is to be found in his 1870 paper 'On hills and dales', in which he discusses in effect the topology of the contours of a scalar field in two dimensions. He does this in the context of the height of land above sea level; but the arguments are equally valid for any scalar field, for example the streamfunction $\psi(x, y)$ of any two-dimensional incompressible steady flow. Maxwell's thinking was topological in character: flexible, widely adaptable, and of great generality; in this as in other areas he was decades ahead of his time.

Postscript

While Maxwell's scientific achievements were truly phenomenal, it must also be recognized that he was not afraid to admit incomprehension in relation to some of the natural phenomena that he addressed. One of these concerned the origin and evolution of the Earth's magnetic field, which he discussed in Part IV, Chapter VIII of his great *Treatise on Electricity and Magnetism.*[28] Having commented on the secular variation of the field (by which he meant the slow wandering of the magnetic poles), he writes with an appropriate sense of wonder:

What cause, whether exterior to the earth or in its inner depths, produces such enormous changes in the earth's magnetism, that its magnetic poles move slowly from one part of the globe to another? When we consider that the intensity of the magnetisation of the great globe of the earth is quite comparable with that which we produce with much difficulty in our steel magnets, these immense changes in so large a body force us to conclude that we are not yet acquainted with one of the most powerful agents in nature, the scene of whose activity lies in those inner depths of the earth, to the knowledge of which we have so few means of access.

The means of access were in due course provided by the penetrating power of seismology (see for example Bullen and Bolt[29]) which established the liquid-metal character of the outer core of the Earth, capable of sustaining the electric currents $\mathbf{j}(\mathbf{x}, t)$ that are the source of the magnetic field $\mathbf{B}(\mathbf{x}, t)$ whose variations we observe at the surface. Gravitational convection and the rotation of the Earth conspire to provide a turbulent (or at least random) flow $\mathbf{u}(\mathbf{x}, t)$ in the core with the property of non-zero helicity oppositely signed in north and south hemispheres, and it is now known that this property alone (implying a degree of knottedness of the vortex lines of the flow) in a body of fluid as large as the Earth's outer core is sufficient to guarantee the growth of a large-scale magnetic field by a dynamo-instability mechanism.[30]

The dynamo instability depends on what are sometimes known as the pre-Maxwell equations (with displacement current filtered out):

$$\frac{\partial \mathbf{B}}{\partial t} = -\nabla \times \mathbf{E}, \quad \nabla . \mathbf{B} = 0, \quad \mathbf{j} = \nabla \times \mathbf{B}$$

coupled with Ohm's law in a moving medium of conductivity σ:

$$\mathbf{j} = \sigma(\mathbf{E} + \mathbf{u} \times \mathbf{B}).$$

Maxwell did have these essential ingredients at his disposal, so that this *most powerful agent in nature* was already imbedded in the theoretical structure that he provided in his 1873 treatise. The fact that a full century had to pass before the dynamo mechanism was finally teased out of the structure may perhaps be seen as yet further indication of just how far James Clerk Maxwell was ahead of his time.

Acknowledgements

I thank David Forfar, Alex Craik, Renzo Ricca, Julian Hunt and Raymond Flood for their most helpful comments on an earlier draft of this chapter.

PART III

Poetry, Religion and Conclusions

CHAPTER 13

Boundaries of Perception: James Clerk Maxwell's Poetry of Self, Senses and Science

STELLA PRATT-SMITH

Maxwell's prominence in the development of nineteenth-century physics so obscures his literary creativity that his poems occupy only the third and final position of his posthumous biography.[1] The physically marginal position scales down their significance, yet they can also be considered as a coherent body of literary work that provides insights into the unique imagination of an exceptional scientist. In describing Maxwell's approach to science and to life, the poems embrace a range of disciplines in ways that emphasize the profound reverberations he experienced between literature, philosophy and physics. As Gillian Beer has pointed out, in the nineteenth century, 'not only *ideas* but metaphors, myths and narrative patterns could move rapidly and freely to and fro, between scientists and non-scientists'.[2] Maxwell's poetry has the characteristic 'hybridity' that scholars have increasingly come to recognize as a central aspect of certain texts.[3] Indeed, if there is one feature Maxwell's poems share, it is their continual questioning of how we know what we come to know. Again and again, the poems address epistemological and disciplinary boundaries, divergences and parallels between individual and collective consciousness, the nature of perception as sensory, rational and non-rational and, last but not least, the interactivity of different forms and forces of matter.

My discussion seeks to introduce the great many poems that Maxwell composed beyond those that are regularly analysed, and it posits Maxwell's investigative impulse as a connecting and fundamentally cross-disciplinary strand between them. The focus is predominantly on his

Fig. 13.1 The young author and illustrator. The last page of a youthful Maxwell's manuscript of 'The Vampyre' (Campbell & Garnett, *The Life of James Clerk Maxwell*, Macmillan & Co. 1882, opposite p. 71).

post-juvenilia, although we might note in passing how Maxwell's earlier works, such as 'The Vampyre' also indicate his deep-seated and early attraction to poetry as a form (see Fig. 13.1).[4]

Maxwell's poems are considered here in three broad, chronological stages, beginning with his compositions in the early 1850s when he was undertaking early explorations of light and colour as a student at Cambridge and, at the same time, using poetic forms to express distinctive fusions of reality, science and fantasy, as well as refine the techniques by which to represent the world around him. After this, we consider the increasingly sharp contrasts apparent in his poetry, between the intellectual, private and public: the intellectual challenge of blending poetry with abstract mathematics; his private dreams and meditations around his marriage in 1858; and his more public concerns as an educator and scientist. In the third and final section, the discussion turns to what is often considered one of the core paradoxes of Maxwell's life, his juxtaposition of scientific practice and religious faith, in relation to contemporary issues such as scientific materialism, evolutionary theory and spiritualism.

Appreciating Maxwell's literary bent is essential for gaining a proper understanding of him. His active interest in literature is evident from his own excited reports of reading Charlotte Bronte's

Villette in 1853 and George Eliot's *Middlemarch* in 1873, and his reported enjoyment of poetry by Tennyson, Browning, and Burns.[5] His poems epitomize a profound delight in language as a means of exploring both literary and scientific perceptions, and the interactions between the two. As Barri Gold suggests in her study of his writings, his 'equations of electricity and magnetism are rather more elegant and timeless than his occasional verse', but 'the two together suggest how analogous, intertwined, and mutually productive, poetry and physics may be'.[6] By reading the two interests in parallel, we gain an extraordinary and often very personal perspective of a figure positioned at the epicentre of the birth of modern physics. The 'small collection of Maxwell's fugitive poems' have a renegade quality at odds with the weighty complexity of his illustrious, public reputation.[7] As several chapters in this volume elucidate, Maxwell's success as a scientist was won through work on a range of topics including electricity and magnetism, the theory of gases, and colour theory – achievements that justly made him famous. By the late nineteenth century, Maxwell was recognized as 'a lover of poetry, a poet' but it was also contended that 'he saw, however, that his true calling was science'.[8] The fact that none of Maxwell's 'fugitive' lines were published in his name during his lifetime suggests that they were largely covert, within the century's shifting dynamics between poetry and science that ranged from Wordsworth's championing of 'the Poet' over 'the Man of Science', Thomas Love Peacock's dismissal of poetry as a 'specious indolence' not fit for grown men, and Herbert Spencer's declaration that 'science excites poetry rather than extinguishes it'.[9] Maxwell's own contention, meanwhile, was that the 'the aim of physical science is to observe and interpret natural phenomena', a purpose strikingly comparable to the main occupation of poetry.[10]

In reading Maxwell's verse more or less chronologically, we witness his gradually developing confidence and maturity, as a writer, as a man, and as a leading figure in the scientific world. By doing so, we avoid the 'excessive concentration on discovery' that, as Geoffrey Cantor, David Gooding and Frank James propose, 'distorts our understanding of both scientists and the activity of science'.[11] We also avoid the danger pointed out by Frank Turner of writing about Victorian intellectuals 'as if they existed outside time', whereby 'we rarely imagine them as young or changing or as developing'.[12] Understanding Maxwell's poetry is an opportunity to achieve more nuanced understandings of his scientific approach and to locate this more wholly, in relation to his time and his life.

An Experimental Youth

Maxwell's earliest mature poems date from his undergraduate years at Cambridge between 1850 and 1854. Three key poems written during this period are 'A Vision, Of a Wrangler, of a University, of Pedantry, and of Philosophy', 'A Student's Evening Hymn' and 'Lines written under the conviction that it is not wise to read Mathematics in November after one's fire is out'.[13] Their candid, almost confessional tone presents a clear impression of how the young Maxwell saw himself, his scientific studies and his doubts about them at the time. The speaker of all three poems is a Cambridge mathematics student at Maxwell's own college, Trinity, which makes it tempting to read them as autobiographical; however, this assumption may be slightly problematic. In his biography of Maxwell, Ivan Tolstoy describes him as 'an intensely private person', evidenced by the 25 year old Maxwell's acerbic description of autobiography as 'the stomach pump of the confessional'.[14] Yet, as Maxwell did not publish his poems and knew no one would read them, they were entirely private works and may, therefore, be autobiographical. With appropriate caution, the most that can be asserted confidently is that Maxwell's poetic speaker is an anonymous figure whose sentiments he might have shared. As such, the first-person narrative voice offers an immediate and revealing portrait of the mid-Victorian student of science, battling with uncertainty and self-doubt as he sets out in the world.

Maxwell's literary skills may appear immature in the early poem 'Hymn', where he strives to rhyme 'off the valley' (5) with 'musically' (7), and 'Thy glory' (42) with 'transitory' (44); equally, however, they may be attempts to represent an individual accent and emphasis, as well as illustrations of a mischievous sense of humour.[15] The sincere or comic intention of the poem's mixed metaphors is similarly ambiguous, in 'Thou that fill'st our waiting eyes / With the food of contemplation' (17–18), and books with 'fresh stacks of buried fire / lurking in their ample pages' (51–2). More distinct in tone are the poems 'A Vision' and 'Lines', which align the key features of Maxwell's life – his scientific studies, poetry and religion. There is humour here too, as the midnight church bells call an end to the mathematical poet's evening studies in 'A Vision' and the frustrated speaker dismisses his 'confounded hydrostatics' (7), with a wish to 'sink it in the deepest sea!' (8), as an 'unblest' subject (24). Rather than the 'pleasure' experienced by Wordsworth's idealized poet-student, 'which accompanies him through the whole course of his studies', Maxwell portrays more realistically the hard grind of intellectual endeavour.[16]

'A Vision' and 'Lines' depict apparitions of women that function, initially at least, to guide the student-speaker away from his doubts. The dream plots bring to mind not only the Romantic models of Keats or Coleridge, but much older classical and medieval traditions of female hallucinations who offer embattled Christians moral guidance, in works such as Boethius's *Consolation of Philosophy* (*ca.* 480–525) and Langland's *The Pearl* or *Piers Plowman* (*ca.* 1330–87). Maxwell's figures are more carefully defined and disturbing, though, particularly the nightmarish 'Hag' of 'A Vision':

> Angular in form and feature,
> Unlike any earthly creature
> Hair of pens and skin of paper;
> Breath, not breath but chemic vapour;
> 'A Vision', 89–94.

A cyborgian figure, in whom any soft physicality is replaced by the rigid materials of education and science, she has a grating 'voice of iron' (122) that seems to 'proceed from metal, broken' (119) and later, as a 'withered beldam', she slinks away 'whining' like a dog (151). The image complicates the better-known and earlier nineteenth-century contention by Thomas Carlyle that 'men are grown mechanical in head and in heart, as well as in hand' with a mechanical grotesque of a woman who dismisses intellectual endeavour in favour of mid-century pragmatism and commerce, and who urges the speaker to concentrate on mathematics rather than poetry, because at least he will always be able to make a living from it.[17] Immediately afterwards, she encourages him to 'seem to have read what he hasn't, / And to do what can't be done' (135–6), creating an association between the mercenary and the fraudulent. The poem equates what is easily attained with shoddy principles and false financial rewards but, perhaps tellingly, in terms of Maxwell's subsequent renderings of women, it is his female spectres that commodify the world around them and exhibit the most limited visions of intellectual capitalism and consumerism.[18]

In Victorian questioning of 'where the boundaries of the self lay', Martin Willis proposes, 'society was also considering the boundaries of science'.[19] Maxwell visualizes a dynamic interaction between self, society and science when he conveys the Hag's Medusa-like stare and visually confusing presence by deploying the language and imagery of optical science: 'Eyes of glass, with optic axes / Twisting rays of light as flax is / Twisted, while the Parallax is / Made to show the real size' (97–100). Read aloud, the lines exemplify Maxwell's keen aural sensitivity, in the sibilance that conveys their 'twisting' contortions. However, the images also reflect the period's fascination with the visual when, as Bernard Lightman reminds us, 'conceptions about vision itself were in the process of being transformed' by spectroscopy and camera technology, resulting in 'the recognition that the human eye was fallible and limited'.[20] Also evident is the direct influence of Maxwell's study of optics and geometry, which he presented the following year in *On the Transformation of Surfaces by Bending* (1854) and, in the apparition's 'eyes of glass' and the 'real size' image, the Real Image Stereoscope he designed.[21]

The composition dates for 'A Vision' and 'Lines' are given as 10 November, one year apart in 1852 and 1853, and the poems represent something of a progression from fierce satire to despondency.[22] In the title pun of 'Lines', intellectual endeavour is reduced to a punishing burden, and the speaker's disillusionment with studying has degenerated to the point of simply 'Moping on from day to day, / Grinding up what will not "pay"' (13–14). The two poems represent the student's angry exhaustion, sleeping fitfully before a dying fire that is, itself, a symbol of the flagging spirit. The cynicism of 'Lines' disparages the costumes of society's educated sages as 'articles not worth wearing' (52) and sets success and integrity in opposition; while excelling in examinations might gain one 'place and pension' (33), the poem suggests, it also requires that 'every scruple is rejected' (37). By the mid-point of the verse, the 'seeming wisdom' of intellectual pursuits are almost entirely dismissed, in the style of William Cowper's declaration that 'if this be learning, most all deceiv'd'.[23] Maxwell's speaker begins to conclude, hopelessly, that a lack of intellect might be preferable to his own struggles, inexorably 'sinking' towards the 'annihilation' of the abruptly shortened final line:

> Happier those whom trifles please,
> Dreaming out a life of ease,
> Sinking by unfelt degrees
> Into annihilation.
> 'Lines', 33–6.

In a message to universities as well as students, the poem warns against pursuing, 'by labour, not by thought' (58–9), the pedantic 'trappings' of intellect, rather than its wiser substance. The apparition of a misleadingly 'learned-looking maid' promotes the earlier Hag's consumerist ethos towards education by mimicking the brash, new tone of attention-grabbing advertising and headlines: 'Pedantry is in demand!/Buy it up at second-hand' (69–70). The distinction Maxwell sets up is between those who are innately and genuinely wise, and those whose learning is merely an imitation of others or their work. The result of ignoring the distinction is, the poem argues, students who may be proud initially but, when their false intellectual guises fall away, are left 'clustering close in blank despair, / Nakedness, cold, and terror' (79–80), exposed as 'learned fools they are indeed, / Learned in the books they read' (85–6). The poem's final plea to 'raise their purblind eyes / To the opening mysteries / Scattered around them ever' (90–3) emanates from a second first-person speaker who, like Maxwell, believed that the attainment of wisdom, rather than just learning, might save them from an otherwise final and futile ending.

Maxwell's motivation, in the three poems, is grounded in a recurrent thread of Christian morality, in the conclusion to 'A Vision' that 'Nature more than symbols prizing, / Learn to worship as we ought' (183–4), in the assertion in 'Hymn' that 'aright to trace/*Thine* to everything created' (72–3; Maxwell's emphasis), and in the resolution to 'Lines' that 'Right should then at last be found/Joining what none may sever' (95–6). What begins to emerge in his early poems is the way in which Maxwell 'was strikingly *whole*' (author's emphasis), as Ivan Tolstoy suggests, and their intermingling of religion, morality and scientific study.[24] Maxwell's commitment to Christian natural theology, as a natural philosopher, was far from unusual for scientists in the 1840s and 1850s; indeed, the combination frequently determined scientific contexts, answers and questions.[25]

The poem 'Reflex Musings: Reflection from Various Surfaces' (1853) is a particularly engaging yet simple example of Maxwell's outlook. Essentially, it describes three successive experiments, each of which demonstrates different correspondences between different forms of matter and how they are perceived. What the poem also illustrates is the vital interactions between science and language, the visual and the imaginative. The ostensibly simple title 'Reflex Musings' is simultaneously oxymoronic and synonymous: the word 'reflex' has associations with the instinctive and immediate, particularly bodily or physical responses, while 'musings' are, by definition, meditative and constructed in the mind. The two words might be diametrically opposed but for the absence of firmness in the notion of 'musings', which might have been offered by an alternative word such as 'thought'. Instead, 'musings' conveys a sense of ideas not confined by rationality, while 'reflections' emanate from intuitive sources more akin to those of reflex. Maxwell uses two apparently different, opposed concepts as reflections of one another in the mirroring form of the poem: the ABABBABA rhyme scheme creates a reflection within each stanza, whereby each half mirrors the other around a centre, reinforcing the self-containment of each experience.[26]

The poem's opening envisages a relatively new phenomenon in the nineteenth century, that of the hurried and anonymous urban crowd. The shifting formations assume an ant-like quality of 'small designs' and 'restless feet', watched over by the speaker, like a scientific observer, from above.

> In the dense entangled street,
> Where the web of Trade is weaving,
> Forms unknown in crowds I meet
> Much of each and all believing
> Each his small designs achieving
> Hurries on with restless feet,
> While, through Fancy's power deceiving,
> Self in every form I greet.
> 'Reflex Musings', 1–7.

Instead of a conventional town–country opposition, the 'street' and the 'dell' portray the operation of similar processes; in the presentation of their 'various surfaces', it is parallels that dominate. The 'web of Trade', for example, meets its equivalence in the 'birches' shadow', whereby both are identifiable presences without actual substance. The true extent of a shadow cannot be distinguished properly from within, just as the grander scale of the economy, currency, and trade cannot be understood by viewing only the strands of individual transactions: fully comprehending either relies on its being conceptualized from a distance.

The reference to 'self', rather than '*my*self', dissipates the potential arrogance of any omniscient viewpoint; instead, the speaker is replicated in the common self of 'every form I greet'. The common, replicated self recurs in the second stanza, with the speaker's image alternately advancing and retreating in the well water's surface, just as it has passing the faces on the street. Maxwell explores the similarities of motion in fluids and optics, sound and space. In the second stanza, he writes

> Oft in yonder rocky dell
> Neath the birches' shadow seated,
> I have watched the darksome well,
> Where my stooping form, repeated,
> Now advanced and now retreated
> With the spring's alternate swell,
> Till destroyed before completed
> As the big drops grew and fell.
> 'Reflex Musings', 9–16.

The well can be read as a crude form of 'wave tank', the glass chamber commonly used to demonstrate the properties of waves in physics and engineering. The poem illustrates Maxwell's interest in both 'imaginary' fluids and in juxtapositions of surfaces and forces, such as those conceptualized just a year previously in Faraday's 'lines of force', which Maxwell's paper 'On Faraday's Lines of Force' (1855–86) would subsequently explore.[27]

On a further historical, literary, and philosophical scale, Maxwell's poem explores how different perspectives underline the mutable nature of the self, with the next stanza's sudden ascent to mountain heights reminiscent of Matthew Arnold's *Empedocles on Aetna* (1852):

> By the hollow mountain-side
> Questions strange I shout for ever,
> While the echoes far and wide
> Seem to mock my vain endeavour;
> Still I shout, for though they never
> Cast my borrowed voice aside,
> Words from empty words they sever –
> Words of Truth from words of Pride.
> 'Reflex Musings', 17–24.

In a Tennysonian 'hollow echo of my own' the speaker is humbled by his voice being 'borrowed', as it exposes the 'empty' futility of his 'words of Pride' and the 'vain endeavour' of explaining cosmology, the mechanics of time and space.[28] Religion surpasses science as he realizes 'words of Truth' beyond technical explanations, with the two concepts of divine Truth and sinful Pride opposed and capitalized to assert their authority.

The single, unqualified 'yes' (25) that launches the final stanza suggests the speaker's acceptance of a Truth more profound than that which he can offer, an answer to his own earlier 'questions strange'.[29]

> Yes, the faces in the crowd,
> And the wakened echoes, glancing
> From the mountains, rocky browed,
> And the lights in water dancing –
> Each, my wandering sense entrancing,
> Tells me back my thoughts aloud,
> All the joys of Truth enhancing
> Crushing all that makes me proud.
> 'Reflex Musings', 25–32.

Interestingly, the speaker refers repeatedly to himself rather than presenting a more universal man, whose 'wandering sense' (29) is too easily entranced. Empirical knowledge is presented as illusory and misleading in that each experience only 'tells me back my thoughts aloud' (30). The poem's conclusion asserts that the sole purpose of sensory experience is the explication of Christian 'Truth', alongside which there is no room for individual vanity or pride.

The poem traces a progression from science towards religion but is also Maxwell's translation of the *experience* of science into poetic form, encapsulating the use of empirical knowledge, experiments and self questioning. It is no coincidence either that Maxwell refers to 'forms' three times in the first two stanzas, in a poem of shadows, caves, and echoes. The poem likely references Plato's 'Myth of the Cave' in *The Republic*, where the world is but appearances and false shadows upon the wall of a cave, and only by leaving the cave and stepping into the light can the true world of forms be perceived. As a verbal expression of Maxwell's private 'reflection', the poem moves from the interior 'cave' of the speaker's mind into what is asserted to be the light of truer understanding about the nature of the world. Maxwell's 'Reflex Musings' seems to reject the validity of experimentation grounded solely in seventeenth-century Baconian empiricism, but it also refuses to reify nineteenth-century positivism. Neither are the seemingly Romantic sublime of pastoral and mountainous locations idealized; instead, they represent distinctly nineteenth-century 'laboratories', where the observance of phenomena leads to intellectual, creative, and spiritual illumination.

Mathematics, Fantasies and Women in Science

Just before graduating from Cambridge, Maxwell composed the poem that might be viewed as his most scientific, 'A Problem of Dynamics' (1854).[30] While it can be argued that 'scientific facts in literary texts need to be understood primarily as a rhetorical ploy' and that 'the literary context evacuates them of their content', 'A Problem of Dynamics' is distinctive because it goes beyond the metaphorical.[31] By using mathematical notations, Maxwell retains the scientific content within a literary context and engages with its workings, rather than giving only an appearance of science. The poem describes the various forces and tensions involved in pulling a horizontal chain from one end and how the curve of the chain changes shape. To effect this, Maxwell uses a poetic form but he also deploys authentic mathematical formulae, notation and terms to explore particular, and very real, scientific interests. Instead of the common subordination of the scientific to the poetic, he makes the science equal the poetry, a reassignment that signals his authorship as a practising scientist and the increasing status and authority of scientific endeavour at the time. While such a technical poem is doubtless enhanced by understanding the mathematical concepts to which it refers, a comprehensive explication of these would go beyond the scope of this study. Nonetheless, it is worthwhile considering the context of Maxwell's scientific work before discussing the poem as a framework for diverse modes of description.

In the mid nineteenth century, as J. T. Lloyd points out, 'times were becoming difficult for the old strain of physicist'.[32] Mathematical methods such as vector calculus needed to be developed in response to the demands of scientists. Accurate explanation of abstract concepts required more advanced descriptive instruments to progress beyond identifying and viewing events in nature, and to explore and prove how they worked. As David Gooding proposes, experimental and theoretical types of knowledge may be more similar than they appear and Maxwell's equations were certainly instrumental in moving science beyond its dependence on physical and visual modelling, to representing abstract concepts.[33] As its title suggests, 'A Problem of Dynamics' focuses on a particular problem; it is both relevant and necessary, therefore, to consider briefly what that is. As well as being a 'problem' in the sense of a puzzle to be solved, what is addressed is the modelling of a chain's movement. The problem's resolution lies in the 'chain' of interrelated equations and, in particular, the 'chain rule' that determines how to change the variable when differentiating, referred to by Maxwell's pun 'when the chain forms a circle' (43). The dilemma for literary analysis is that any discussion of the poem's content is hampered by the hurdle of verbalizing its content.[34] Mathematical notation and formulae have the capacity to communicate abstract information, but literary analysis relies on words, less precise devices that depend on shifting real world referents and associations. These issues make 'A Problem of Dynamics' stimulating for at least two reasons: first, it represents a stage in scientific and intellectual history when the information to be communicated had become so advanced and abstract that it almost defied verbal expression. Second, despite these obstacles, Maxwell still attempts to use the specifically *verbal* vehicle of poetry to communicate theories whose essence is virtually beyond words. The extent to which he succeeds is debatable and largely determined by the mathematical or literary capabilities of its readers.

It is exactly the complexity of Maxwell's subject that makes poetry such a suitable means of expression. Just as mathematical symbols are a method of shorthand, the condensed poetic form can signify the elusive nuances of matters ostensibly beyond its means. The consistency of poetic form and its brevity makes it a good vehicle for scientific explication; as Alan Rauch suggests, 'the very structure of poetry, which can be governed by "rules" that guide the practice of poetic form, is itself "scientific" and thus, like so many other genres of writing, is empathetic to the rule-governed practices of scientific enquiry'.[35] In just 64 lines of deceptively naïve, rhyming couplets, Maxwell executes a complex mathematical study that, using only words, would have required more lines in prose. More than this, the poetic form affords a level of musicality and imaginative possibility that might be absent otherwise. The tumbling cascade of tightly-packed syllables carries an incessant, mesmeric quality that shapes it into a scientific chant. In doing so, it draws together not just poetry and science but also the technical and artistic, the practical and enchanting.

Maxwell pre-empts the difficulties of scanning mathematical notation by using the rhythm as a built-in guide:

> Then $\frac{V_n}{V_t}$ the tangent will equal,
> Of the angle of starting worked out in the sequel.
> 'A Problem of Dynamics', 13–14.

To maintain the rhythm, 'V_n over V_t' has to be read exactly that way, despite the word 'over' not being physically present. Meanwhile, the struggle to maintain a consistent rhyme scheme is treated with humour; for example, in the lines 'Now joyfully leaving *ds* to itself, a-/Ttend to the values of T and of α' (47–8). Initially, the enjambment of the interrupted word 'attend' appears clumsy but, on the contrary, it creates a deliberate rhyme between '(it)self, a' and the symbol 'alpha', that ends the subsequent line (α). Maxwell's skill in rhyming is further evident in his expert matching of 'velocities' (10) with the phrase 'across it is'; 'the first thing of course is' (15) with 'effectual forces'; 'serve at your' (21) with 'curvature'; and 'to do it, he' (27) with 'Continuity' (28). On a thematic level, the poem's subject of motion and tension is emphasized

A PROBLEM IN DYNAMICS.

19th Feb. 1854.

AN inextensible heavy chain
Lies on a smooth horizontal plane,
An impulsive force is applied at A,
Required the initial motion of K.

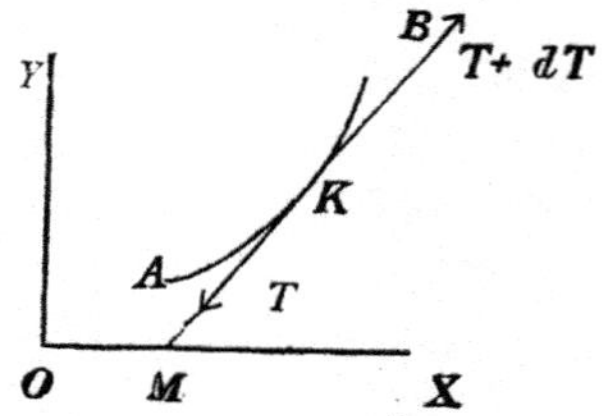

Let *ds* be the *infinitesimal* link,
Of which for the present we've only to think;
2 s

Fig. 13.2 A diagram to illustrate the poem 'A Problem in Dynamics', where Maxwell takes a problem relating to the impulsive force applied to a heavy chain and translates it into poetry (Campbell & Garnett, *The Life of James Clerk Maxwell*, Macmillan & Co. 1882, p. 625).

by the vigorous movement of 'tugged' (17), 'distorting convulsion' (l.48), and 'the force of impulsion to beat to a jelly' (29). As an extended mathematical joke, 'A Problem of Dynamics' reflects Maxwell's humour, even as its dextrous rhythms and patterning reveal his awareness of 'the old alliance between music and mathematics' (see Fig. 13.2).[36]

Despite the mathematical realities of 'A Problem of Dynamics', it seems from Maxwell's subsequent poem 'Recollections of Dreamland' (1856) that he was far from done with dream-like qualities and shifting perspectives that worried the boundaries of the sustained omniscient viewpoint. His maturing poetic skills allowed him to depict connections between scientific creativity, dreams and fantasy, and re-deploy the Romantic trope of the dreaming-genius poet. The night hours are less troubled that in his earlier poems and become a time when the 'inner world of dreams' illuminates the poet's 'outer world of action' (6). Again, internal and external worlds repeatedly switch places, imitating the disorientation of falling asleep.

> For the inner world grows wider as the outer disappears,
> And the soul, retiring inward, finds itself beyond the spheres.
> Then, to this unbroken sameness, some fantastic dream succeeds,
> Vague emotions rise and ripen into thoughts and words and deeds.
> Old impressions, long forgotten, range themselves in Time and Space,
> Till I recollect the features of some once familiar place.
> Then from valley into valley in my dreaming course I roam,
> Till the wanderings of my fancy end, where they began, at home.
> 'Recollections of Dreamland,' 9–16.

Sleep and dreaming become vehicles for travels through 'Time and Space', to the 'treasure caves' of 'departed days' (26). The dreamer's meetings with the dead assert their imaginary nature, depicting the experiences as fictitious time travel in ways that forestall their interpretation as spiritual, ghostly or supernatural. The depiction differs from 1850s scientific studies of dreaming; unlike William Carpenter's 'Somnambulist', who is 'without any controlling or directing power on the part of the mind', the speaker here is very much in control, not merely dreaming but lucid dreaming.[37] Instead of John Addington Symonds's 'dubious regions, dimly-lighted, phantom-peopled', Maxwell's speaker confidently asserts: 'I know the forms that meet me are but phantoms of the brain' (30).[38] Maxwell's dreaming speaker is also conscious, the stress on 'know' strengthening his emphatic determination to remain objective in the highly emotional encounters with memories of the dead.

The meetings are remembered positively, as ways to 'link the past and present into one continuous life' (38) and examine the nature of thought. He posits the possibility of the mind as a liquid medium, where dreams are 'empty bubbles, floating upwards through the current of the mind?' (47) The image seems to reference Thomas Nashe's dismissal of the physiological basis of dreams in *Terrors of the Night* (1594): 'nothing else but the bubbling scum or froth of the fancy, which the day hath left undigested'.[39] It is more likely that they also reflect Maxwell's research on electricity at the time, in which he refers to heat as 'an imaginary fluid', thin tubes of continuous, incompressible liquid that travels along routes akin to Faraday's 'lines of force'. The adaptation of the physical model of fluidity to describe dreaming reflects the exchange of ideas between Maxwell's science and his poetry, and illustrates his earlier reference to the 'guiding radiance' (5) of dreams gleaming upon his 'outer world of action' (6).

Maxwell's metaphors retain the idea of electrical force and light, so that dreams represent unrecognized inner potential, precious parcels that exist separately from the waking conscious.

> There are powers and thoughts within us, that we know not, till they rise
> Through the stream of conscious action from where Self in secret lies.
> But when Will and Sense are silent, by the thoughts that come and go,
> We may trace the rocks and eddies in the hidden depths below.
> 'Recollections of Dreamland', 48–51.

Maxwell extends the metaphor of liquidity to depict dreams as rising from a deeper, river-like self, through which it connects to the earlier pastoral scenes. The mysteries of what actually constitutes the self are not pursued; instead, they are relinquished with the abrupt declaration that, 'what is ours we know not, either when we wake or when we sleep' (44). Maxwell's reticence about unearthing the self, nature, or the universe pronounces, like Carlyle's Scandinavian, that 'the thing is larger than man, not to be comprehended by him; an Infinite thing!'[40] It is a feature that becomes more pronounced in Maxwell's later poems, reflecting the conservative strain of his Protestant faith, and its reliance upon biblical and divine, rather than human, explanations.

In Daniel Brown's investigation of science in Victorian poetry, he notes Tennyson's proposition that 'science and poetry "feel" Nature in different ways; they have different "dreams" of Nature'.[41] If that was so, Maxwell's poem about dreaming combines the feelings of both – refusing to mystify Nature, yet also defying propositions of scientific objectivity with a speaker who is much more than an industrious bystander.[42] The recurrence of dreams, imagination and perception in Maxwell's poems is as personal as it is subjective, and engages with the period's growing fascination with dreaming and the nature of consciousness. Brown explains the ramifications of this link: 'the dream – a state of consciousness in which products of the imagination are experienced with the conviction [that] they are in fact actual sense-impressions – critiques the positivist faith in sense-data.'[43] While Maxwell's poems portray imaginative experience and dreams as sense-impressions, his speakers repeatedly reject the 'truth' or reliability of sensory perceptions, empiricism and positivism. For Maxwell, 'Truth' is ultimately divine and beyond human comprehension. Even in 'Recollections of Dreamland', where an overt religious conclusion is absent, imagination and dreams are kept entirely distinct from waking, sensory experience, and yet they are also valuable products of memory as well as future abilities.

Maxwell's engagement with different forms of representation, perception and perspective are illustrated further by the importance of musicality in verses such as 'To the Air of "Lorelei"' (1858) and the operatic origins of 'Tune, *Il Segreto per esser felice*' (1858).[44] However, new poetic influences are also evident in both these poems and others from the same year, such as 'I've Heard the Rushing', 'Will you come along with me?', 'Why, when our sun shines clearest?' and '(To his Wife)'.[45] Maxwell and Katherine Mary Dewar were engaged in February 1858 and married on 2 June 1858 so that, as Campbell and Garnett propose, his writings 'contain the record of feelings' relating to his married life.[46] As might be expected, the witty academic word-play of 'A Problem of Dynamics' is replaced by a newly personal and emotional quality. The dream-like inspiration of how Maxwell composed 'To the Air of "Lorelei"' is decidedly Coleridgean; as he writes:

> Last week I was walking with the tune of the "Lorelei" running through my head, and it set itself into a "kind of allegory". The words are very crude, but that is the way they came together. I profess myself responsible for most things I say, but less for this than for most.[47]

The inspired nature of the poem's origin is duplicated in the title reference to 'Lorelei', the name of a beautiful, legendary siren whose singing lured navigators on the Rhine to their doom. At the time, Maxwell was a suitor on the brink of proposing marriage and the poem reflects his awareness of women's captivating capacities. The opening locates the speaker in an idyllic, pastoral setting:

> Alone on a hillside of heather,
> I lay with dark thoughts in my mind,
> In the midst of the beautiful weather
> I was deaf, I was dumb, I was blind.
> 'To the Air of "Lorelei"', 1–4.

Maxwell is known to have read Robert Browning, and the fourth line brings to mind the 'water-beetle with great blind deaf face' of Browning's poem 'Sibrandus Schafnaburgensis' (Browning, 1844: 51–3). However, the context of nature's profusion is quite different for the two writers. As Pauline Fletcher suggests, Browning is 'celebrating the wonderful fecundity and energy of a natural world in which . . . God may be either present or absent'.[48] Maxwell's speaker is saved from such waking oblivion:

> I knew not the glories around me,
> I counted the world as it seems,
> Till a spirit of melody found me,
> And taught me in visions and dreams.
> 'To the Air of "Lorelei"', 5–8.

Music is portrayed as the redemptive 'spirit of melody' that rescues the speaker from 'wilful confusion' (15). Once he hears the natural world 'mingling its notes with our own' (17–18), he is able to rejoice in it with an excitement reminiscent of Andrew Marvell's pilgrims finding 'a temple where to sound His name' ('Bermuda', 32).[49] Nowhere is Maxwell's joy more evident than in the convergence of poetry and music, nature and divinity.

'Will You Come Along With Me?' is similarly joyous, with rhythms that imitate a foot-stamping *ceilidh* or country dance.[50] Its folkloric character is reinforced by the Scottish colloquialisms of 'bonny', 'braes' and 'ain', conveying Maxwell's affection and excitement about his home country, just as his known fondness for the poetry of Robert Burns indicated his appreciation of its rustic traditions. The natural world is figured solely in terms of what it offers the poem's speaker. The themes of harmony and divine influence remerge again in 'Tune', which dates from immediately after his engagement. Its operatic subtitle, 'Il Segreto per esser felice', translates as 'the secret of happiness'; however, its tone differs sharply from the carefree exhilaration of 'Will You Come Along With Me?' Instead, it focuses upon the tensions of the Aristotelian 'good life', which are neatly summarized by Mark Young's question, 'if my highest aim is to seek only to further my own well being, what is to prevent me from doing that at the expense of my neighbour?'[51] As such, the poem begins with warnings about thinking too much and about becoming overly attached to other people: 'Do not love them, nor hate them, nor care for their fate / But keep a look out for your own'; it recommends, instead, to 'eat your dinner in quiet, / And live like a sensible man' (9–14). However, such solitary tranquillity is presented as stultifying:

> While we boast of our tranquil enjoyments,
> The means of enjoyment are flown,
> Both our joys and our pains, till there's nothing remains,
> But the tranquil repose of a stone.
> 'Tune, *Il Segreto*', 21–4.

Perhaps surprisingly, the speaker concludes that the pursuit of individual happiness is an unnatural impulse:

> In Nature I read
> Quite a different creed,
> There everything lives in the rest;
> Each feels the same force,
> As it moves in its course,
> And all by one blessing are blest.
> 'Tune, *Il Segreto*', 29–34.

What these poems illustrate is Maxwell's continued fascination with interconnectedness and the place of the individual in relation to the whole. His letters, like his poetry, reveal that attraction too, in seamlessly incorporating the language and imagery of science. When he expresses interest in the protagonist of Max Müller's *Deutsche Liebe*, for example, he describes him as someone who is

> conscious of the shell that surrounds him and divides him from others, and able neither to live in it nor to break it. But the shell is only the outer surface of a minute drop of fluid, impenetrable because minute, and arising from molecular forces in himself.[52]

Despite discussing a fictional character's sense of physical boundaries, Maxwell's description is that of a scientist peering down a microscope. He examines beyond the 'shell' of appearances, to scrutinize the 'minute' and invisible world of molecular forces. At the same time, he merges conchology with chemistry, and molecular physics with biology. Later in the same letter, he analyses social relations in a similar way, discussing the human 'transfusive tendency' and the 'fusion' of marriage and friendships.[53] In doing so, he exhibits the 'uninterrupted chain of deduction from the great principle that governs the universe', such as that envisioned by Mary Somerville in *On the Connexion of the Physical Sciences* (1834).[54]

The amalgamation into poetry of Maxwell's perceptions of the natural world, alongside both science and philosophy, is further evident in 'I've Heard the Rushing' (1858). Despite its pastoral, waterside setting, it has none of the immediate merriment and optimism of his earlier poems; instead, it illustrates David Forfar's comment that, in Maxwell's poetry, 'the charm on the surface often concealed a quite contrary meaning for those able to see beneath'.[55] The poem opens with the deceptive vitality and grandeur of ferocious 'mountain torrents', but these quickly dissipate, 'lost in the desert' (5) with 'high resolves growing weaker every day' (7). The speaker's dejection reaches its height in the following lines:

> . . . cares, springing round me,
> With creeping tendrils bound me,
> And all I once hoped for was wearing fast away.
> 'I've Heard the Rushing', 8–10.

The poem is one of Maxwell's saddest, conveying as it does the entrapment of a man, 'bound' by 'creeping' yet unspecified 'cares'. Although it declares that 'all now is righted' by the union of 'our souls' (18–19), the years to come are portrayed in terms of the 'sorrows' they entail (20). C. F. W. Everitt suggests that the Maxwells' marriage 'was not without its strains' and that 'Mrs Maxwell was jealous and difficult'; however, it is not possible to deduce from the poem itself precisely what the source of the speaker's disappointment might have been.[56] Both 'I've Heard the Rushing' and 'Why, When Our Sun Shines Clearest' are dated only for the year of 1858, leaving it unclear at what point in the year they were composed and whether it was before or after June, when the couple were married. What is evident, though, is that the sorrow of the first poem barely touches the utter desolation of the second, which opens with persistent questioning.

> Why, when our sun shines clearest,
> Why, when our hopes seem nearest,
> Why, when our life feels dearest,
> Rises a secret pain—
> 'Why, When Our Sun Shines Clearest', 1–4.

Against a happy backdrop is a cry against tragedy, with happiness marred by a catastrophe that results in 'Hope's perfect mirror broken–' (6); yet the source of the speaker's distress remains indistinct. Instead, the poem represents the physical experience of emotional pain, in the succession of interrupted lines that emulate attempts to articulate a suffering that remains repeatedly incomplete. Again, Maxwell's focus is boundaries and exchanges – this time between the physical, the emotional, and the verbal. While the central mystery around which the poem rotates is never revealed, the spiritual resolves all human conflicts of emotion and thought:

> Nothing of ours is pure.
> Still must such thoughts upbraid us,
> Seeking our own to aid us;
> God, not ourselves, hath made us;
> Trusting in Him we're sure.
> 'Why, When Our Sun Shines Clearest', 12–16.

By not specifying the cause of the poem's anguish, 'Why, When Our Sun Shines Clearest' offers a curious blend of modern narrative and traditional ethos. Identifying the source of the 'pain' at the poem's centre is left to the reader, rather than delineated by its author, while religion is offered as a panacea that no difficulty is beyond the reach of. The verses portray hope and fear, happiness and disappointment, to represent the constantly changing peaks and troughs of man's universal experience. Their intimacy and vulnerability is not something which occurs so clearly again, especially in his subsequent poems about women.

Women tend to feature in Maxwell's early poems as fantastic or grotesque figures like the 'Hag'. His perceptions of women in the world of science are apparent particularly in 'Lectures to Women on Physical Science I' (undated) and 'Lectures to Women on Physical Science, II' (July, 1874).[57] The poems reflect Maxwell's attention to the rotational kinetics of molecules in the 1870s; however, they also convey the era's contention that women were considered unsuited to the rigours of academic study and their education not just unnecessary but, as the 1868 article 'A College for Women' in *John Bull* suggests, detrimental to society:

> She cannot equal man on this ground, but in her own sphere – so difficult to define – so easily felt – her loss will be irreparable. Is it needful? Cannot we look back to the old models of womanly virtues, instead of forwards to a sort of *she*-man?[58]

While women were present at Cambridge, they were not permitted to attend lectures at the Cavendish Laboratory during Maxwell's time as Cavendish Professor of Physics, partly due to his ruling that all 'students should be male'.[59] As Paula Gould cautions, however, 'we should be wary of removing the Cavendish from a more general context of late Victorian Cambridge' where 'Maxwell could almost be regarded as progressive in allowing female footsteps across the threshold at all'.[60] The issue of women's education is forefront in the two 'Lectures' poems, the first of which deals with female learning, the second with female teaching. Neither fares well under Maxwell's hand. Several critical studies have focused on the history of women's education from the female perspective; what these poems represent is the male perspective, the dismissal with which women had to contend, and the defensive derision their participation prompted.[61]

Set in 'a small alcove with dark curtains' (prelude), 'Lectures to Women on Physical Science I' is located in a deceptively cosy space. The opening proceeds to contrast women's new appearance in the scientific laboratory with their traditional role in poetry, in a parody of Tennyson's 'the splendour falls on castle walls' (1):[62]

> The lamp-light falls on blackened walls,
> And streams through narrow perforations,
> The long beam trails o'er pasteboard scales,
> With slow-decaying oscillations.
> 'Lectures, I', 1–4.

The lone female student in the laboratory appears as confined as a dungeon with prison bars. However, cutting across the gloom like an incantation, is each stanza's final couplet, with its incitement to 'Flow, flow . . .', then to 'Swing, swing . . .', and finally to 'Break, break . . .' Rather than being simply an experiment in electricity, it becomes a hypnotic spell that conjures up the immediacy and speed of the forces of nature, in a 'flashing, quivering, and dying' (6) that makes them seem organic and alive. The dramatic excitement is built up further by the sudden interruption of 'O look! How queer!' (7) What is portrayed is not the cool objectivity of scientific investigation; rather, it is electricity made sensational, a thrilling entertainment of 'gliding fire!' (9) This is scientific experimentation as Maxwell thinks it is seen through a woman's eyes; however, a dismissive man's voice reasserts its dominance in the final stanza, halting the pathetically misguided female effort to understand science.

O love! You fail to read the scale
Correct to tenths of a division.
To mirror heaven those eyes were given,
And not for methods of precision.
'Lectures, I', 13–16.

Maxwell's speaker articulates, albeit light-heartedly, many contemporary views of women in science, views that, as Gould observes, 'raised many fundamental questions, including the nature of intellectual identity and the physical suitability of women for study'.[63] The poem goes further than that, too, by suggesting that the female student in question is physically incapable of precision, let alone the composure needed for scientific objectivity.

The other poem of the pair, 'Lectures to Women on Physical Science, II', provides evidence of Maxwell's interest in how social boundaries were changing around him.[64] The title mixes the factual with the elusive, in seeming to connect with the first poem but withholding any further information. Gender is a primary concern but the poem's key figure, the farcically-named 'Professor Chrschtschonovitsch, Ph.D.', remains without gender throughout, not least because the adjective 'prim' with which it opens can refer to either sex (OED). The 'women' to whom the lecture is addressed remain entirely silent throughout and, although a lecturer and a student are said to be present, at various points, it is unclear which one is speaking. Ambiguities such as these are marked and continuous enough to suggest they are deliberate, as a way representing the poem's turmoil of identities, genders and roles, and mirroring the disordered scientific principles about which the characters argue. The speaker decries work that is 'stiff with scientific shoddy' (26) and barely fit for a music lesson, let alone the precision required of science, expressing objections similar to those against 'second-hand' learning, in the earlier poem 'Vision' (70). Lack of precision was anathema to Maxwell and, in his poems at least, it is closely related to women's attempts at scientific endeavour (see Fig. 13.3).

The poem appears to be addressed to 'Professor Chrschtschonovitsch, Ph.D.', who can be traced, I propose, to the Russian mathematician Sofia Kovalevskaya (1850–91), one of the first

Fig. 13.3 Sofia Kovalevskaya – perhaps the real Professor Chrschtschonovitsch, Ph.D (*The Illustrated London News*, 1884, 85, p. 121).

women to gain a Ph.D. and the first woman student at Heidelberg University, which is referred to in the opening lines: 'Prim Doctor of Philosophy / From academic Heidelberg!'[65] The poem was written in the same month and year that Kovalevskaya gained her doctorate from Göttingen, where Maxwell was elected to the academy the following year; their research topics also overlapped so it is likely that they met at some point, or he at least knew of her.[66] The originality of Kovalevskaya's doctoral thesis was sharply questioned, making her even more probable as the target of Maxwell's derision.[67] As the Russian mathematicians Egorov and Yubin suggest, Kovalevskaya's key theorem only 'applies to equations of a very general form and is widely used in the modern general theory of partial differential equations'.[68] Despite the cumulative evidence that the poem refers to Kovalevskaya, Maxwell never makes explicit his target's identity and thus allows the figure to personify a broader resistance to what he considered the degenerating scientific and educational standards of his time. The depiction of physical violence towards a possibly female target uses the scientific language of forces; thus we have ostensibly light-hearted lines such as 'You seized my hand – you gave me pain, / By torsion of a wrist so supple' (33–4), and 'Were every hair of every tress /. . . (Which you, no doubt, imagine mine), / Drawn towards you with its breaking stress–' (36–8). We also have, albeit in playful format, an extreme and highly physical expression of Maxwell's frustrations about women and their presence in the sciences.

The locations Maxwell explores in his poems in this period could hardly be more different, ranging from the worlds of abstract mathematics, inspirational dreams, and the Scottish countryside, to the claustrophobic interiors of the scientific laboratory, domestic life, and the classroom. While they are often localized, they compare to the equally diverse and compelling emotional states that the poems express, from reflective meditations, playful delight, and extremes of sadness, to derision and irritation. Whether he was writing about mathematics, marriage or method, his poetry was invariably thoughtful and passionate. The combination of the local and the diverse indicates Maxwell's determination to be his own man, with views not bound by conventions, a valuable insight that remains largely absent from investigations of his scientific practice or his religion. His subsequent poems demonstrate the immense significance of that characteristic, in terms of how he saw his role within the scientific community and its public presentation.

Sensation, Passion and Evolution

In the final decade of his life, Maxwell's poems engaged more directly with the people and the wider concerns of the scientific community. By this time, he was an established authority, in receipt of honorary doctorates from the universities of Edinburgh, Oxford and Pavia, and a member of academies as far-flung as Amsterdam, Göttingen, Vienna, Boston, Philadelphia and New York.[69] His illustrious position as an established figure in Victorian science might make unexpected the dramatic and sometimes sensual tone of 'To the Chief Musician upon Nabla: a Tyndallic Ode' (1871), but it is explained by the context of relationships Maxwell had with some of the day's leading scientists.[70] In keeping with the 'fugitive' nature of Maxwell's poetry mentioned earlier, the 'Tyndallic Ode', as it is commonly known, was published anonymously in the journal *Nature* in 1871. The poem's spectacular first-person opening is surely one of Maxwell's most powerful:

I come from fields of fractured ice,
Whose wounds are cured by squeezing,
Melting they cool, but in a trice,

Get warm again by freezing.
Here, in the frosty air, the sprays
With fern-like hoar-frost bristle,
There, liquid stars and their watery rays
Shoot through the solid crystal.
'A Tyndallic Ode', 1–8.

The simple rhymes disguise the conceptual and verbal expertise of the poem's sibilant oppositions, where hot meets cold, constraint holds back against explosion, and 'fern-like' frailty (6) is set against the harshness of 'bristle' (6). Initially, it gives an appearance of fantasy, but the poem also relates to specific contemporary figures. As Campbell and Garnett explain in a footnote, the 'Chief Musician' to whom the poem is addressed is Peter Guthrie Tait (1831–1901), Maxwell's fellow physicist and friend, who was the first person to discuss fully the mathematical 'nabla' symbol.[71] Conspicuously not footnoted is the poem's allusion to John Tyndall (1820–93), with whom Maxwell found himself increasingly at odds, in terms of theoretical physics, religious belief and personally. Gowan Dawson has drawn attention to the widespread disapproval of Tyndall's 'unseemly hankering for the opposite sex'.[72] Tyndall's reputation for having several girlfriends before he married in 1876 prompts Dawson to read the passionate and sensual tone of 'A Tyndallic Ode' as both a parody of Tyndall's 'proclivity for breaking out into excitably effusive and ardent scientific language' and an attack upon this other, more personal propensity.[73] Regardless of Maxwell's motivations at the time of writing, however, the extraordinary verbal sensuality of the poem certainly defies any notion that Maxwell lacked poetic skill.

The second stanza proceeds to envelop even greater contrasts, of time and scale:

I come from empyrean fires—
From microscopic spaces,
Where molecules with fierce desires,
Shiver in hot embraces.
'A Tyndallic Ode', 9–12.

Ranging from ancient Greece to personified minutiae, Nature is embodied with 'wounds' that open and close erotically in a veritable orgy of the atomic, electrical, and chemical. The ode's frenzied chaos of flashing lights and lurid colour subsides abruptly in the slow, controlled movement in the third stanza, when 'our eye' (17) is placed to a lens, drawing the reader into voyeuristic collusion. The experiment with electricity and sound waves travelling through a vacuum is translated by cloaks, crystal, and 'dark rays' (17) into an act of sinister magic, drawing on a dangerous 'source of power' (20) to conjure up the 'air-drawn form' of a ghostlike 'mystic fish' (31). It is an exuberant, almost bacchanalian performance, where the boundaries between man and chemical matter dissolve into a dance of synchronicity, accompanied by a siren-like flame, which 'sweetly sings' (35). The screech of an actual siren signifies not only modern industrial technology but perhaps also Maxwell's impatience with the superstitious paraphernalia of the dramatized Faustian contract. Re-asserting his own agenda, he draws the scientist back to his real purpose where 'grasp of mind' is more important than 'fancy scientific' (53–4) and truth requires 'thought' as well as 'sights and sounds' (55–6). The shift suggests Maxwell's earnest commitment to the scientist's vocation as systematically observant and rigorous, rather than simply entertained and enchanted.

The disappearance of the observer's decorum and distance was abhorrent to Maxwell, as was the degeneration of science into spectacle or superstition. He saw Tyndall's florid sensationalism as compromising scientific integrity, and entering what Simon Schaffer describes as a 'sanctuary of minuteness and of power where molecules obey the laws of their existence, clash together in fierce collision, or grapple in yet more fierce embrace, building up in secret the forms of visible things'.[74] Again, it is the boundary between what can be seen and what is really going on that intrigues Maxwell. Yet the poem still defends his traditionalist perspective, as did his public work during this period, attempting to counter 'the materialist implications of Tyndall's molecular physics and Huxley's evolutionism', as a stand 'against the derivation of intelligence from molecular motions alone'.[75] For Maxwell, what was central was harmony, grounded in natural theology, both as a poetic theme and as a reflection of the divine at work in the world.

'A Tyndallic Ode' demonstrates how, in these last stages of his life, Maxwell saw what Schaffer refers to as 'molecular uniformity as a sign of divine purpose and free will'.[76] The poem's final stanza highlights this fundamental opposition between Maxwell and his opponent.

Go to! prepare your mental bricks,
Fetch them from every quarter,
Firm on the sand your basement fix
With best sensation mortar.
'A Tyndallic Ode', 57–60.

For Maxwell, the construction of Tyndall's work is flawed not just because it is based on unproven 'sensation' theories, but also because of it has no proper foundation in religious belief. Tyndall's 'wall' is prone to collapse, on the basis of Jesus's warning that 'everyone that heareth these sayings of mine, and doeth them not, shall be likened unto a foolish man, which built his house upon the sand' (Matthew 7:26). The poem reacts to the 'picture which science draws of the world' being increasingly coloured by evolutionary theory and atomistic materialism, in the light of Darwin's *Origin of Species* (1859).[77] The tensions between religion and science as 'the two mightiest antagonists' became a 'battleground', wherein there was no possibility of either one accommodating the other.[78] Tyndall and Maxwell became embroiled, whether they liked it or not, especially after Maxwell publicly asserted the incompatibility of such thinking with his own molecular theories, in his 1873 Bradford address 'Discourse on Molecules':

No theory of evolution can be formed to account for the similarity of molecules, for evolution necessarily implies continuous change, and the molecule is incapable of growth or decay, of generation or destruction.[79]

Tyndall responded to Maxwell's proposition by defending the new theories in his 'Belfast Address' (1874), which was published in *The Times*. As well as attacking Roman Catholic dogma for its negativity towards scientific progress, Tyndall questioned 'the legitimacy of Maxwell's logic', comparing it to the primitive belief of Pierre Gassendi (1592–1655) that 'God, who created earth and water, plants and animals, produced in the first place a definite number of atoms, which constituted the seed of all things'. He mocked as naïve Maxwell's contention that atoms 'remain unbroken and unworn', and even more so his conviction that this constituted proof of the divine, saying:

In his 'manufactured articles,' as he calls the atoms, Professor Maxwell finds the basis of an induction which enables him to scale philosophic heights considered inaccessible by Kant, and to take the logical step from the atoms to their Maker.[80]

As if to emphasize the contrast between his own character and Tyndall's, Maxwell did not respond publicly but privately, with the poem 'British Association 1874 – Notes on the President's Address' (1874).[81] His restraint is replicated in the uniformity of the poem's rhyming couplets, but his resentment is clear nonetheless. He ridicules Tyndall's zeal for 'nothing but atoms and void' (8) and castigates his assumptions about creation and the beginning of time:

> So treading a path all untrod, the poet-philosopher sings
> Of the seeds of the mighty world – the first-beginnings of things;
> How freely he scatters his atoms before the beginning of years;
> How he clothes them with force as a garment, those small incompressible spheres!
> 'British Association, 1874', 14–16.

Maxwell dismisses Tyndall as a melodic but self-indulgent 'poet-philosopher' rather than a true scientist, in assuming ownership of atoms as 'his' and casting them about capriciously. To Maxwell, Tyndall's assumptions about the world's creation treat science as a game, like the chessboard described by Huxley where 'the rules of the game are what we call the laws of Nature'.[82] In clothing atoms in his own theories, Tyndall is depicted as a clever but misguided tailor, stitching together an external appearance for the material world but thereby masking its divine nature.

Maxwell suggests that the evolutionist narrative is based on flawed and opportunistic intuitionism:

> Thus the pure elementary atom, the unit of mass and of thought,
> By force of mere juxtaposition to life and sensation is brought;
> So, down through untold generations, transmission of structureless germs
> Enables our race to inherit the thoughts of beasts, fishes, and worms.
> 'British Association 1874', 31–8.

Maxwell objects to Tyndall bringing the atom to life by 'the force of mere juxtaposition' because it reduces man to a temporary and incidental vehicle, rather than a divinely designed, purposeful creation. For him, the sole agency of change was God; nature was not capable of changing itself, a view resembling that of William Paley and James Paxton at the turn of the century.[83] As physicist Philip Marston highlights, by the 1870s Maxwell's commitment to biblical commentary tended to be shared not by scientists but instead by theologians such as fellow Eranus Club members Brooke F. Westcott and J. B. Lightfoot.[84]

Although Tyndall repeatedly proclaims his own commitment to Christian beliefs, he also expounds Epicurus's contention that 'as long as we are, death is not; and when death is, we are not' and the idea that 'the acceptance of mere authority led, as it always does in science, to intellectual death'– assertions that implicitly deny both the overarching authority of a Christian God and the doctrine of resurrection. Maxwell's poem accentuates Tyndall's idolatrous egotism and his alignment with pre-Christian and atheist thinkers; 'First, then, let us honour the atom, so lively, so wise, and so small; / The atomists next let us praise, Epicurus, Lucretius, and all' (37–8). Maxwell resists arguing explicitly for the role of the divine in the universe, and the poem's final reference to the futility of 'endless discussions' (45) suggests a battle-weariness with secular opinion. Although Peter Bowler reminds us that 'most Victorian physicists did not share the materialistic opinions articulated in Tyndall's notorious Belfast address of 1874', Maxwell's continued deep faith meant that there was ever less common ground between him and the scientific community of the day.[85]

Engaging further with the scientific community's debates over science and religion, and written in Belfast during the same year, is the poem 'Molecular Evolution'.[86] The analytical approach adopted by David Strauss's *The Life of Jesus Critically Examined* (1846), Ludwig Feuerbach's *The Essence of Christianity* (1854) and Ernest Rénan's *The Life of Jesus* (1863) indicates the way in which, as Jane Irwin has pointed out, science was increasingly 'competing in areas conventionally thought to be the preserve of religion'.[87] Maxwell refers to evolution not merely as fundamental to the century's topical developments, but the role of Christianity in scientific study; as a result, Lewis Feuer reports, he was 'angered that his own students repudiated the Scriptural account of the Creation and Geology' and 'angered to the highest pitch' by their greater faith in Darwinism than the Bible.[88]

The poem targets the 'evolution' of British Association members from their formerly wise, yet humble and self-deprecating gravity as 'British Asses' (11) to the newly ferocious 'Red Lions' (12), who set upon scientific discovery with reckless fervour. In classical style, Maxwell sarcastically hails the power of sage 'nonsense' to undermine all 'towers of Truth' (25) and derides the acceptance of Tyndall's atomism and Huxley's scientific naturalism as the end of reason:

> Yield, then, ye rules of rigid reason!
> Dissolve, thou too, too solid sense!
> Melt into nonsense for a season,
> Then in some nobler form condense.
> Soon, all too soon, the chilly morning,
> This flow of soul will crystallize,
> Then those who Nonsense now are scorning,
> May learn, too late, where wisdom lies.
> 'British Association, 1874', 26–34.

With the deft assurance of a mature and skilled poet, Maxwell follows the chaotic collapse of reason from the dawn of death to the Day of Judgement, in a sober warning to his over-zealous fellow scientists. The destiny of materialist theories is portrayed in terms of a laboratory experiment, with 'sense' as a substance melting into nonsense and purifying intellectual heat into the 'nobler form' of spiritual awareness.

The four brief stanzas reveal the hidden hurt and righteous anger Maxwell felt about the rejection of biblical explanations by scientists and intellectuals. His concern about secularization extended to its effect on the next generation of scientists. The 'Song of the Cub' (1874), for instance, depicts the repercussions of flawed mentoring upon a young aspiring scientist.[89] Maxwell traces the trajectory of his speaker's enthusiastic yet mistaken pursuit of empirical science, towards its ultimate disillusionment:

> O where are those high feasts of Science?
> O where are those words of the wise?
> I hear but the roar of Red Lions,
> I eat what their Jackal supplies.
> I meant to be so scientific,
> But science seems turned into fun [. . .]
> 'Song of the Cub', 17–22.

The blame for these dashed hopes is laid squarely upon the 'Jackal' Tyndall. The label characterizes him as a carnivorous scavenger feeding off others' intellectual spoils, against the biblical assertion that non-believers would become food for the jackals (Psalm 63: 9–11). It is also possible that Maxwell refers to Thomas Huxley's contention that 'small dogs are descended from jackals, and the big dogs ditto wolves', as evidence of evolution.[90] Although Huxley's research was not published till 1880, after Maxwell's poem was composed; he had, however, spent March 1872 observing and sketching jackals in Egypt. It is not unlikely, therefore, that Maxwell had heard of his work well before writing the poem and uses the evolutionists' own theories against its key proponents.

Maxwell's strife with fellow scientists continues further in the final poem in Campbell and Garnett's collection, 'A Paradoxical Ode' (1878), the best known and, arguably, Maxwell's most poetically accomplished work. It is addressed to 'Hermann Stoffkraft, Ph.D., the Hero of a recent work called "Paradoxical Philosophy"', sequel to *The Unseen Universe* (1875) by physicists Balfour Stewart and Peter Guthrie Tait, in which the fictional Stoffkraft is converted to their way of thinking. A further subtitle notes the poem as 'After Shelley', in reference to Percy Bysshe Shelley's representation of Faustian ambition in *Prometheus Unbound*.[91] It opens with the memorable lines: 'My soul is an entangled knot, / Upon a liquid vortex wrought' (1–2), with Campbell and Garnett's footnote offering the alternative to 'entangled' of 'amphicheiral'.[92] One of the paradoxes of 'A Paradoxical Ode' is the different tones in which it can be read. Brian Stapleford, for example, sees it as 'a poetic celebration of the perversities of emergent analytical science' while Gillian Beer understands it to be 'Maxwell's sardonic survey of his contemporaries' (unsuccessful) attempts to dislodge the soul from matter and energy'.[93] Beer's interpretation is supported, particularly, by the line indicating that Klein and Clifford only '*think* the infinite is now at last destroyed' (13).[94] Maxwell goes on to condemn evolutionism as a paltry culmination of man's efforts to explain his existence:

> But when thy Science lifts her pinions
> In Speculation's wild dominions,
> We treasure every dictum thou emittest,
> While down the stream of Evolution
> We drift, expecting no solution
> But that of the survival of the fittest.
> 'A Paradoxical Ode', 14–19.

The metaphor of the new, speculative science as a winged creature implies its lofty, yet misguided distortion into mere flights of fancy and, again, Maxwell turns its proponents' own ideas against them by incorporating Herbert Spencer's legendary phrase.[95] The poem epitomizes Maxwell's protest that the speculations of evolutionism, materialism, and spiritualism went beyond competencies and scientific proofs that could be rightfully claimed. As he states in 'Discourse on Molecules'

> Science is incompetent to reason upon the creation of matter itself out of nothing. We have reached the utmost limits of our thinking faculties when we have admitted that because matter cannot be eternal and self-existent it must have been created.[96]

Beer says of 'A Paradoxical Ode' that it 'has a semantic vigour and precision not unlike Hopkins' own' but that it is 'metrically more conservative'.[97] The replicated form of each stanza might be viewed as conservative; however, there is still considerable complexity and innovation in the poem's structure, in its unusual 13-line stanzas and the refusal of the rhyme pattern to conform to any traditional scheme. Rather than in terms of form, however, where Maxwell stands out is in his rejection of Balfour and Tait's theories about 'aether' (23, 25) to which, as Peter Bowler explains, several leading scientists were acquiescing at the time: 'the ether was a tenuous material spread through the whole of space, capable of linking and energizing material structures. Lodge also followed William Crookes in open support of spiritualism, firmly linking his views into a philosophy designed to challenge scientific naturalism and reinstate a role for religious belief.'[98] Maxwell saw these connections as logically and technically flawed, and questioned them in his essays, for example, in 'Psychophysik' of 5 February 1878, when he writes

> Let us suppose that a thinking man is built up of a number of thinking atoms. Have the thoughts of the man any relation to the thoughts of atoms or of one or more of them?[99]

Stewart and Tait's *Paradoxical Philosophy* encapsulates both the parallels and the divergences between Maxwell and certain colleagues. Maxwell, guided in all things by biblical teachings, asserted the harmony of science and scripture, in accord with the 1865 'Scientist's Declaration on Science and Scripture', a tract claiming that 'apparent conflicts arise from misinterpretations of Scripture and from presumptuous certainty about one's own scientific conclusions', which was circulated widely by a small, anonymous group of scientists.[100] Stewart and Tait, on the other hand, proposed that the visible universe was only a portion of the whole; its energy did not dissipate and this was, they proposed, the basis of immortality. Their writings about the after-life gained a tremendous following, laying the foundations for later fascination with the psychic, occult and spiritualist. Despite writing against Tyndall's materialistic determinism, Stewart and Balfour's concept of what Maxwell encapsulates as 'Unending Continuity!' (28) was at odds with Maxwell's beliefs. Maxwell believed in the total division between the material and spiritual worlds, the first being transitory, the other eternal and spiritual. With Maxwell's recurrent interest in the nature of boundaries, it was a one-way relationship; indeed, in an 1862 letter to Rev. Lewis Campbell, he proposes that

> There is action and reaction between body and soul, but it is not of a kind in which energy passes from the one to the other,— as when a man pulls a trigger it is the gunpowder that projects the bullet, or when a pointsman shunts a train it is the rails that bear the thrust. But the constitution of our nature is not explained by finding out what it is not. It is well that it will go, and that we remain in possession, though we do not understand it.[101]

Like the earlier 'Recollections of Dreamland', Maxwell continued to believe in there being a 'point at which Science must stop' which, rather than being a form of censorship, was due to the limited capacities of science itself. As he comments in 'Discourse on Molecules',

> . . . Science is arrested when she assures herself, on the one hand, that the molecule has been made, and, on the other, that it has not been made by any of the processes we call natural.[102]

Maxwell's hesitation about interrogation is surely one of the core paradoxes of his life. When he wrote 'A Paradoxical Ode', he knew that he was already feeling the symptoms of the cancer that would kill him at only 48, and it expresses the alignment between his role as a physicist and as a Christian. That being so, its pleading tone is surprising, showing as it does a degree of fearfulness about death and the breakdown of the partition between the physical and non-physical worlds.

Oh never may direct Creation
Break in upon my contemplation;
Still may thy causal chain, ascending,

Appear unbroken and unending,
While Residents in the Unseen—
Aeons and Emanations—intervene,
And from my shrinking soul the Unconditional screen.
'A Paradoxical Ode', 33–9.

In begging for his 'contemplation' or conscious mind never to encounter his soul directly – that part of man which may be considered closely 'God's direct creation' – it is as though he knows he could not tolerate a meeting of the two. He asks that the Christian tenets continue to 'appear unbroken and unending'; however, the word 'appear' suggests a flicker of doubt, implying in his request for his 'shrinking' soul to be screened a dread of the 'Unconditional' divine retribution. The trepidation of facing that final connection links back to the poem's initial reference to Shelley, in a warning about the promethean overreaching of boundaries inscribed by higher powers and its regretful consequences.

Demonstrating the consolation of such fears, the subsequent 'Addendum' counteracts the earlier desperate pleading with the calm surety of a prayer.

And where that chain is lost to sight
Let viewless fancies guide my darkling flight,
Through atom-haunted worlds in series infinite.
'A Paradoxical Ode', *Addendum*.

The 'darkling flight' into 'atom-haunted worlds' suggests a sphere of multiple possibilities where the material of the physical world itself is spectral. In 'The Scientific Use of the Imagination' (1870), Tyndall advises scientists to let their imaginations 'rise above the physical limitations of vision'.[103] The two men's concern about what can be seen, physically and mentally, reflects the Victorian fascination with what Kate Flint describes as 'the slipperiness of the borderline between the visible and the invisible'.[104] Tyndall believed the 'seeking intellect' would inevitably reveal the mechanics of existence; in contrast, the 'viewless fancies' of Maxwell's after-life humbly concede the limitations of empiricism and even the imagination itself.[105]

The power and gravity of these last few lines fittingly resolves the paradoxes of the preceding ode and acts as an obituary. If the comic bent of certain earlier verses prompted uncertainty about how seriously Maxwell took his vocation and faith, the 'Addendum' and these more mature works show his earnestness. Meanwhile, the ode itself reveals his ultimate recognition of pastoral obligation, the influence of the present upon the future, and the creative convergence between faith, poetry and scientific practice.

*

Maxwell's poems touch upon many different subjects and span a period of several decades; however, they make virtually no mention of the century's political or social change, let alone the Crimean War or the expansion of the British Empire. Clearly, they were never intended by Maxwell to be a comprehensive or even representative reflection of his times, just as his own position in science was not necessarily so; as Peter Harman suggests, Maxwell 'should not be regarded as a professional scientist himself. His way of life was that of a Scottish laird, who often relied on inherited property in pursuing his own scientific interests.'[106] What Maxwell's poetry does represent is how he negotiated the labyrinthine private and public paths of mid-century scientific science and secularization. His poetry shows an increasing concern about how others thought, exemplifying George Levine's suggestion that 'ideas live in culture not disembodied, but as actions, attitudes, assumptions, moral imperatives'.[107] The poems' gradual shift of focus reflects the century's re-evaluations of established viewpoints and margins, texts and authorities, over many decades. In 1844, Leigh Hunt proposed that 'poetry begins where matter of fact or science ceases to be merely such, and to exhibit a further truth; that is to say, the connexion it has with the world of emotion, and its power to produce imaginative pleasure'.[108] Maxwell's poems indicate that, for him, facts and science were never 'merely such'; instead, they provided the true basis of poetry and were as genuinely moving as any form of imagination. The lack of distinction between arts and sciences at the time meant, as Renee Bergland suggests, that 'often poets and scientists were one and the same: writers fiddled with microscopes and telescopes while scientists read, recited, and even wrote poetry'.[109] For Maxwell, imaginative pleasure did not have to be based purely on fantasy; the workings of the surrounding world offered the very material of awe and meditation. An illustration of this relational viewpoint is offered by one of Maxwell's shortest, yet best-known poetic efforts, a ditty called 'Gin a Body'. Ostensibly a parody of Robert Burns's ditty on sexual encounter 'Comin Thro' the Rye', Maxwell's composition can be read just as easily as a description of molecule interactions, and a metaphor for the influence of Maxwell's life and work on others.

Gin a body meet a body
Altogether free,
How they travel afterwards
We do not always see.
'Gin a Body', 9–12.

With characteristic concision, in just 17 words Maxwell conveys worlds that are profoundly literary and scientific, individual and universal, and simultaneously personal and abstract. Maxwell may not always have agreed with his colleagues, but his theories and innovations provided a unique impetus for the efforts of his contemporaries and others in the future. In that sense, his poetry represents the lively responsiveness of his contributions, and its enduring reach across the boundaries of literature and science.

CHAPTER 14

Maxwell, Faith and Physics

PHILIP L. MARSTON

Though he never entered into theological controversy, and only occasionally in his scientific writings indicated in a sentence or two the side he took in questions which have recently been brought prominently before the public by some of the more popular men of science, those who had an opportunity of seeing into his home-life knew him to be an earnest Christian. About three weeks ago he remarked that he had examined every system of Atheism he could lay hands on, and had found, quite independently of any previous knowledge he had of the wants of men, that each system implied a God at the bottom to make it workable. He went on to say that he had been occupied in trying to gain truth, that it is but little of truth that man can acquire, but it is something to 'know in whom we have believed.' William Garnett (*Nature*, 13 November 1879)[1]

Introduction

During Maxwell's era there was increasing public attention to religious scepticism, at times associated with the proclamations of 'popular men of science'. From all accounts, however, Maxwell's Christian faith remained steady during his adult life and was unaffected by criticism or controversy over the fundamentals of the Christian faith. Furthermore, Maxwell's approach to physics conveyed no inconsistencies with his faith. General public interest in Maxwell's positions on matters of faith is evident as early as 1872 and remained evident for half a century following his death. By the mid 1870s, Maxwell was aware of diverse reactions to his public statements. To appreciate Maxwell's faith, it is appropriate to begin with the values conveyed by his parents and his Scottish culture.

Early Years, Adolescence and Studies in Edinburgh (1831–50)

Parental guidance and background: Maxwell was the beneficiary of thoughtful but limited, loving Christian instruction from his parents. Maxwell's early education at the remote estate of Glenlair in Kirkcudbrightshire was largely supervised by his mother (Frances Clerk Maxwell, *née* Cay, 1792–1839) until the time of her death when Maxwell was only 8 years old. By this age he had memorized the whole of the 119th Psalm, and it is said 'His knowledge of Scripture, from his earliest boyhood, was extraordinarily extensive and minute'.[2] Over three decades later, Maxwell reported that his mother was 'guided by religious thought and very independent of the exhortations of acquaintances, clerical or lay. Religion was a forbidden subject in her father's [Robert Hodshon Cay] family as his mother [Frances Cay *née* Hodshon] was a Roman Catholic and all discussion was avoided religiously.'[3] While the Cay family had non-conformist roots in north-eastern England traceable back to the period of persecution during the Restoration, Maxwell's mother's family had become Episcopalian.[4] In addition to Bible memorization, Maxwell's mother reportedly encouraged him to 'look up through Nature to Nature's God'.[5] Maxwell's father (John Clerk Maxwell, 1790–1856) provided some companionship and a steady example, supporting the central tenants of the Church of Scotland, and serving as an elder in a church near Glenlair. John showed concern for the spiritual as well as the intellectual nurture of his son. Their relationships within their extended family were highly valued throughout each of their lives.

In 1841, Maxwell moved to Edinburgh to continue his education. This was practical because his aunts from each side of the family resided there, each providing a home during extended periods when his father needed to reside in Glenlair. With his move came a broadening of the religious elements of his education and an exposure to events that helped to form his ecclesiastical views. The broadening arose in part because his Episcopalian aunt, Jane Cay, took young Maxwell with her to St. John's Chapel on Sunday afternoons. Maxwell also attended St. Andrew's Church (Church of Scotland) in central Edinburgh with his father. This church was near his Aunt Wedderburn's residence where Maxwell often stayed.

John Knox and the Reformation: When considering Maxwell's religious environment associated with his Scottish heritage, the residual influence of John Knox (*c.*1514–72) and other Scottish reformers on the Church of Scotland is unavoidable.[6] Daniel Dewar (*c.*1787–1867), Maxwell's future father-in-law and a teacher of church history, was highly favourable towards Knox's influence. In his 944-page *magnum opus* on *The Church* (1845),[7] Dewar lamented recent criticism of Knox, writing, 'When we reflect that he was the unflinching advocate of true religion and sound morality, it cannot surprise us that infidel historians, and especially Hume, should never lose an opportunity of vilifying his character'. Knox returned to Scotland in 1559, following a period of close association with the Swiss reformer, John Calvin. This preceded a pivotal period of the Scottish reformation when Knox wrote or co-wrote the major documents of the Scottish Protestant Church. Knox was concerned not only with articles of faith, but also with church practice and responsibility, some of the latter areas remaining unresolved long after Knox's death. Issues affecting the church during Maxwell's life included obligations of the church to educate and assist the poor, the means of financing the church and selecting and ordaining pastors.[8]

The Disruption of 1843 and aftermath: In May 1843 an important event occurred that influenced Maxwell's attitudes and those of his father concerning religious 'ferment'. The primary event commonly referred to as 'the Disruption' of the Church of Scotland was the culmination of a decade of tension associated with church–state relations; issues included the approval or rejection of proposed ministers and the financial support or patronage of local churches.[9] A substantial number of ministers favouring the evangelical leader Thomas Chalmers (1780–1847), the Non-Intrusionist Party, chose to leave the Established Church of Scotland. This decision on the part of the ministers and the parishioners who followed them to the newly formed *Free Church of Scotland* carried with it major financial implications because the ministers were vacating their livings, and the parishioners their places of worship. The break with the established Church occurred at the general Assembly, which in 1843 was meeting in St. Andrew's Church, the building where young Maxwell worshiped with his father. Associated with the Disruption was a procession out of St. Andrews of hundreds of ministers seceding from the Church of Scotland. While it is not known if Maxwell himself viewed this procession, accounts have survived by two of Maxwell's schoolmates recalling the procession approximately six decades later. One of those schoolmates became Maxwell's lifelong friend and biographer: Lewis Campbell.[10,11]

Shortly before his death, Maxwell recalled some consequences of the Disruption in conversation with Fenton Hort, a friend from his college years. According to Hort's account Maxwell said,

> The ferment, about the Free Church movement had one very bad effect. Quite young people were carried away by it; and when the natural reaction came, they ceased to think about religious matters at all, and became unable to receive fresh impressions. My father was so much afraid of this, that he placed me where I should be under the influence of Dean Ramsay, knowing him to be a good and sensible man.[12]

Looking back on the disruption and its aftermath, some leaders of the Free Church also expressed concern with negative spiritual reactions in some who sided with the movement. Dean Ramsay, however, was affiliated with St. John's Scottish Episcopal Chapel in Edinburgh and was the author of a standard catechism likely used by Maxwell when he was a member of Ramsay's catechetical class.

As previously noted, Maxwell sometimes also attended St. Andrew's church and was reportedly equally acquainted with the catechism of the Church of Scotland. While it is not known what the specific teaching at St. Andrew's was like during Maxwell's attendance, his pastor, Thomas Crawford, eventually became a highly regarded Professor of Divinity at the University of Edinburgh.[13] He was especially known for his biblically based exposition of doctrine in areas that remained of interest to Maxwell in his adult life.[14] Crawford's teaching stood in opposition to the hyper-Calvinism of the period and emphasized the idea that the unresolvable mysteries of Christianity are not incompatible with truth.[15] The opening lines of his final major work display the priority of scripture in Crawford's thinking: 'If we receive the witness of men, the witness of God is greater' (I John 5:9).[16]

Schooling and friendships with Lewis Campbell and Peter Guthrie Tait: Maxwell's attendance at Edinburgh Academy initially resulted in harsh treatment by his peers and reason to doubt his father's wisdom in placing him there. Eventually, however, his enrolment gave rise to two lifelong friendships, and those friendships are important in the consideration of Maxwell's religious views. His friendship with Campbell is especially relevant because Campbell (1830–1908), at the request of Maxwell's widow, became the main author and editor of much of the relevant material in the standard biography. Shortly after the Disruption, Campbell moved with his mother and brother to a residence on Heriot Row in Edinburgh, near where Maxwell resided with Mrs Wedderburn. Though Tait (1831–1901), like Maxwell, became a physicist, and like Maxwell was concerned about the growing religious scepticism of the Victorian era, their reactions and involvement differed during their adult lives. Maxwell, Campbell and Tait each benefitted from the classical education at Edinburgh Academy that included Latin, Greek and mathematics. Maxwell and Tait subsequently built on their mathematical training, while Campbell became a classical Greek scholar after serving briefly as an Anglican minister following his college education.

Maxwell entered the University of Edinburgh in November 1847, continuing his education at Cambridge in the Autumn of 1850. Campbell left for the University of Glasgow in 1846, moving on to Oxford in 1849, and Tait left for Cambridge in 1848. While both Tait and Maxwell included natural philosophy and mathematics in their studies in Edinburgh, Maxwell's studies were significantly broader and included lectures on logic and metaphysics from Sir William Hamilton. Hamilton's contributions to philosophy included the editing of the earlier work of Thomas Reid, one of the founders during the prior century of the Scottish school of 'Common Sense Philosophy'. Reid's thinking was in part a response to Hume's epistemological scepticism. Writing to Campbell in 1847, Maxwell indicated that '. . . [Hamilton's] Logic lectures are far the most solid and take most notes', and, in a subsequent letter, noted his familiarity with 'Reid's account' of logic.[17] Maxwell learned of Hamilton's distinction between 'Knowledge and Belief', beliefs including inexplicable convictions or convictions falling short of certainty, and Hamilton's discussion of 'Natural Realism', our consciousness of mind and of the external world.[18] Maxwell probably also heard Hamilton discuss Freewill and Necessity; Hamilton lectured that '. . . he who disbelieves the moral agency of man must, in consistency with that opinion, disbelieve Christianity'.[19] Hamilton aimed not 'to teach *philosophy* but [how] to *philosophize*'.[20] Following Hamilton's death, in 1856, as his views became better known, the response of Christian theologians to Hamilton varied widely, though it has been reported that Hamilton himself maintained a sincere respect for the Bible.[21] Hamilton became physically impaired by 1847, and this probably curtailed his popular hospitality of students in his home on Great King Street, only a short distance from Maxwell's residence. In spite of Hamilton's famous scorn of mathematics, Maxwell regarded him highly as a teacher, while being sometimes critical of Hamilton's philosophy or how it was portrayed.[22]

Concerning other people who impressed Maxwell during his time in Edinburgh, Campbell remarked that Maxwell hung the likeness of James Forbes and George Wilson in his dwelling in Trinity College, Cambridge.[23] Forbes was his natural philosophy professor, and allowed Maxwell to readily access laboratory equipment and recommended Maxwell for Cambridge admissions. Much less has been documented of the early aspects of the relationship between Maxwell and Wilson, at the time a chemistry lecturer for the Edinburgh Veterinary College and School of Arts.[24] Eventually, in 1856, Wilson (1818–59) had grounds for professional interaction with Maxwell when he included a letter from Maxwell on colour vision near the end of his monograph on *Colour Blindness*. What made Wilson unusual were his physical trials and his reputation. His left foot was amputated in 1843, prior to the use of anaesthesia. By reputation he displayed from about that time 'a dedication of life to Christian ends and uses, which henceforth made every place and work sacred'.[25] To appreciate the likely nature of Wilson's early relationship with Maxwell, one need only study the intensely caring letters by Wilson to an unidentified young male written between 7 January 1847 and April 1848.[26] The letters not only include discussions of the known scientific interests of young Maxwell, they also include evangelical expositions of the Bible.[27] Consider also a letter in 1860 to Wilson's sister by an unidentified person, similar to Maxwell in age and circumstance, indicating concerning Wilson that 'I count it a blessing to have known such a man, such an example'.[28]

By October 1849, Maxwell had become sufficiently aware of modern attacks on the historicity of the Biblical gospel to give his attention to a critical reply. The particular occasion concerned David Strauss's attempt to portray the gospels as mythologized accounts, *Das Leben Jesu* (1835–36), which prompted a parody by an American minister, Theodore Parker, that Maxwell repeated in a letter to Campbell.[29] After having been initially impressed by Strauss's fictionalization, various English observers eventually concluded that Strauss could not account for the perfection of Christ as recorded in the gospels.[30]

Cambridge: College and Teaching Fellowship (1850–56)

Following his first term at Peterhouse, Maxwell transferred to a much larger college, Trinity, which brought him into contact with a diverse group of friends. Charles Hope Robertson, a friend from Edinburgh, roomed near Maxwell in Trinity and suffered occasionally from difficulties with his vision. Years later, Robertson recalled how Maxwell often gave up an hour of his recreation time to read aloud assignments for him. He also recalled examples of where either he aided Maxwell in defending scripture, or Maxwell offered his own scriptural insight. He summarized his views on Maxwell as:

> If shortly described, he might be said to combine a grand intellect with childlike simplicity of trust. He was too deep a thinker to be sceptical, but too well read not to feel for others' difficulties. All his experiments led him to greater reverence for the Great First Cause, heartily agreeing with Young's *Night Thoughts*, 'An undevout astronomer is mad'.[31]

George Tayler, a new acquaintance at Trinity who first met Maxwell early in 1852, much later recalled the exceptional breadth of Maxwell's reading:

> Grave and hard-reading students shook their heads at his discursive talk and reading . . . he read a good deal in other lines of study than natural philosophy. Sir Thomas Brown[e]'s *Religio Medici* was one of his favourite books.[32]

When considering Maxwell's interest, Browne's declarations such as the following concerning 'the general beauty in the works of God' are relevant:

> . . . yet this rule of His He doth sometimes pervert, to acquaint the world with His prerogative, lest the arrogancy of our reason should question His power, and conclude He could not. And thus I call the effects of Nature the works of God, whose hand and instrument she only is; and therefore, to ascribe His actions unto her is to devolve the honour of the principal agent upon the instrument; which if with reason we may do, then let our hammers rise up and boast they have built our houses, and our pens receive the honour of our writings.[33]

Maxwell summarized his own beliefs in a letter to Lewis Campbell in November 1851 as 'I believe, with the Westminster Divines and their predecessors ad Infinitum that "Man's chief end is to glorify God and to enjoy him for ever".'[34] Maxwell, however, was generally critical of the quality of preaching he received from the Church of England in Cambridge during his years as a student, and this discomfort may have helped to motivate the breadth of his reading.[35] Maxwell's poetry of this period occasionally displayed a reverence to his creator perhaps most clearly evident in the following stanzas of his *Student's Evening Hymn* from April 1853:

Teach me so Thy works to read
That my faith,—new strength accruing,—
May from world to world proceed,
Wisdom's fruitful search pursuing;
Till, thy truth my mind imbuing,
I proclaim the Eternal Creed,
Oft the glorious theme renewing
God our Lord is God indeed.

Give me love aright to trace
Thine to everything created,
Preaching to a ransomed race
By Thy mercy renovated,
Till with all thy fulness sated
I behold thee face to face
And with Ardour unabated
Sing the glories of thy grace.[36]

During 1853, there were other developments that are especially revealing of Maxwell's religious nature. Maxwell accepted an invitation to relax early in June in Otley (near Ipswich) at the home of Reverend Charles Benjamin Tayler, the uncle of Maxwell's student friend, George Tayler. Maxwell had been working diligently during the spring in preparation for the Mathematical Tripos examination scheduled for January 1854. Early during his visit Maxwell became seriously ill, suffering from what was described as 'brain fever' requiring several weeks of recovery. Maxwell long remembered the kindness rendered, recalling over a decade later how C. B. Tayler had 'carried me about when I could not move myself . . .'.[37] Concerning this difficult time, Campbell wrote,

He referred to it long afterwards as having given him a new perception of the Love of God. One of his strongest convictions thenceforward was that 'Love abideth, though Knowledge vanish away.' And this came to him at the very height of the intellectual struggle.[38]

Maxwell himself, in a section of his letter of 8 July 1853 to Rev. Tayler, sheds some light on his 'intellectual struggle'. In reply to a letter from Tayler, Maxwell wrote after returning to Cambridge,

. . . I maintain that all the evil influences that I can trace have been internal and not external, you know what I mean—that I have the capacity of being more wicked than any example that man could set me, and that if I escape, it is only by God's grace helping me to get rid of myself, partially in science, more completely in society,—but not perfectly except by committing myself to God as the instrument of His will, not doubtfully, but in the certain hope that that Will will be plain enough at the proper time.[39]

The events of June 1853 are sometimes viewed as an Evangelical awakening for Maxwell.[40] At the very least, it was a time where Maxwell gained a heightened awareness of God's grace pertaining to his own wrongdoing.

During the prior academic year, Maxwell had broadened the nature of his college discussions by accepting an invitation to join the select Cambridge essay group commonly known as *The Apostles*. His membership over four years brought him into contact with an extremely diverse group, many of whom found reasons to disagree with Maxwell privately or publicly. The group included C. J. Monro (who, with R. B. Litchfield, had prolonged correspondence with Maxwell),

H. M. Butler,[41] F. J. A. Hort (Maxwell's neighbour when they both returned to Cambridge in the early 1870s), V. Lushington (who eventually became a proponent of the antitheist French philosopher Auguste Comte), R. H. Pomeroy[42] and F. W. Farrar. Maxwell's participation was perhaps unusual in that, though willing to contribute a thoughtful essay, he was reluctant to argue with opponents.[43] Writing to Monro early in 1855 he said 'It is a fearful thing to answer when a man tackles you with arguments. I won't argue at all, least ways not with them as tries to argue me.'[44] During discussions at meetings of the Apostles in which each member had to speak in turn, Farrar described how Maxwell's remarks 'were often hardly intelligible to anyone who did not understand the characteristics of his mind' but that if pursued, the relevance of his first remark 'became abundantly manifest'.[45]

During the final months of his preparation, in 1853, for the Tripos examination, Maxwell wrote his father a detailed essay concerning the main parties within the Church of England, also asking him about a prior family acquaintance, Thomas Erskine. By 1882, only his father's reply survived,

(Glenlair, 16 December 1853)

I knew Thomas Erskine of Linlathen very well long ago. He . . . is author of various religious books.

Your dissertation on the parties in the Church of England goes far beyond any knowledge. I would need an explanatory lecture first, and before I can follow the High, Broad, and Low, through their ramifications.[46]

Young Maxwell's interest in Erskine may have been because of Erskine's theological influence on Rev. F. D. Maurice, a member of The Apostles from an earlier generation who became an influential Christian activist during Maxwell's generation. Following Erskine's own Evangelical awakening at about the time he knew Maxwell's father, Erskine wrote a series of books revealing his progression from Evangelical to Liberal positions. Maxwell's lost essay to his father was almost certainly brought about by an a lengthy anonymous essay devoted to 'Church Parties' published in *Edinburgh Review* (October 1853). Eventually the author was revealed to be Rev. W. J. Conybeare (1815–57). Here, it is sufficient to note Conybeare's main divisions as *High Church*, *Low Church* and *Broad Church* and that from Conyebeare's perspective, the parties had several doctrines in common including: Christ as the Incarnation and Atonement for sin, Conversion by Grace and Justification by Faith. He also noted that terminologies used to describe each party and their subdivisions depended on the perspective of a given writer. It is revealing that Maxwell devoted his attention to such issues at such a pivotal time during his mathematical studies. Campbell, when considering Maxwell's interest in Evangelical (or Low Church) expressions of Christianity following his crisis of June 1853, emphasizes the breadth of Maxwell's interests.[47] He acknowledges Maxwell's interest in the Bible and 'Scottish Calvinism' and his study of the evangelical authors John Owen and Jonathan Edwards, but he goes on to note Maxwell's regard for the written sermons of (Broad churchman) J. C. Hare[48] and his concern over the ecclesiastical persecution of F. D. Maurice.

By late January 1854, the Tripos examination was over and Maxwell was placed second behind E. J. Routh. During February Maxwell tied Routh in a relatively specialized examination, the Smith's Prize.[49] All things considered, Maxwell's father was pleased with his placement in these examinations. In spring 1854, Maxwell received his Bachelor's Degree, staying on until 1856 at Trinity. That period was important for his continued independent development as a scientist, if for no other reason than his paper late in 1855 'On Faraday's Lines of Force'.

Return to Scotland: Aberdeen and Marriage (1856–60)

The health of Maxwell's father, John, was deteriorating by the winter of 1854–55 and James was giving his father more attention. Consequently, when in February 1856 James learned that the Chair of Natural Philosophy at Marischal College, Aberdeen, was vacant, it was appropriate for him to apply. The Scottish college schedule would enable him to spend more time in Glenlair assisting his father. Unfortunately, John died at noon on 3 April, about a month before James was offered the Marischal professorship. James was present in Glenlair at the time of John's passing and he quickly wrote his uncle Robert Cay for assistance. In subsequent correspondence to Maxwell's Aunt Cay, Robert noted:

It is beautiful to see . . . when he wants to make a change, his anxiety, lest people should think he disapproves of the former customs. For instance, he wished to have the servants in for prayers every evening, instead of our reading by ourselves and reading to them separately; he was quite afraid they should think his doing so would look as if he thought it was wrong, it not having been done before.[50]

Since James now had full responsibility for Glenlair and needed the greater flexibility permitted by the Scottish schedule, he accepted the professorship, and upon assuming responsibilities delivered his Inaugural Lecture on 3 November. The lecture clearly and repeatedly indicates his views as to the wisdom of the Creator: '. . . that the laws of nature are *not* mere arbitrary and unconnected decisions of Omnipotence, but that they are essential parts of one universal system in which infinite Power serves only to reveal unsearchable Wisdom and external Truth.'[51] He spoke of the mystery 'that every atom of creation is unfathomable in its perfection' and that, '. . . we ought to admire the wisdom of Him who has so arranged these mysteries that we find first that which we can *understand* at first and the rest in order so that it is possible for us to have an ever increasing stock of *known* truth concerning things whose nature is absolutely incomprehensible'.[52] He indicates the 'reverent study of the order of Creation' was beautiful to students of theology. As it turned out, one of his students from Aberdeen, present 23 years later at Maxwell's graveside service, devoted his life to the Christian ministry.[53]

Maxwell likely found the geographic isolation from friends and family in Aberdeen challenging during the six months the college was in session, Campbell describing Maxwell's experience in Aberdeen in 1857–58 as the 'crisis of his life'.[54] Maxwell was sufficiently pleased with his Inaugural Lecture that he mailed a newspaper clipping summarizing the text to relatives.[55] One of his first new friends in Aberdeen was the Professor of Moral Philosophy, William Martin, who walked with Maxwell '*everyday*' for some time, until Martin had a period of poor health.[56] It may be because of Martin's expertise concerning *Butler's Analogy* that Maxwell resumed his interest in that Christian apologetic while in Aberdeen.[57] In the eighteenth century, Bishop Butler developed an analogy between the visible course of nature and the biblical system of order. In 1858, Martin became a major supporter of an Evangelical Revival in Aberdeen.

During his first year in Aberdeen, Maxwell had visited on occasion the family of Principal Daniel Dewar of Marischal College. In September 1857, he visited Ardhallow near Dunoon, Argyll, during the annual visit there by Dewar's family. In February 1858 he became engaged to the Principal's daughter, Katherine Mary Dewar, emphasizing in correspondence to friends the importance of their mutual understanding. To Litchfield he included remarks on their mutual elements of faith:

You will easily see that my 'confession of faith' must be liable to the objection that Satan made against Job's piety. One thing I would have you know, that I feel as free from compulsion to any form of compromised faith as I did before I had any one to take care of, for I think we both believe too much to be easily brought into bondage to any set of opinions.[58]

When considering Maxwell's observations concerning the strength of Katherine's faith, it is appropriate to notice the level of stress likely placed on Principal Dewar and his family following the Disruption of 1843. From early in his career, Dewar was viewed as supportive of Thomas Chalmers' evangelical leanings. When Chalmers shifted his Glasgow ministry to St. John's, in 1819, to expand his ministry to the lower classes, Dewar was called to replace him at the Tron Church (until 1833) and at various times during Chalmers' expanding leadership of the evangelical wing of the Church of Scotland, Dewar occasionally appeared on platforms with him or supported Chalmers' positions.[59] But when the Disruption came, Dewar did not shift to the *Free Church*, perhaps to avoid jeopardizing his faculty position at Marischal College.[60] It may also be relevant that, in 1843, Dewar was likely in the midst of writing his work on *The Church*.[61] It is plausible that Dewar's family was aware of criticism of him for failing to join the *Free Church*.[62]

Prior to their wedding in June, Maxwell made a trip south for nearly a month. The occasion of this trip was to serve as the best man at Lewis Campbell's wedding in Brighton and to visit friends in London. The trip is important because of the revealing nature of the correspondence with Katherine. The fragment available from his letter of Sunday, 2 May shows Maxwell establishing a pattern of correspondence based on Bible reading:

Now you must remember that all I say about texts and matters of that sort is only a sort of help to being together when we read, for I am not skilful to know what is the right meaning of anything so as to tell other people, only I have a right to try to make it out myself, and what I say to myself I may say to you.[63]

Subsequent letters concern their mutual reading of specific chapters of Scripture: 6 May (Isaiah 51, Galatians 5 on fruits of the Spirit), 10 (Ephesians 3), 13 (Ephesians 6), 16 (Philippians 3). His discussion of Phil. 3:21 on Sunday 16 May was followed by an expression of concern for one of his friends he had visited that weekend near London:

What a description of the power of Christ in the last verse, over 'all things,' and our vile bodies among the rest, and what a day it will be when He has done all His work and is satisfied.

I think the more we enter together into Christ's work He will have the more room to work His work in us. For He always desires us to be one that He may be one with us. Our worship is social, and Christ will be wherever two or three are gathered together in His name.

I have been vexed that I could not speak better to—. I had a long walk with him, talking of what people have believed, and what was necessary to be believed. I hope we may come to understand each other, but more that he may come to the clear light. I wish I could speak to him wise words. He is so anxious to hear, and I to speak, and then the words are all wind after all.[64]

Following the wedding of Maxwell to Katherine, in Aberdeen on 2 June 1858, they eventually made their way to Glenlair, both for a period of rest and to introduce Katherine to the joys and responsibilities of life at Glenlair. Maxwell needed to return to Aberdeen by early November to resume his professional duties. Not long after returning, the Maxwells would have become aware of a shift in the religious climate and an expansion of evangelical preaching. Following other religious services in Aberdeen, William Martin, Maxwell's former walking partner from 1857, invited a lawyer from Liverpool, Reginald Radcliffe, who had become an effective evangelist. Radcliffe arrived in Aberdeen by late November 1858 and soon began services for children in a Congregational mission room in a low-income region. The impact of those services was such that James Smith, the minister of Greyfriars Parish Church (of the Established Church of Scotland) took the controversial step of allowing Radcliffe to preach on Sunday evenings and provided other forms of support.[65] Radcliffe's ministry was especially effective among college and working men of Aberdeen. Concerning these events Campbell noted:

Amongst the human phenomena surrounding him, one which genuinely interested him was the religious 'revival' which took place about that time in Scotland. His intercourse with evangelical friends in England had prepared him to sympathise with such 'experiences,' and his Calvinistic reading had familiarised him with the language used. And he was less jealous of Antinomianism than of a cut and dried morality. But he was in no wise distracted from his professional duties by this or anything else, and although he referred to it in conversation, it has left no trace in any of his remaining letters which I have seen.[66]

While Campbell's perspective of 'no trace' may be applicable, a thoughtful account of these events is given in a series of anonymous letters to an anonymous recipient, published as *Five Letters on the Religious Movement in Aberdeen.* The letters are dated from March 1859 and are followed by an Appendix: *Nature, Probability, and Necessity of a Religious Renewal.* The author, using the pseudonym *Omicron*, indicates that he looks forward to seeing the recipient during the upcoming meetings of the British Association for the Advancement of Science (BAAS) meeting (held in Aberdeen in September 1859). He has not only a high regard for scripture and familiarity with the Established Church, but also makes repeated reference to writings of Jonathan Edwards and has 'read not a few' religious biographies. Omicron discusses charges made against James Smith by the Presbytery and Smith's defence of his openness to Radcliffe. Not only was lay preaching against the rules of the Established Church, it could be taken as contrary to the advice of Knox, nearly three centuries earlier. Omicron examines both sides of various issues making repeated allusions to Christ's Parable of the Sower, and notes that 'divinity students . . . became but dry preachers'.[67] He ends with a perspective possibly paralleling Maxwell's:

While we are to discourage everything that partakes of a spurious revival and to abjure every attempt to create a mere religious excitement, it is surely the duty of all true Christians to exert and purify all the ordinary agencies within their power—of our ministers to preach with increased faithfulness and labour with renewed energy, and of our people to welcome them as God's ambassadors—to cultivate a holy and blameless walk—to dwell in unity and peace, and above all to join as one man that the voice of earnest prayer may ascend as one vast petition to the throne of God for a manifestation of the Spirit's power, a great awakening, and an abundant increase to the church in our beloved land.[68]

In February of 1859, a scientific paper by Clausius appeared on the path length of a particle of gas between collisions. This captured Maxwell's attention and led to his now famous predictions of the probability distribution for the velocity of gas particles and the independence of the shear viscosity on the gas pressure.[69] He presented that research at the 1859 BAAS meeting, along with selected work on colour vision and on Saturn's rings. In spite of his popularity with serious students, he lost his position owing to the merging of universities in Aberdeen. After the completion of the academic year in the spring of 1860, and the retirement and relocation of Principal Dewar, the Maxwells had little reason to return to Aberdeen. While travelling in April 1860, Maxwell sent Katherine letters concerning chapters of scripture, 1 Corinthians 12 and 13 on spiritual gifts and charity.[70]

London, Glenlair and Scientific and Religious Developments (1860–71)

In late June and early July 1860, Maxwell attended the now famous meeting of the British Association in Oxford, reporting on recent progress on his theory of gases and measurements related to colour vision. In July he was offered the professorship of Natural Philosophy at King's College, London, with teaching duties beginning in October. He began the session having only recently recovered from small pox. Through early 1861 the Maxwells resided at 7 Palace Gardens Villas, Kensington,[71] moving to 8 Palace Gardens Terrace, which remained their London address at least through April 1868. During the summer of 1861, Maxwell formulated his initial model relating wave motion in ether to 'the numerical relation between statical and magnetic effects of electricity', which appears to have been mentioned first in a letter to Michael Faraday in October.[72] While Maxwell had corresponded occasionally with Faraday since 1857, moving to London offered the possibility of some personal contact, though by then Faraday's health was declining. Maxwell, on the occasion of lecturing on colour vision at the Royal Institution in May 1861, appears to have dined with Faraday. His admiration of Faraday's insights on electric and magnetic fields is evident in his preface to his *Treatise on Electricity and Magnetism* (1873). His essay on Faraday written for the *Encyclopedia Britannica* in the late 1870s also suggests a sympathy with Faraday in broader issues: 'When, on rare occasions, he was forced out of the region of science into that of controversy, he stated the facts, and let them make their own way.' Maxwell went on from there to quote a section of Faraday's 1854 lecture on mental education:

> Yet even in earthly matters I believe that 'the invisible things of Him from the creation of the world are clearly seen, being understood by the things that are made, even His eternal power and Godhead'; and I have never seen anything incompatible between those things of man which can be known by the spirit of man which is within him, and those higher things concerning his future which he cannot know, by that spirit.[73]

Maxwell ended his essay by quoting the way Dr. Bence Jones, the Secretary of the Royal Institution, ended his own biographical essay: 'His standard of duty was . . . formed entirely on what he held to be the revelation of the will of God in the written Word, and throughout all his life his faith led him to act up to the very letter of it.'[74]

Maxwell's move to London also enabled him to interact with people of his generation who subsequently became involved in various controversies. These included Fleeming Jenkin and Balfour Stewart[75] who, beginning in 1863, assisted Maxwell in developing a standard of electrical resistance, measurements of interest to Maxwell because they influenced the relation of electrostatic and magnetic units important in his prediction of the velocity of electromagnetic waves. Jenkin became involved in controversy when he perceived scientific conjectures as being accepted by faith. He wrote two anonymous essays. One, written about the time Jenkin was assisting Maxwell, concerned scientific and mathematical criticisms of Charles Darwin's book *On the Origin of Species* (1859). The other examined 'The atomic theory of Lucretius' and related recent developments.[76] Jenkin's criticism of the efficiency and limitations of natural selection attracted much more attention. Maxwell's relationship with Jenkin may have been relatively significant, given that Maxwell served as god-parent at the baptism of one of Jenkin's sons, Bernard Maxwell Jenkin, in 1867.

When the Maxwells, in 1861, moved to their dwelling at Palace Gardens Terrace, they were situated about 100 m north of a small Anglican Church, St. Paul's, Vicarage Gate.[77] It was not this church that Maxwell mentions, however, when he wrote in February 1866 from London to his Anglican friend from 1853, Rev. C. B. Tayler:

Many people's minds seem to be shut up with solemn charms, so that though they seem Christians, and know what they mean to speak about, they can say nothing. At Cambridge I heard several sermons from excellent texts, but all either on other subjects or else right against the text. There is a Mr. Offord in this street, a Baptist who knows his Bible, and preaches as near it as he can, and does what he can to let the statements in the Bible be understood by his hearers. We generally go to him when in London, though we believe ourselves baptized already.[78]

The pastor mentioned here was John Offord of Palace Gardens Chapel. This chapel was built in 1861–62, situated near the intersection with Bayswater Road.[79] It was about 400 m north of Maxwell's residence. John moved late in 1862 to London from Plymouth to serve in the new chapel.[80] During his time in London he worked hard at basic Bible teaching and evangelism, being well respected by Charles Spurgeon and other evangelical leaders. Offord died from overwork in June 1869, having published several small books of scripture-filled sermons. By 1869, however, the Maxwells were only occasionally in London, having given up their Palace Gardens Terrace residence by mid 1868. They had retained that residence long after Maxwell's retirement from Kings College in Spring 1865, continuing the practice of splitting time between Glenlair and London.

In June of 1864, Maxwell stayed in London working on electrical measurements and presumably developing his theory of electromagnetic waves. Katherine had gone to the healthier climate in Glenlair. This was the occasion of a series of letters by Maxwell to Katherine on basic teaching from the New Testament Gospels (Mark and John) and Epistles (Romans, 1 and 2 Corinthians and 1 John):

[Thursday] 23d June 1864.

Think what God has determined to do to all those who submit themselves to His righteousness and are willing to receive His gift. They are to be conformed to the image of His Son, and when that is fulfilled, and God sees that they are conformed to the image of Christ, there can be no more condemnation, for this is the praise which God Himself gives, whose judgment is just. So we ought always to hope in Christ, for as sure as we receive Him now, so sure will we be made conformable to His image. Let us begin by taking no thought about worldly cares, and setting our minds on the righteousness of God and His kingdom, and then we shall have far clearer views about the worldly cares themselves, and we shall be continually enabled to fight them under Him who has overcome the world.

[Sunday] 26th June 1864.

. . . I have come from Mr. Baptist Noel. The church was full to standing, and the whole service was as plain as large print. The exposition was the Parable of Talents, and the sermon was on John iii. 16. The sermon was the text writ large, nothing ingenious or amusing, and hardly any attempt at instruction, but plain and very serious exhortation from a man who evidently believes neither more nor less than what he says.[81]

The pastor mentioned was Baptist W. Noel of the Baptist John Street Chapel near London's Holborn district, north of King's College.[82]

During the late 1860s, when in Glenlair, Maxwell's leadership included leading the household prayers described as being '*extempore*'. The following example, soon after having been printed in the *Life* was quoted in *Does science aid faith in regard to Creation?*[83] and was subsequently included in a *Prayer Book*.[84]

Almighty God, who hast created man in Thine own image, and made him a living soul that he might seek after Thee and have dominion over Thy creatures, teach us to study the works of Thy hands that we may subdue the earth to our use, and strengthen our reason for Thy service; and so to receive Thy blessed Word, that we may believe on Him whom Thou hast sent to give us the knowledge of salvation and the remission of our sins. All which we ask in the name of the same Jesus Christ our Lord.[85]

The partial success of Maxwell's formulation of the kinetic theory of gases, especially his verification (with Katherine's assistance) of the only slight dependence of viscosity on gas pressure by the mid 1860s, turned his interests toward atomic theory, such that Tait could write in 1880 'During the last ten years of his life he had no rival ... in the whole wide domain of molecular forces ...'.[86] As early as 1860, Maxwell had realized that his formulation of kinetic theory failed to predict the observed ratio of the specific heat of air at constant volume to that at constant pressure. Late in 1867, however, his attention turned to the second law of thermodynamics, and by April 1868 he had gained an appreciation of its broader implications. He had read in the 4 April *Saturday Review* an anonymous article 'Science and Positivism' and made some remarks in a letter to Mark Pattison that he wished to communicate to the author. The article examined criticism of Comte's Positivism and recent developments, given by E. M. Caro of the Sorbonne in a recent book, and noted the 'slow but sure advance towards extinction' of the Cosmos.[87] Expanding on this issue, Maxwell noted in letters to Pattison '... our experience of irreversible processes ... leads to the doctrine of a beginning & an end instead of cyclical progression for ever' (7 April) and,

I think we are likely to arrive at physical indicators of a beginning & an end. This end is not a destruction of matter or of energy but such a distribution of energy that no further change is possible without an intervention of an agent who need not create either matter or energy but only direct the energy into new channels.[88]

In December 1867, Maxwell had conceived of the consequences of 'an agent' that eventually became known as a superhuman demon.[89] While to Maxwell the theological implications of the second law pertaining to 'a beginning & an end' were obvious, he may have also viewed his superhuman 'agent' as an avenue for fulfilling the 'new heaven and a new earth' eschatology of Revelation 21:1.[90]

At the British Association meeting in Liverpool in 1870, Maxwell addressed the Mathematical and Physical Section as its president. In this address Maxwell examined some limitations of science and touched on issues of faith. In 1867, Maxwell had become interested in stellar spectroscopy, communicating an analysis of the Doppler shift to the astronomer William Huggins.[91] In his address of 1870, Maxwell examined the remarkable similarity between terrestrial and stellar spectra and applications to the establishment of measurement standards. Concerning the stability of molecular properties he concluded,

. . . when we reflect that no power in nature can now alter in the least either the mass or the period of any one of them, we seem to have advanced along the path of natural knowledge to one of those points at which we must accept the guidance of that faith by which we understand that 'that which is seen was not made of things which do appear.'[92]

The phrase at the end that Maxwell placed in quotation marks is from Hebrews 11:3 which reads: 'Through faith we understand that the worlds were framed by the word of God, so that things which are seen were not made of things which do appear.'[93] Writing in 1872, Murphy noted this section in a book for broader audiences.[94] Maxwell also examined the irreversible dissipation of energy and the cosmic implications:

This idea of a beginning is one which the physical researches of recent times have brought home to us, more than any observer of the course of scientific thought in former times would have had reason to expect.[95]

At about this time Maxwell had interrupted his progress in writing sections of his *Treatise* so as to prepare his *Theory of Heat*, a volume that used only elementary mathematical methods. The first edition was published late in 1871. Some of the themes mentioned in the aforementioned 1870 address were noted, including the implication of the dissipation of energy concerning 'a superior limit to the antiquity of the observed order of things'. Maxwell ends with an essay in which he summarizes why 'the kind of speculation with which we have become so familiar under the name of theories of evolution is quite inapplicable to the case of molecules'.[96] In contrast to Descartes' theory of the evolution of molecules, Maxwell finds, 'Our molecules . . . are unalterable by any of the processes which go on in the present state of things . . .' and concludes,

But if we suppose the molecules to be made at all, or if we suppose them to consist of something previously made, why should we expect any irregularity to exist among them? If they are, as we believe, the only material things which still remain in the precise condition in which they first began to exist, why should we not rather look for some indication of that spirit of order, our scientific confidence in which is never shaken by the difficulty which we experience in tracing it in the complex arrangements of visible things, and of which our moral estimation is shown in all our attempts to think and speak the truth, and to ascertain the exact principles of distributive justice?[97]

The phrase 'distributive justice' was likely used by Maxwell in the sense it was used in Thomas Hobbes' *Leviathan*: 'the act of defining what is just'.[98] An undated handwritten manuscript of this section suggests that Maxwell worked with his wording to extend the meaning conveyed in various sentences: an example is the conclusion of 'to think and speak the truth'.[99] Evidently, in 1870, Maxwell publically supported the importance of a Creator in a scientific context *and* he viewed the Creator as the source of morality.

Return to Cambridge (1871–79)

Maxwell was offered the new Professorship in Experimental Physics in March of 1871 and took up residence in Cambridge by November. In the 1860s and 1870s, informal conversation remained a major part of a Cambridge student's education. By the 1870s, there was growing interest in the exclusive constraints on knowledge imposed by Comte's Positivism. Reactions varied, but a particularly influential scholar was Henry Sidgwick (1838–1900) who, after an extended period of leadership of the Cambridge Apostles, felt compelled to resign his Fellowship at Trinity in 1869 because he no longer was willing to affirm belief in the doctrines of the Church of England.[100] In June 1871, Parliament facilitated greater diversity in Cambridge by passing the Universities Test Act, abolishing religious tests as a requirement for fellowships. The growing emphasis on constraining knowledge, however, was such that by 1879 the future Prime Minister Arthur Balfour, a Cambridge graduate of 1869, was motivated to publish his cautionary *Defence of Philosophic Doubt*, critical of the narrow epistemological constraints of Positivism. In his better known *Foundations of Belief* (1895), he also thought it appropriate to examine evidence raised by Maxwell in the 1870s concerning what Balfour termed an 'Argument from Design': the similarity between molecules and 'manufactured articles'.

For his own part, Maxwell was frequently cautious concerning the limitations of contemporary scientific understanding. In his Cambridge *Introductory Lecture on Experimental Physics* (1871) he emphasized 'new scientific ideas' and the 'unsearchable riches of creation'. In one of his closing paragraphs he indicated that,

> ... our molecular hypothesis may, in its turn, be supplanted by some new theory of the constitution of matter; but the idea of the existence of unnumbered individual things, all alike and all unchangeable, is one which cannot enter the human mind and remain without fruit.[101]

Three years later, in his review of Plateau's work on soap bubbles, Maxwell concluded '... when a physicist is pushing his way into the unknown regions of molecular science. In such matters everything human, at least in our century, must be very imperfect . . .'.[102] Similarly, in his review of H. W. Watson's small book on kinetic theory, in 1877, Maxwell welcomed mathematical propositions pertaining to the constitution of molecules,

> ... likely to startle us out of our complacency, and perhaps ultimately to drive us out of all the hypotheses in which we have hitherto found refuge into that state of thoroughly conscious ignorance which is the prelude to every real advance in knowledge.[103]

While reviews of Maxwell's *Theory of Heat* made it clear that his mode of presentation provided a much needed textbook, the most qualified reviewer, Professor Clausius of Bonn, pointed out in January 1872 that Tait (in an earlier text) and Maxwell had given misleading presentations of entropy and the history of thermodynamics.[104] In response, Tait entered into a series of defensive rebuttals, while Maxwell humbled himself by correcting his errors in revised editions. During his early period in Cambridge, however, the limited time Maxwell had for writing must have been devoted to completing his two-volume *Treatise on Electricity and Magnetism*, the Preface being dated 1 February 1873. These volumes reveal aspects that Maxwell could describe with confidence, while displaying caution in other areas. An example of the former is his diagram, Fig. 14.1, of the spatial relationship between electromagnetic fields 'at a given instant' along a ray of 'plane-polarized light', perhaps the most important diagram in nineteenth century theoretical physics.[105] Examples of the latter are his caution concerning an 'atom of electricity' and avoidance of field singularities such that 'a point charged with a finite quantity of electricity cannot exist in nature'.[106]

In Cambridge, the Maxwells resided at 11 Scroope Terrace in a building adjacent to Trumpington Street. The location was convenient, the walk to the Cavendish Laboratory being about 900 m, and the immediate neighbourhood providing opportunities for stimulating conversation. Nearby, at 6 Scroope Terrace, resided Brooke Foss Westcott (1825–1901), the Regius

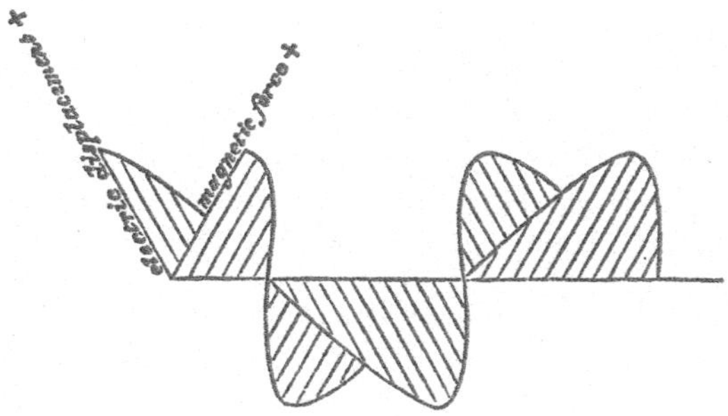

Fig. 14.1 Maxwell's diagram of the spatial relationship between electromagnetic fields 'at a given instant' along a ray of 'plane-polarized light'. (Maxwell, J. C. (1873). *Treatise on Electricity and Magnetism*, Article 791. Oxford University Press.)

Professor of Divinity. After 1872, Maxwell's former acquaintance from the Apostles, Fenton Hort, resided in an adjacent building, St. Peter's Terrace. The vicar William H. Guillemard of the nearby church that the Maxwells attended (Little St. Mary's) also resided nearby.[107] Westcott, a prolific publisher of collected sermons,[108] was especially popular with students for informal discussions at the Sunday suppers he held in Scroope Terrace.[109] Westcott and Hort are often remembered as scholars who systematized New Testament textual criticism as a method for reconstructing the original Greek text. With their Cambridge associate, J. B. Lightfoot (1828–89), they are also widely remembered as influential members of the committee that released a translation of the New Testament in 1881, the *Revised Version*. Lightfoot is also highly regarded as a scholar of New Testament Greek and history.[110]

Some of what is known about the breadth of Maxwell's interests during the 1870s is associated with his participation in the informal *Eranus Club* started by Westcott in May 1872 'for the purpose of hearing and discussing essays prepared by the members'.[111] Among the members were some former Apostles and others having diverse religious perspectives. During the Maxwell era, these included (not likely all present at any given meeting): Westcott, Hort, Lightfoot, Maxwell, Sidgwick, Stokes, Henry Jackson, Alfred Marshal, J. R. Seeley, V. H. Stanton, James Stuart and Coutts Trotter.[112] The meetings, ordinarily scheduled for Tuesday evenings, with approximately six meetings a year, provided the opportunity 'for an interchange of ideas somewhat more serious and methodical than is suitable at an ordinary social gathering'.[113] Some members went so far as to print a title and outline of their presentation in advance of the meeting.[114] Trotter, a Trinity College physics lecturer, was deeply respectful of Maxwell's faith, while Seeley, a Professor of Modern History, had in 1866 published anonymously *Ecce Homo*, controversial as an unorthodox presentation of Christ. The general topic of Maxwell's Eranus Club essay on 'Free Will' (11 February 1873) will be noted in a subsequent section. How Maxwell managed to find the time for this essay is unclear: late in January he published in *Nature* an anonymous review of William Thomson's reprint of papers on *Electrostatics and Magnetism*; on 1 February he had completed the preface to his *Treatise*; and on 21 February he read a lengthy essay at the Royal Institution in London on 'Action at a Distance'. After giving a non-mathematical summary of his electromagnetic theory of light, in one of its closing paragraphs, the Christian context of his thinking is clearly stated: 'The vast interplanetary and interstellar regions will no longer be regarded as waste places in the universe, which the Creator has not seen fit to fill with the symbols of the manifold order of His kingdom.'[115]

The growing demands on Maxwell's time, while establishing the Cavendish Laboratory, and his summer responsibilities at Glenlair were such that after he had committed to giving a prestigious invited lecture at the September 1873 Bradford meeting of the British Association, he endeavoured to be relieved of this task.[116] His invited lecture 'On Molecules' turned out to be his most controversial public pronouncement concerning the relationship between faith and physics. While the majority of the lecture is devoted to summarizing molecular properties and procedures for measuring those properties, his controversial material appeared in the closing paragraphs expanding on themes from his 1870 lecture and the section on molecules in his *Theory of Heat*:

None of the processes of Nature, since the time when Nature began, have produced the slightest difference in the properties of any molecule. We are therefore unable to ascribe either the existence of the molecules or the identity of their properties to the operation of any of the causes which we call natural.

On the other hand, the exact quality of each molecule to all others of the same kind gives it, as Sir John Herschel has well said, the essential character of a manufactured article, and precludes the idea of its being eternal and self existent.[117]

While the similarity with *manufactured articles* had been previously noted by John Herschel[118] and by Maxwell's former Trinity College Master, William Whewell,[119] this perspective had been generally forgotten, and certainly had not been not tied with the remarkable spectroscopic stability of molecules discussed in Maxwell's lecture.[120] In a later paragraph of his lecture, Maxwell continues this creation theme:

Science is incompetent to reason upon the creation of matter itself out of nothing. We have reached the utmost limit of our thinking faculties when we have admitted that because matter cannot be eternal and self-existent it must have been created . . .

That matter, as such, should have certain fundamental properties—that it should exist in space and be capable of motion, that its motion should be persistent, and so on, are truths which may, for anything we know, be of the kind which metaphysicians call necessary. We may use our knowledge of such truths for purposes of deduction but we have no data for speculating as to their origin.

But that there should be exactly so much matter and no more in every molecule of hydrogen is a fact of a very different order. We have here a particular distribution of matter—a collocation —to use the expression of Dr. Chalmers, of things which we have no difficulty in imagining to have been arranged otherwise.[121]

The concept here of a favourable *collocation* of matter was emphasized years earlier by Thomas Chalmers in his 1833 *Bridgewater Treatise*, but had also been largely forgotten by Maxwell's generation of scientists.[122] Maxwell closes by contrasting the stability of molecules with inherent decay of astronomical systems and observes:

But though in the course of ages catastrophes have occurred and may yet occur in the heavens, though ancient systems may be dissolved and new systems evolved out of their ruins, the molecules out of which these systems are built—the foundation stones of the material universe—remain unbroken and unworn.

They continue this day as they were created—perfect in number and measure and weight, and from the ineffaceable characters impressed on them we may learn that those aspirations after accuracy in measurement, truth in statement, and justice in action, which we reckon among our noblest attributes as men, are ours because they are essential constituents of the image of Him Who in the beginning created, not only the heaven and the earth, but the materials of which heaven and earth consist.[123]

Maxwell lectured on Monday evening, 22 September. The lecture was given brief but favourable press coverage, and even after five decades the physicist Oliver Lodge (1851–1940) could recall how the lecture aroused his 'utmost admiration'.[124] Joseph Hooker, a botanist, expressed a low opinion of the lecture, however, in a letter to Charles Darwin, indicating that the lecture lacked significance.[125] By the following Saturday, Darwin had been able to read Maxwell's lecture and suggested that Hooker would view it more favourably by reading it.[126]

Maxwell's next significant public pronouncement came in the form of a book review, published anonymously in *Nature* in January 1874. The book under review was Balfour Stewart's *Conservation of Energy*. We can be confident that Maxwell wrote the review because Westcott attributed the review to Maxwell[127] and because some peculiar sentences of the book review are identical to text in Maxwell's (at the time unpublished) Eranus essay on *Free Will*.[128] Maxwell's remarks in the review are important because he examines the consequences of irreversible processes in much broader terms than in prior discussions. He had previously emphasized the singularities encountered in Fourier's analysis of heat flow caused by reversing the direction of time. To Maxwell, the cosmological implications were clear:

> Here, then, we have an irreversible process always going on, at a greater or less rate, in the universe. If, therefore, there was ever an instant at which the whole energy of the universe was available energy, that instant must have been the very first instant at which the universe began to exist. If there ever shall come a time at which the whole energy of the universe has become unavailable, the history of the universe will then have reached its close. During the whole intervening period the available energy has been diminishing, and the unavailable increasing by a process as irresistible and as irreversible as Time itself. The duration of the universe according the present order of things is therefore essentially finite, both *à parte ante* and *à parte post*.[129]

Probably the first significant philosophic criticism of Maxwell's lecture on 'Molecules' was given by the mathematician William Kingdon Clifford (1845–79) in April 1874 at the London Sunday Lecture Society.[130] In his lecture, Clifford objected to Maxwell's opposition to applying evolutionary thought to molecules and strongly objected to the broader implications of a beginning inferred by various writers from singularities in the backwards analysis of heat flow.[131] If Clifford, a gifted essayist, had noticed the more general discussion by Maxwell in *Nature* (January 1874) he was unaware of Maxwell's authorship. Maxwell, who was busy preparing for the formal opening of the Cavendish Laboratory in June of 1874, was probably unaware at the time of Clifford's criticism. Maxwell's position was clear from a Latin verse above the Cavendish entrance, which in English is 'Great are the works of the LORD, studied by all who have pleasure in them'.[132]

Public awareness of Maxwell and his lecture on 'Molecules' grew significantly following John Tyndall's controversial *Belfast Address* as the President of the British Association in August 1874. Oliver Lodge recalled this address, lasting nearly two hours, as 'the chief pronouncement of the materialism of the nineteenth century'.[133] Concerning the address and the local responses, Lodge remarked, 'I remember feeling the atmosphere grow more and more sulpherous as the materialistic utterances went on . . .'.[134] The paragraphs of the *Belfast Address* devoted to Maxwell were prefaced by remarks on Pierre Gassendi (1592–1651) who, instead of asserting that atoms were eternal, had allowed that God created atoms. After complementing Maxwell, Tyndall continued:

> There seems to be this difference, however, between Gassendi and Maxwell. The one *postulates*, the other *infers* his first cause. In his 'manufactured articles,' as he calls the atoms, Professor Maxwell finds the basis of an induction which enables him to scale philosophic heights considered inaccessible

by Kant, and to take the logical step from the atoms to their Maker. Accepting here the leadership of Kant, I doubt the legitimacy of Maxwell's logic; but it is impossible not to feel the ethic glow with which his lecture concludes . . .[135]

Since the Belfast Address and the associated controversy have been widely studied[136] it is necessary here to note some aspects of the controversy that are frequently overlooked. It is especially helpful to consult local accounts, suggesting that the address *as delivered* differed at times from the early published text[137] and that Tyndall had made significant historical errors.[138] Discussions of the address during the subsequent half-dozen years sometimes referred to Maxwell in contrast with Tyndall, hence raising awareness of Maxwell's British Association addresses of 1870 and 1873.[139]

Maxwell responded with a lengthy serio-comic poem, 'Notes of the President's Address', one form of which was eventually published anonymously in *Blackwood's Magazine*.[140] This included a line critical of Tyndall's understanding of *Butler's Analogy*. He penned two revealing related short poems: 'Molecular Evolution' and 'Song of the Club', referring to the 'Red Lions' club of established scientists. The opening lines of the latter poem made clear the overreaching domain of the *Belfast Address*:

I KNOW not what this may betoken,
That I feel so wondrous wise;
My dream of existence is broken
Since science has opened my eyes.
At the British Association
I heard the President's speech,
And the methods and facts of creation
Seemed suddenly placed in my reach.[141]

Writing to his uncle Robert Cay in October 1874, he included a humorous note on evolutionary thought he had penned also at the Belfast Meeting,

Tell her [Aunt Jane] to look out in the November number of Blackwood for notes on Professor Tyndall's speech. Here is an observation which occurred to me at Belfast and which I submitted to the Apostle of Evolution—Herbert Spencer.

Domestic Evolution:

The ancients made foes, whom they saved from the slaughter
  Into hewers of wood and drawers of water.
  We moderns, reversing arrangements so rude,
  Prefer ewers of water and drawers of wood.[142]

In the poem Maxwell contrasts the attitude of God-fearing Israelites with modern materialists. The Israelites honoured an oath made to God and preserved the lives of the Gibeonites even though they had been deceived by them (Joshua 9). In contrast, the materialist is satisfied with the ewers and drawers in hand.

At the meeting in France of the Association for the Progress of Science, the President opened with an address that was much closer to Maxwell's views.[143] The seriousness of the controversy during the months following the Belfast Address is evident, because scholarly journals such as *Nature* tolerated open discussion of the limitations of science. In October, Maxwell's old friend C. J. Monro published a letter critical of his assertion that molecules are like 'manufactured articles', Monro evidently forgetting Maxwell's desire (expressed to him in 1855) to avoid 'arguments'.[144] In opposition to Tyndall, P. H. (Philip Henry) Pye-Smith (physician and physiologist, Guy's Hospital, London) published in November a translation of a critique by the German physiologist Rudolf Virchow (1821–1902) from 20 years earlier:

> Of all kinds of dogmatism, the materialistic is the most dangerous, because it denies its own dogmatism, and appears in the garb of science; because it professes to rest on fact, when it is but speculation; and because it attempts to annex territories to natural science, before they have been fairly conquered.[145]

Following his trip to Belfast, Maxwell returned to Glenlair to work on his article on 'Atoms' for the third volume (published in 1875) of the 9th edition of the *Encyclopedia Britannica.* Maxwell gave greater attention to the relationship between atoms and molecules and a detailed discussion of William Thomson's vortex ring theory of atoms. He retained some of the themes related to creation and the beneficial collocation of matter associated with molecules, expanded the discussion of reasons for viewing atoms as manufactured articles and strengthened his attribution to Herschel. The most important new theme taken up in this essay was discussion relating the finite size of molecules to constraints on theories of genetics and heredity:

> Thus molecular science sets us face to face with physiological theories. It forbids the physiologist from imagining that structural details of infinitely small dimensions can furnish an explanation of the infinite variety which exists in the properties and functions of the most minute organisms.[146]

One of the theories rightfully criticized by Maxwell was the hereditary theory of Pangenesis, a 'provisional hypothesis' originally published by Charles Darwin in 1868.[147] Maxwell's concern about the limitations imposed by the finite size of molecules related to the way in which Darwin had proposed virtually complete genetic information was passed through *many* subsequent generations in a dormant form. Maxwell realized, however, that there must be some molecular mechanism for retention of ancestral peculiarities, but that the amount of information retained was limited. He noted this in his letter to Campbell in September 1874.[148] A few scientists of the period attempted to give attention to Maxwell's constraints, however some failed to appreciate the depth of Maxwell's concerns.[149] Ultimately the resolution required the abandoning of Darwin's Pangenesis in favour of Mendelian inheritance and the molecular structure of DNA. The discovery of the electron gave early hope that Maxwell's dilemma would be resolved.[150]

The most widely read response to Tyndall's Belfast Address written by *physical scientists* appeared anonymously early in 1875: *The Unseen Universe or Physical Speculations on a Future State.* Within a few months the second edition appeared and numerous subsequent editions followed, the sixth appearing in 1876. By that year, Maxwell's friends Balfour Stewart and P. G. Tait were listed as the authors. There are elements of their presentation that were in general agreement with Maxwell's thinking, such as '. . . the argument is in favour of the production of the visible universe by means of an intelligent agency residing in the invisible universe' with a reference to Herschel and Maxwell concerning the 'marks of [atoms] being manufactured articles'.[151] Other examples are the notion of a 'Great First Cause' and that the 'transformable energy' of the 'visible universe' must 'come to an end'. There were, however, aspects central to their presentation not in agreement with Maxwell's position.[152] A major emphasis (explained relatively succinctly in the Preface to the Second edition, and reprinted in subsequent editions) concerns the extreme priority of the 'principle of Continuity' . . . 'which leads us, whatever state of things we contemplate, to look for its antecedent in some previous state of things also in the Universe'. They apply this principle to the problem of immortality by making the supposition that upon death there is 'a transference from the visible universe to some other order of things intimately connected with it'.[153] They go on to speculate concerning the nature of the connection. One of the possibilities is that, since consciousness appears to be connected somehow with the ether, from the principle of Continuity there should be another ether, connected with the first, connected also with consciousness in the immortal state. While the expression was not used by Stewart and Tait, for various critics (including Clifford, one of the most forceful) the other ether became known as the 'second ether'. Christians holding orthodox views of scripture also frequently expressed their displeasure with the book.[154] Tyndall, for his part, issued a series of replies to critics and saw no reason to avoid continuing mention of Maxwell; in the *Fortnightly Review* of 1875, he refers in a derogatory way to 'the Power whom Gassendi and Clerk Maxwell present to us under the guise of a "manufacturer" of atoms . . .'.[155]

By 1875, Maxwell was aware of the growing interest in his own views concerning the relation between physical sciences and religion. In March he received a special invitation to join the Victoria Institute, an organization sponsoring a journal and annual meetings devoted to reconciling science and issues related to faith. Unlike his Cambridge friend Professor G. G. Stokes (who eventually served as the Institute's president), Maxwell declined to join, indicating,

Among the objects of the Society are some of which I think very highly. I think men of science as well as other men need to learn from Christ, and I think Christians whose minds are scientific are bound to study science that their view of the glory of God may be as extensive as their being is capable of. But I think that the results which each man arrives at in his attempts to harmonise his science with his Christianity ought not to be regarded as having any significance except to the man himself, and to him only for a time, and should not receive the stamp of a society. For it is of the nature of science, especially of those branches of science which are spreading into unknown regions to be continually—[156]

At this point the draft of his reply ends. In his personal collection of published book reviews of the various editions of his *Theory of Heat*, are copies of reviews from 1876 of the fourth edition.[157] In each case an anonymous reviewer went out of his way to examine Maxwell's views concerning atoms bearing the stamp of 'manufactured articles' and his difficulties in accounting for the state of molecules by 'natural selection'. In November 1876, the Anglican Bishop C. J. Ellicott was preparing for publication some lectures he had prepared on *Modern Unbelief: its principles and characteristics* in which he had noted the similarity between molecules and manufactured articles. He used this occasion to write Maxwell for additional information and to ask him about his understanding of 'the creation of light' in the beginning of Genesis. Maxwell's thoughts on science and Scripture were revealing. Concerning the specific aforementioned question related to Genesis he emphasized again the changing nature of science:

If it were necessary to provide an interpretation of the text in accordance with the science of 1876 (which may not agree with that of 1896), it would be very tempting to say that the light of the first day means the all-embracing æther, the vehicle of radiation, and not actual light, whether from the sun or from any other source.[158]

Maxwell's concluding remarks are especially important for appreciating his perspective:

But I should be very sorry if an interpretation founded on a most conjectural scientific hypothesis were to get fastened to the text in Genesis, even if by so doing it got rid of the old statement of the commentators which has long ceased to be intelligible. The rate of change of scientific hypothesis is naturally much more rapid than that of Biblical interpretations, so that if an interpretation is founded on such an hypothesis, it may help to keep the hypothesis above ground long after it ought to be buried and forgotten.

At the same time I think that each individual man should do all he can to impress his own mind with the extent, the order, and the unity of the universe, and should carry these ideas with him as he reads such passages as the 1st Chap. of the Ep. to Colossians (see Lightfoot on Colossians, p. 182), just as enlarged conceptions of the extent and unity of the world of life may be of service to us in reading Psalm viii.; Heb. ii. 6, etc.[159]

The reference to Colossians concerns the active role of Christ in Creation. By consulting the referenced commentary (by his Cambridge friend J. B. Lightfoot) the noted page emphasizes what Lightfoot termed 'the creative and administrative work of Christ . . .'.[160] By holding to this scripture, Maxwell was avoiding the demiurge view of the creator held by some Platonists and Deists of an earlier generation. Ellicott replied to Maxwell's letter, thanking him for the 'packet of pamphlets' that included the encyclopedia article on 'Atoms'. Eventually, Ellicott appears to have given Maxwell a copy of *Modern Unbelief* since a copy was listed in his personal library along with another item building on Maxwell's perspective.[161]

During his tenure as a Cambridge professor, Maxwell remained an Elder in a local Church of Scotland near Glenlair. His obligations in Cambridge were such that he could return to Glenlair in time to 'officiate at the midsummer communion'.[162] Some years he was also able to return to Glenlair for a brief period near Christmas. In December 1873, when an occasion required him to be separated from Mrs Maxwell, he wrote her a letter similar to those of earlier years:

I am always with you in spirit, but there is One who is nearer to you and to me than we ever can be to each other, and it is only through Him and in Him that we can ever really get to know each other. Let us try to realise the great mystery in Ephesians v., and then we shall be in our right position with respect to the world outside, the men and women whom Christ came to save from their sins.[163]

When in Glenlair he continued to lead devotions with the workers and families affiliated with the property. He was well liked in the local community, having a reputation of caring for families in need going back to the late 1860s. A biographer writing at the centenary of Maxwell's birth noted the favorable recollections of some living near Glenlair:

The fragrance of his memory still lingers. In some of the cottages a pair of faded eyes will light up at the mention of his name, and an old figure, hunched over the fire for warmth, will turn to exclaim fervently:—'Eh, he was a richt nice mon!' Which in Galloway is praise indeed.[164]

Even as late as 1877, Maxwell was asked to attend meetings concerned with the welfare of those in the region.[165] Adding to Maxwell's cares, his wife's health was deteriorating in the 1870s. For a while her sister, Mrs McCunn from Dunoon, resided with them in Cambridge.[166] Then, by spring 1877, his own health began to deteriorate, markedly so (according to Maxwell) by summer 1879.

Published criticism of Christian thought was sufficiently widespread that the 1870s editor of the liberal *Fortnightly Review*, John Morley, noted in his *Recollections* that a historian had identified 1877 as the year English rationalism was most intensely expressed.[167] Maxwell's examination of aspects of the issues is evident from two essays he wrote in 1878: 'Psychophysik', for the Eranus meeting of 5 February 1878,[168] and a review published in *Nature* in December 1878 of Stewart and Tait's latest book, *Paradoxical Philosophy*.[169] Maxwell opens the first of these essays with a rhetorical 'Whence came we? Whither are we going? And what should we do now? . . .' followed by a recognition of recent changes concerning even the nature of such questions. While he never mentions Milton in this essay, he could well assume all of his listeners were familiar with words Milton put in the mouth of Adam: '. . . how came I thus, how here? Not of myself; by some great Maker then, in goodness and in power preeminent. . .'.[170] Concerning the changing view of such issues, Maxwell narrows his theme and proceeds to examine contemporary perspectives concerning consciousness at the cellular level.

In his review of *Paradoxical Philosophy*, Maxwell went out of his way to draw on material from his 'Psychophysik' essay, much like he had previously drawn on material from his 'Free Will' essay in his anonymous review for *Nature* of Stewart's book *Conservation of Energy*. It is thus appropriate to combine the discussion of his 1878 Eranus essay with his review for *Nature*. As background information on the novel *Paradoxical Philosophy*: in comparison with *The Unseen Universe* this account of the proceedings of the fictitious learned Paradoxical Society makes informal reading.[171] The novel may be taken as less offensive than W. H. Mallock's recently published *New Republic* which contained what Maxwell termed 'echoes of well-known voices'.[172] By contrast Maxwell viewed Dr. Stoffkraft of the Paradoxical Society as one who 'makes it his chief care so to brandish his materialistic weapons as to not hurt the feelings of his friends'.[173] The tone and substance of Maxwell's review make it clear that while he doesn't *endorse* Stewart and Tait's version of a reconciliation of science and religion, he is also not in agreement with the materialists. Maxwell pokes fun at the 'conviction of the perpetual validity of the "Principle of Continuity" enforced by the tremendous sanction [of Stewart and Tait as well as the Materialists] that if at any place or at any time a single exception to that principle were to occur, a general collapse of every intellect in the universe would be the inevitable result'. His closing paragraph of his review gives a gentle condemnation of the reasoning of *The Unseen Universe*:

> The progress of science, therefore, so far as we have been able to follow it, has added nothing of importance to what has always been known about the physical consequences of death, but has rather tended to deepen the distinction between the visible part, which perishes before our eyes, and that which we are ourselves, and to shew that this personality, with respect to its nature as well as to its destiny, lies quite beyond the range of science.[174]

His disagreement with Stewart and Tait, as well as the position supported by Stoffkraft, are also evident from his comic poem, 'To Hermann Stoffkraft' that he sent to Tait late in 1878, cautioning Tait that it was not for publication in *Nature*.[175] As in several other poems following Tyndall's Belfast Address, Maxwell gives a *reductio* of the materialist position.

In his book review as well as his Eranus essay Maxwell examines conflicting scientific opinions concerning the theory of consciousness and its origin, described as 'plastidule souls'. A letter to Tait, in December 1877, suggests that Maxwell became aware of this speculative theory by reading recent essays in *Nature*; he mentions the discussion of Professor Nägeli from October, though it is likely that he also read one (also in October) by Ernst Haeckel, a German biologist. Haeckel defines 'plastidule' as a 'molecule of protoplasm'.[176] Haeckel extended the usage to 'psychic activities of the protoplasmic molecule' and also introduced the terms 'plastidule soul' and 'atom-soul'. With this background, Nägeli went on to develop a theory of consciousness, and it is his theory that Maxwell dismisses. Nägeli appears to have been at least partly motivated by the desire to overcome a difficulty of evolutionary theory brought up by the German physiologist, Emil du Bois-Reymond, a well-known contributor to the field of electrophysiology. Du Bois-Reymond had pointed out in 1872 that 'in the first trace of pleasure which was felt by one of the simplest beings in the beginning of animal life on the earth, an insuperable limit was marked. . .'.[177] Professor Nägeli, however, asserted 'there exists a continuous gradation without definable limit' in consciousness and his argument extends down to, and including, molecular mechanics. Making light of Nägeli's presentation of dynamics and the associated pleasure *felt by molecules*, Maxwell summarizes 'to attribute life, sensation, and thought to objects in which these attributes are not established by sufficient evidence is nothing more than the good old figure of Personification'. He also criticized Herbert Spencer's trivialization of the problem of the origin of consciousness.

In his Eranus essay, Maxwell also examines some positions of Liverpool physician John Drysdale, whom he describes as a 'Christian materialist'. While his reaction to Drysdale is in keeping with his general support of free will and opposition to mechanical determinism, his mentioning of Drysdale is important because of the broader aspects of Drysdale's recent essay that Maxwell appears to have studied. In that essay, Drysdale takes issue with Stewart and Tait's discussion of *The Unseen Universe*; he also dismisses Maxwell's concern over the incompatibility of estimates of molecular size with the prevailing thought about mechanisms of heredity. Drysdale's essay does, however, capture the bias of some scientists against Christians, when he summarizes the way in which a recent disciple of Haeckel dismisses a critic.[178]

W. K. Clifford continued to contribute to the public discussion on science and religion in the late 1870s nearly up to his death on 3 March 1879. By mid 1876 he had started having significant respiratory problems, likely aggravated by his residing in London. The biologist, T. H. Huxley, and the scholar and lawyer, Frederick Pollock, arranged, without Clifford's knowledge, to take up a collection from his friends to send him to a warmer climate (and better air) for part of the summer. Maxwell contributed £5, writing to Pollock that he hoped for Clifford's recovery and (what may be interpreted as) a calming of Clifford's thoughts.[179] Clifford had been active in the Metaphysical Society, a selective London discussion club with a renowned list of members. Unlike Maxwell's essays for the Eranus, however, Clifford's essays were quickly published in widely circulated periodicals.[180] Shortly after Clifford's death, his *Lectures and Essays* on general and philosophic topics were collected and republished. Appropriately, Maxwell was asked to write a review, and he studied the essays during the summer of 1879 but was unable to submit a review due to his declining health. Hort, in a letter to Campbell of February 1882, recalled Maxwell's remark 'there were many things in the book that wanted trouncing, and yet the trouncing had to be done with extreme care and gentleness, Clifford was such a nice fellow'.[181] According to Hort, up to the summer of 1879, Maxwell had not apprehended for himself 'anything worse than transient ill-health'.[182]

For Maxwell, 1879 began with a sad note: The house of the Attendant for the Cavendish Laboratory had been constructed in an unsanitary way such that, in January, three of the Attendant's children contracted typhoid. One of them died.[183] The enrolment in Maxwell's lectures was small[184] and by autumn he was unable to lecture. The accounts given of Maxwell's final months recorded by Campbell and Garnett, and available supplemental material, give no indication of a weakening of Maxwell's biblical faith so evident in his letters to Katherine from the late 1850s, early and mid 1860s and early 1870s. When William and Mrs Garnett visited the Maxwells in Glenlair in September, they were startled by the changes in his physical appearance, and perhaps only then fully appreciated the gravity of his situation. Prior to Maxwell's return to Cambridge in early October, he was able to attend for a final time his local Scottish Kirk in Corsock, and an important account of the visit has survived. It happens that George Walker, one of his former students from Aberdeen, became the minister at a neighbouring Scottish

Kirk and thus renewed his acquaintance with Maxwell and became acquainted with the local minister (probably George Sturrock). Walker passed on the following account from the local minister:

... the text happening to be, 'Eye hath not seen nor ear heard,' etc, he was more than usually attentive; and afterwards, in great pain and weakness, opened out his mind to his friend in some wonderful thoughts upon the probably nearness to us, even in this life, of the Unseen Universe, to him then near indeed. In broken and difficult utterances he spoke of the invisible world as encompassing us about, heaven bordering upon earth, the human touching the divine; that it is not easy to fix on the boundary-line between, that the darkness is within ourselves, that the light is very great and very near, but that the Spirit of God comes and dispels our darkness, and then all becomes light, God filling the soul with a flood of light. For what the eye cannot see is spiritually discerned.[185]

Two accounts have survived from Rev. Guillemard (of Little St. Mary's in Cambridge) which are of value, even though he indicates Maxwell was not his 'Intimate Friend'. They primarily concern visits with Maxwell during his final weeks. One of these letters was originally intended for publication and notes:

He ... joined fervently in all acts of prayer, listened with a most intelligent interest to all I read, either out of the Bible (which he knew well-nigh by heart) or out of any of our great devotional writers in prose or poetry; was especially fond of any new hymns, and frequently capped such by reciting from his wonderful memory some parallel passages of his favourite old authors, specially George Herbert.[186]

Once, as Guillemard was putting on his liturgical garments to offer Maxwell the Lord's Supper, Maxwell took it upon himself to recite George Herbert's poem *Aaron* in which the central reflection is,

Christ is my only head,
My only heart and breast,
My only music, striking me, e'en dead;
That to the old man I may rest,
And be in Him new drest.

The poem suggests that Maxwell was looking beyond the garments to the core of his faith. Guillemard wrote an account, late in 1879, to a friend of Maxwell from the early 1850s, Rev. Isaac Bowman. Here, Guillemard included more personal information:

... he was never downcast or overburdened, and yet he was the humblest and most diffident of men, with the deepest sense of his own unworthiness, of his many short-comings, of his neglected opportunities. 'But he loved much, and love had cast out fear.' ... his illness drew out the whole heart and soul and spirit of the man: his firm and undoubting faith in the Incarnation and all its results; in the full sufficing of the Atonement; in the work of the Holy Spirit.[187]

The most biographical account was provided in F. J. A. Hort's letter to Campbell of February 1882 (previously quoted) that includes Maxwell's summary of his life, ending with the testimony:

I have been thinking how very gently I have been always dealt with. I have never had a violent shove in all my life.

The only desire which I can have is like David to serve my own generation by the will of God, and then fall asleep.[188]

Hort also noted Maxwell's observation:

My interest is always in things rather than in persons. I cannot help thinking about the immediate circumstances which have brought a thing to pass, rather than about any will setting them in motion. What is done by what is called myself is, I feel, done by something greater than myself in me. My interest in things has always made me care much more for theology than for anthropology . . .[189]

Hort, who lived very near the Maxwells, needed to rush off from one of his visits to attend the opening services of the new Divinity School in Cambridge.[190] Hort's own perspective on Maxwell included the observation:

The testimony of his unshaken faith to Christian truth was, I venture to think, of exceptional value on account of his freedom from the mental dualism often found in distinguished men who are absorbed chiefly in physical inquiries. It would have been alien to his whole nature to seclude any province of his beliefs from the free exercise of whatever faculties he possessed; and in his eyes every subject had its affinities with the rest of the universal truth.[191]

Maxwell died at noon on Wednesday 5 November, and Hort's account above is consistent with what he wrote his daughter four days later: 'He was one of the greatest men living, and, I believe it is not too much to say, one of the best; his rare powers being united to a high type of pure and simple Christian character. There is no one who can take his place here.'[192] Another early account notes,

So far from possessing aught of the spirit of the recluse, his greatest object in life seemed to be the happiness of those about him. During the nine years that he held the chair of Experimental Physics in Cambridge he never uttered a hasty word even to his servants, but treated all alike as brethren and equals. If mistakes were made or accidents occurred which caused him trouble, expense, or delay, he never complained, but tried to smooth the way for those who were to blame . . . For twenty-one years his devotion as a husband was almost without parallel, while he held the noblest and most exalted views of the married tie. Only a few days before his death he remarked that the love between husband and wife taught us much respecting the higher love of God. Professor Maxwell seems to have been a diligent biblical student, as well as to have been intimately acquainted with the works of many of the Christian hymn writers. He took great pleasure in repeating from memory some of the hymns of George Herbert, Richard Baxter, and others. Baxter's hymn beginning—

Lord, it belongs not to my care
Whether I die or live,
And terminating—
My knowledge of that life is small,
The eye of faith is dim;
But 'tis enough that Christ knows all,
And I shall be with Him.

seemed to be a special favourite. One great feature in his religion was that he held no doctrine at second hand, but proved all things. A few days before his death he said that he had investigated every system of atheism he could meet with, and that, independently of any prior knowledge of the needs of man, he found that every system involved a God at the foundation to make it workable. He then added—'It is little of the truth we can gain here, but it is something to know in whom we have believed!'[193]

According to *The Cambridge Review*, a funeral over Maxwell's remains was held on Monday in the Chapel of Trinity College. At the close the choir sang the hymn, 'Jesus Lives! thy terrors now can no longer, death, appall us'. Hort, Westcott and Stokes attended and quite possibly other members of the Eranus, along with Garnett and numerous members of the University. (Lightfoot was unable to attend, having gone to Durham as Bishop, as Westcott did a decade later.) Maxwell's remains were buried in Parton (near Glenlair) with George Walker at the graveside.

Rev. Henry Montagu Butler remembered Maxwell at the Sunday service on the 16 November at Great St. Mary's Church (near Trinity College). Butler, who like Hort overlapped with Maxwell's tenure as an 'Apostle', recalled Maxwell's '. . . unfailing reverence, the singular absence of any taint of the breath of worldliness in any of its thousand forms'.[194] A footnote to his sermon included in *The Cambridge Review* (but omitted by Campbell and Garnett) referenced Maxwell's controversial 1873 address on molecules. The controversy could not have escaped Maxwell's associates in Cambridge.[195]

Free Will, Determinism, Morality and Chaos

Beginning in the mid 1850s, through the remainder of his life Maxwell shared (with other Christian thinkers of his generation) a concern over the growing popularity of determinism applied to human actions.[196] As noted by Hamilton, his philosophy teacher from the late 1840s, a shift from free will to determinism could undercut Christian teaching on responsibility. This concern, or related aspects of moral perceptions, can be found in various of Maxwell's essays from the 1870s, including all of his known Eranus essays and some of his book reviews. The full title of his 1873 Eranus essay only hints at his concern: 'Does the progress of Physical Science tend to give any advantage to the opinion of Necessity (or Determinism) over that of the Contingency of Events and the Freedom of the Will?' By that time, the conservation of energy was widely accepted as applicable to living organisms, leaving the question of how could it be possible for the soul to alter behaviour in the absence of a source of energy? To resolve this problem Maxwell drew upon two physical principles: (a) the direction of motion of an object is altered by a force applied perpendicular to the motion without a transfer of energy, and (b) for unstable systems very small changes in the conditions can produce large changes in the outcome. Related to (a) Maxwell noted in his 1873 essay that the 'soul of an animal' functions as a 'steersman of a vessel—not to produce, but to regulate and direct the animal powers'.[197] As early as 1862, in correspondence he had likened the action of the soul to the actions of a railway pointsman who shifts the track taken by a train.[198] Related to (b) Maxwell notes in his anonymous 1874 review of Stewart's *Conservation of Energy* that 'In unstable systems, like antecedents do not produce like consequents; and as our knowledge is never more than an approximation to the truth, the calculation of what will take place in such a system is impossible to us'. Following the corresponding discussion in his 1873 essay, Maxwell noted that 'singular states' associated with instabilities should be relevant to 'human endeavour'. He closes with the profound observation:

If, therefore, those cultivators of physical science from whom the intelligent public deduce their conception of the physicist, and whose style is recognised as marking with a scientific stamp the doctrines they promulgate, are led in pursuit of the arcane of science to the study of the singularities and instabilities, rather than the continuities and stabilities of things, the promotion of natural knowledge may tend to remove that prejudice in favour of determinism which seems to arise from assuming that the physical science of the future is a mere magnified image of that of the past.[199]

Maxwell's essay remained timely even when it was first published in 1882. When Prime Minister W. E. Gladstone read Campbell and Garnett's *Life* early in 1883, he noted several items of interest in the essay on free will.[200] It is perhaps unfortunate that Maxwell neglected to publish his essay in 1873, because the issues raised were relevant not only to Tyndall's 1874 *Belfast Address* but also to an address at the same meeting by the biologist T. H. Huxley on 'Animal Automata'. While it is not known if Maxwell listened to Huxley's address, his knowledge of theology was undoubtedly sufficient to conclude, along with Robert Watts, that Huxley's assertion was incorrect that the doctrines of the eighteenth-century American theologian Jonathan Edwards supported his position that 'man is a conscious automaton'. Watts quotes Edwards:

[I would say] that man is entirely, perfectly and unspeakably different from a mere machine, in that he has reason and understanding with a faculty of Will, and so is capable of volition and choice . . . and by means of these things is capable of moral habits and moral acts . . .[201]

By late in 1878, when Maxwell gave his final observations on these issues in his review of *Paradoxical Philosophy* in *Nature*, he had incorporated in his thinking the related analysis of the Catholic mathematician Joseph Boussinesq. At the time Boussinesq had published an analysis of the solution of systems of differential equations associated with certain complicated systems.[202] Maxwell was sufficiently excited by this development that he discussed it in a letter in February 1879 to Francis Galton, opening the discussion with 'Do you have any interest in Fixt Fate, Free Will &c.'[203] Boussinesq discovered situations for which – in Maxwell's terms – a 'bifurcation of path occurs' in the solutions such that if 'the Will' were coupled with an analogous physical system, the outcome is altered without any expenditure of 'trigger-work' by the soul.

Over a century later, a pioneer of the modern mathematical theory of chaotic motion, James Yorke, noted the insightfulness of Maxwell's essay on 'Free Will' in anticipating some aspects of, and the importance of, the modern theory of chaotic motion.[204] The case for Maxwell's appreciation of the sensitivity to initial conditions is made even stronger by noting a closing paragraph of his 1874 anonymous book review, in which he distinguished dynamically allowed reversibility (based on 'exact reversal') from imperfect reversibility: 'If but one atom were to receive a velocity differing infinitesimally from an exact reversal, that atom would leaven the whole universe with that tendency to dissipation of energy which actually exists . . .'[205]

Remembering Maxwell as a Man of Faith

It is impossible to think of him whom we have so lately lost, to whom first the charge of Cavendish laboratory was committed, Professor Clerk Maxwell, and to recollect his subtle and profound thought, his tender and humble reverence, without being sure that the close connection between physics and theology, is still a living reality among us. I owe a deep debt of gratitude to Professor Maxwell for his researches . . .[206]

In Maxwell's era, issues relating to science and faith were sufficiently polarizing that during his lifetime and in the subsequent half-century his related views were brought up in what appear to be hundreds of publications. It may be surprising in the twenty-first century to realize that some of the most extended of these discussions appeared anonymously. For example, in 1880 a thoughtful book, *The New Truth and the Old Faith*, was published in London by 'A Scientific Layman'. It includes a favourable account of some of Maxwell's views. That book may have been in part a response to books by David Strauss and to the highly critical book *A Candid Examination of Theism* by 'Physicus' (a pseudonym) which appeared in 1878.[207] Physicus described aspects of Maxwell's discussion of molecules having the appearance of manufactured articles as an 'atrocious piece of arrogance'. Even though, by 1879, a review had appeared pointing out the harsh treatment of Maxwell, nothing has survived to suggest that Maxwell was aware of this criticism. The main concern of Physicus was to counter the exposition in Robert Flint's *Theism* (1877). Eventually, Physicus was revealed to be George John Romanes (1848–94) during a period in which Romanes had forsaken his Christian faith.[208] He subsequently returned to faith and in a posthumous publication respectfully mentions Maxwell and Stokes as illustrious scientists on the side of orthodoxy.[209] As an indication of his variability in the 1870s, Charles Darwin's library contained a copy of *Candid Examination* sent to him by Romanes; inspection of this book shows a penned revision of the text in Romanes' hand on one of the pages.[210]

After the improved understanding of radioactive decay in the early 1900s, Professor John Cox was particularly critical of Maxwell's statements about atomic stability.[211] Others have emphasized, however, that from the perspective of classical physics, the atom is remarkably stable. The biologist J. Arthur Thomson remarked, concerning Maxwell's 1873 address on Molecules, in the 1915–16 Gifford Lectures:

> For 'molecules' a modern chemist would read 'atoms', and even then he would remind us of the apparent disintegration of the atom in radio-active substances. But in the main what Clerk-Maxwell said remains true. It is evident that the fulcrum of the inorganic on which organisms rest their lever is one of reliable steadiness.[212]

Some physicists trained in the first third of the twentieth century reflected on Maxwell's positions, such as James Jeans in *Astronomy and Cosmology* (where an analogy of 'astronomical objects' with 'manufactured articles' is noted)[213] and Hendrik Casimir, in his autobiography.[214] If Katherine was concerned over the controversy related to her late husband's manifestations of faith, she didn't discourage Campbell and Garnett from including illustrations related to his faith in their biography. Campbell sent R. B. Litchfield, Maxwell's college-era friend, proofs of the first two-thirds of the *Life*. Litchfield found the time to read it even though his father-in-law, Charles Darwin, had died only a few months prior. Early in August 1882, Litchfield wrote C. J. Monro concerning the proofs: 'The amount of evidence of strong religious belief (besides religious feeling) was a surprise to me.'[215] C. J. Monro died late in November 1882, so that if he ever had the opportunity to read the *Life* his reaction is not known.

After the *Life* appeared, late in 1882, it was given numerous favourable reviews in religious as well as scientific periodicals. An anonymous review in the religious *London Quarterly Review* of April 1883 is so intensely personal that it suggests the reviewer had a direct knowledge of Maxwell. The opening discussion places Maxwell in the context of Francis Bacon's observation 'Slight tastes of philosophy may perchance move to atheism, but fuller draughts lead back to religion'. The reviewer asserts that Maxwell was 'diametrically opposed' to Tyndall's views expressed in the *Belfast Address* and hints that he also opposed the 'nebular hypothesis of Laplace'.[216] The reviewer reflects,

A profound sorrow was universally felt at Cambridge when his death became known, for all who had come into contact with him had been impressed with his worth. He was not only eminent in science, but he was firm and tender in friendship, moving no envy by his success, and showing none at the honours paid to others.[217]

The reviewer summarizes in closing,

We have not many men to lose like Professor Maxwell, and it is pardonable if those who long to see the thinking, throbbing world of science spiritualised by a living faith in God, and sobered by a reverence for revealed truth, should feel that his departure has left a gap which cannot easily be filled. But he lived nobly, and future generations will be the richer for his life.[218]

Prime Minister Gladstone, while lacking direct knowledge of Maxwell, from his notes on the *Life* was impressed by Maxwell's concern over 'Common Christianity', Maxwell's service as a church elder and his correspondence with Bishop Ellicott.[219] Glazebrook, one of Maxwell's associates at the Cavendish, noted in 1895 in the preface of a short biography that Maxwell was 'deeply loved by all who knew him'.[220]

The potentially most damaging personal recollection of Maxwell was given by the statistician, Karl Pearson, half a century after he had visited Maxwell's home in Cambridge to take a section of the Smith's Prize examination in January of 1879. Pearson recalled having spoken 'disrespectfully of Noah's Flood' during lunch after which he asserts: 'Clerk Maxwell was instantly aroused to the highest pitch of anger, reproving me for want of faith in the Bible! I had no idea at the time that he had retained the rigid faith of his childhood, and was, if possible, a firmer believer than Gladstone in the accuracy of Genesis.'[221] Pearson's recollection is also difficult to understand considering Maxwell's statement, noted by Campbell (but lacking date and context): 'I have no nose for heresy.'[222] It appears likely that Pearson lacked a personal relationship with Maxwell to appropriately interpret Maxwell's response.[223] Elsewhere, in a lecture in the 1920s, Pearson remarked that '. . . it is impossible to understand a man's work unless you understand something of his character and unless you understand something of his environment'.[224] Maxwell's usual treatment of controversial issues of his era was such that, long after his death, he retained the admiration of Charles Darwin's son, George Darwin, who had appreciated Maxwell's lectures and advice. George was not only careful to acknowledge his personal scientific debt to Maxwell,[225] he contributed in 1899 to a memorial for Maxwell at the Scottish Kirk in Corsock (see Fig. 14.2 and Table 14.1). By late in 1882, even John Tyndall was able to admire Maxwell's phrase for atoms as 'foundation stones of the material universe' while never endorsing the notion of 'manufactured articles'.[226]

To appreciate the aforementioned statement about 'no nose for heresy', it must be remembered that Maxwell shunned intense arguments. Campbell also made a point of emphasizing Maxwell's tolerance in relating to others.[227] Notice, though, that Campbell's mentor at Oxford, Benjamin Jowett, was one of the most controversial Broad Churchmen of the era. With scientists, however, differences over basic theological issues would have been easier to debate had Maxwell been inclined to get involved. For example, in a widely publicized speech in London, in 1866, T. H. Huxley confused the deeply theological concept of 'justification *by* faith' with the 'justification *of* faith'.[228] In other cases, scientists simply succumbed to zealous Bible critics who did not live long enough to see their theories overturned. Clifford, for example, was favourably impressed by critics who asserted the fourth Gospel (the Gospel of John) was 'late and legendary'.[229] For a refutation, Maxwell could have consulted Lightfoot or his neighbour Westcott.[230]

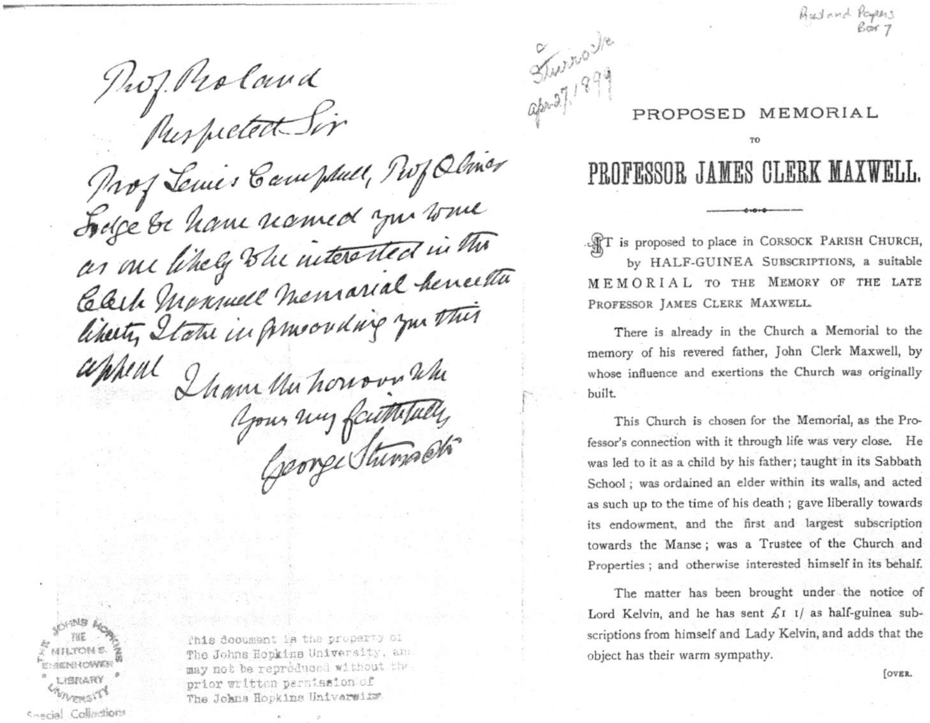

Fig. 14.2 Rev. George Sturrock distributed this request, in 1899, for a memorial to James Clerk Maxwell. This copy had been signed by Sturrock and sent to Professor H. A. Rowland of Johns Hopkins University. The requested amount was ½ guinea (slightly over £½). The memorial was added to the Corsock Parish Church near Glenlair. Supporters, printed on the reverse side (Table 14.1), included scientists, mathematicians and engineers, and some former students or assistants who had interacted with Maxwell during his professorships at Aberdeen and Cambridge. Some supporters resided near, or farmed on, Glenlair and some were Maxwell's relatives. Used with permission from MS 6, Papers of Henry A. Rowland, Box 7, Special Collections, Sheridan Libraries, Johns Hopkins University.

Of Maxwell's own understanding of Scripture concerning Creation in relation to science, his repeated inferences of the subtleness of nature (only some of which have been noted in the present chapter) are important. For example, concerning Romans 1:20 (quoted earlier in Maxwell's discussion of Faraday), 'For the invisible things of him from the creation of the world are clearly seen, being understood by the things that are made, *even* his eternal power and Godhead; so that they are without excuse.' Maxwell would have agreed with Martin Luther's understanding that nature is such that the witness has always been present and will always be present.[231] See also the aforementioned quotation by Maxwell of Hebrews 11:3 in his British Association address of 1870.[232] Thus, Maxwell's mistakes in the light of future discoveries are consistent with the witness of nature during the Victorian era, just as Newton supported nature's witness in his own era in his *Principia* and his *Opticks*. The general similarity of Maxwell's and Faraday's religious views were evident to Maxwell's contemporaries.[233] It appears likely that Maxwell and Faraday would have been amused if they could have known that, as late as 1955, the Astronomer Royal, Sir Herbert Spencer Jones, would open a review article on 'The Origin of the Solar System' with the assertion

Table 14.1 List of those with Lord Kelvin who 'among others, also cordially approve and support the proposal' for the church memorial in Corsock explained in Fig. 14.2. The list was printed on the reverse side of the request, having been compiled by Rev. George Sturrock. The following notes are appended here: (a) Prime Minister of the United Kingdom from 1902–05; (b) Cambridge Eranus club member or visitor; (c) a son of Charles Darwin; (d) alumnus of Maxwell's teaching in Aberdeen.

Lord Armstrong, Bamburgh Castle.	The Family of the late Principal J.D. Forbes, per Professor Geo. Forbes, FRS, M.I.C.E., London
W. A. Watson Armstrong, Esq., Bamburgh Castle	Professor J. A. Fleming, D.Sc., FRS, London
Wellwood Anderson, Esq., Dumfries	R. W. Fergusson, Esq., Kilquhanity
Right Hon. A. J. Balfour, M.P., D.C.L., LL.D., Whittinghame (a)(b)	Francis Galton, Esq., D.C.L., D.Sc., FRS, London
Professor Hugh Blackburn, Fort-William	Thos. Gladstone, Esq., King's Grange
Mrs Blackburn, Fort-William	William Gillespie, Esq., Castle-Douglas
Miss Blackburn, Kirkcudbright	Rev. Thos. Greig, M.A., Edinburgh
Dr. A. Buchan, Edinburgh	Robert Grierson, Esq., Castle-Douglas
Professor A. Barr, D.Sc., Glasgow	Sir William Huggins, London
Professor Campbell Brown, D.Sc., Liverpool (d)	A. Young Herries, Esq., Spottes
Professor Crum Brown, M.D., D.Sc., LL.D., Edinburgh	Professor Oliver J. Lodge, D.Sc., FRS, Liverpool
Professor W. F. Barrett, FRSE, Dublin	Professor J. Larmor, FRS, Cambridge
Professor Lewis Campbell (Clerk Maxwell's biographer), London	Professor T. Lindsay, D.D., and Mrs. Lindsay, Glasgow
Mrs Lewis Campbell, London	Rev. A. Laidlaw, Glasgow
Professor Chattock, Bristol	Sir Henry Mance, London
John Clerk, Esq., London	Mrs Wedderburn Maxwell, Oxford
Mrs John Clerk, London	Major Wedderburn Maxwell, Glenlair
Mrs Campbell, Vevay, Switzerland	Mrs J. Wedderburn Maxwell, Glenlair
John Campbell, Esq., Slagnaw	W. H. Maxwell, Esq., Munches
Rev. W. C. Callendar, M.A., Ladhope	W. J. Maxwell, Esq., Terraughtie
Mrs Cannon, Oxford	Ben. Rigby Murray, Esq., Parton
Rev. Dr. Dallinger, FRS, Lee	George Rigby Murray, Esq., Parton
T. W. Dunn, Esq., Cambridge	Dr. Alex. Muirhead, Shortlands, Kent
Mrs Dunn, Cambridge	Gavin Muirhead, Esq., Glenlair Farm
Professor J. Dewar, FRS, London	Hugh Muirhead, Esq., Glenlair Farm
Mrs J. Dewar, London	Robert Muirhead, Esq., Blackhills
Professor G. H. Darwin, FRS, Cambridge (c)	Professor Minchin, Cooper's Hill
Mrs Murray Dunlop, Corsock	Dr. R. Murray, Stockport
Captain H. Murray Dunlop, Corsock	Jas. Murray, Esq., Disley
John Dickson, Esq., Perth	Miss Maitland, Chipperkyle
Thos. Dickson, Esq., C.A., Edinburgh	Hamish M' Queen, Esq., Crofts
Professor Ewing, FRS, Cambridge	J. Hunter Maxwell, Esq., Castle-Douglas
Professor Everett, D.C.L., FRS, Belfast	Professor Charles Niven, FRS, Greenwich
Professor G. F. Fitzgerald, FRS, Dublin	W. H. Preece, Esq., FRS, London
	Professor J. H. Poynting, D.Sc., FRS, Birmingham
	Rev. J. A. Hunter Paton, Duddingston

continued

Table 14.1 (continued)

Professor T. Preston, Dublin	Jas. Swinburne, Esq., M.I.C.E., London
J. Denison Pender, Esq., London	Colonel Sanderson, Glenlaggan
Professor Ramsay, Ph.D., FRS, London	Rev. W. A. Stark, Kirkpatrick-Durham
Sir G. G. Stokes, F.R.S., Cambridge (b)	J. M'Gavin Sloan, Esq., Courier and Herald, Dumfries
Lady Stokes, Cambridge	Rev. Geo. Sturrock, Corsock
Sir Mark J. M' Taggart-Stewart, Bart., M.P. Southwick	George Sturrock, Esq., Aden
A. Skirving, Esq., Croys	Professor J. J. Thomson, Cambridge
Mrs Skirving, Croys	Professor Tait, D.Sc., Edinburgh
Very Rev. Principal Story, D.D., LL.D., Glasgow	Professor S. P. Thompson, F.R.S., London
Mrs Story, Glasgow	Miss Wedderburn, Oxford
Professor H. Sidgwick, Litt.D., Cambridge (b)	Rev. Geo. Walker, B.D., Castle-Douglas (d)
Mrs Sidgwick, Cambridge	Robert Walker, Esq., M.A., Aberdeen (d)
W. N. Shaw, Esq., M.A., Cambridge	Rev. D. and Mrs Wark, Auchencairn

'The solar system shows so many conspicuous regularities that it could not possibly have been formed by chance'.[234]

In Maxwell's professional papers archived in the Cambridge University Library is a page, written in his own hand, which appears almost to have been left in the collection by accident. On the page Maxwell has written Philippians 4:8, followed by short note, (Fig. 14.3)

> Finally, brethren, whatsoever
> things are true, whatsoever
> things are honest, whatsoever
> things are just, whatsoever
> things are pure,
> whatsoever things are lovely,
> whatsoever things are of
> good report; if there
> be any virtue, and if there
> be any praise, think
> on these things.
> Dr. Abercrombie's book on this
> fesit *[sic]* has been the delight of her life.[235]

While the date and circumstances are not known, the stationary used was from Ardhallow, Dunoon, the home of Katherine's sister where Maxwell first visited with the Dewars in 1857 and occasionally with Katherine in the 1860s and early 1870s. The book referred to is *Think on These Things*,[236] a devotional classic by John Abercrombie (1780–1844), an Edinburgh physician known to Maxwell's family.[237] Though the circumstances are not known, Maxwell's jotting reflects something of his own priorities and habits. Contemplation of Maxwell's Christian experience would be incomplete without also reiterating his intense sense of *reverence* for his Creator evident in some of his own correspondence and serious poetry and commented on by many of his friends and associates.[238] Stokes chose different words in correspondence, in 1895, with someone who

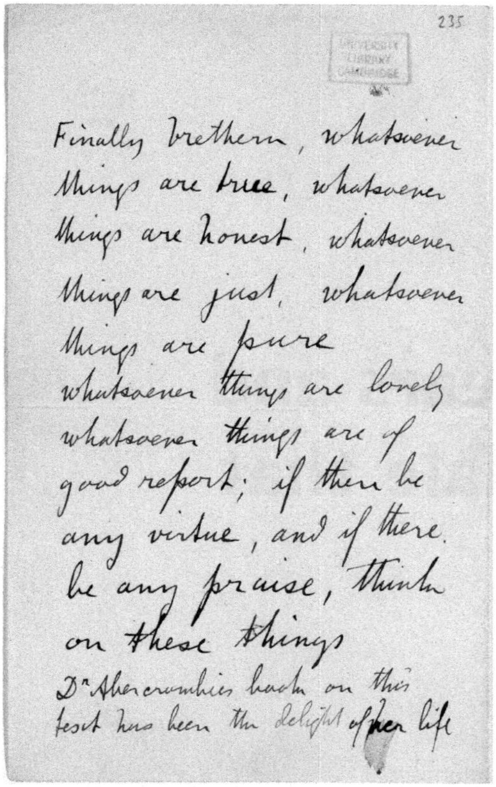
235

UNIVERSITY LIBRARY CAMBRIDGE

Finally brethern, whatsoever
things are true, whatsoever
things are honest, whatsoever
things are just, whatsoever
things are pure
whatsoever things are lovely
whatsoever things are of
good report; if there be
any virtue, and if there
be any praise, think
on these things
Dr Abercrombie's book on this
text has been the delight of her life

Fig. 14.3 Philippians 4:8 followed by short note written by Maxwell on an unidentified occasion. Reproduced by kind permission of the Syndics of Cambridge University Library. Department of Manuscripts classmark: CUL MS.Add.7655./ii/235.

evidently had not known Maxwell. After indicating he had known Maxwell well and had been 'intimate' with him, Stokes described Maxwell simply as a 'deeply religious Christian' man.[239] Other descriptions of Maxwell used by his peers include terms such as *simplicity, modesty, humility, charity, deep earnestness* and *unworldliness*.[240] Using twentieth-century terminology, Maxwell was more than just a 'Christian intellectual'. Concerning religion, he held 'there is nothing like the old thing after all'.[241]

Acknowledgement

I am grateful to Geoffrey R. Treloar for drawing my attention to letters from Maxwell to Lightfoot in the Durham Cathedral Library and to Chris Jakes (Cambridgeshire Libraries) for locating the lists denoted in the references as {FLCL} and {FLEL}.

CHAPTER 15

I Remember Years and Labours as a Tale that I have Read

MARK McCARTNEY

What! Has Time run out his cycle, do the years return again?
Are there treasure caves in Dreamland where departed days remain?
I have leapt the bars of distance - left the life that late I led -
I remember years and labours as a tale that I have read.

James Clerk Maxwell, *Recollections of Dreamland.*

During June 1856[1] Maxwell wrote what is, to this reader at least, his most evocative and powerful poem, *Recollections of Dreamland*. In the poem, Maxwell addresses obliquely and sensitively the death of his father just two months before. Released from a day's work and from 'the midst of facts and figures' he dreams of his idyllic childhood home in Galloway and of hearing and seeing the dead of his past. The poem shows depth of literary skill and psychological insight and given Maxwell's wide reading of literature and poetry, may even be a nod to the genre of medieval dream-poems exemplified by Chaucer or Langland. The poem is a marker that, as Stella Pratt-Smith has shown in Chapter 13 on Maxwell's wider poetic corpus, in James Clerk Maxwell we are dealing with a man of literary substance. Maxwell's biographers, Campbell and Garnett, note that after his election to a fellowship at Trinity 'he found time for a full course of classical English reading. And as all that he read he read critically, and had it thereafter in perfect possession, his literary acquirements were by this time of no mean order.'[2]

Further, as Phillip Marston has delineated in Chapter 14, Maxwell maintained a serious and thoughtful Christian faith throughout his life, which combined both extensive theological reading and practical Christian action. He could also exhibit a concern for philosophical matters that can be traced back to his undergraduate days and Sir William Hamilton's philosophy lectures at Edinburgh University. Campbell and Garnett claim that Maxwell's grasp of philosophy was 'an aspect of his genius of which too little is known',[3] with one of his obituaries commenting that 'he was almost as much at home with students of philosophy and theology as with those of physics'.[4]

One of Maxwell's more recent biographers, Francis Everitt, sees Hamilton's teaching that human knowledge is of the relationships between objects rather than the objects themselves, as a key influence in Maxwell's move from a vortex model of rotating fluid and idler wheel particles (see Figure 1.10 of Raymond Flood's Chapter 1) to a dynamical field theory of electromagnetism devoid of detailed mechanism.[5]

As a child, on wet days Maxwell read 'voraciously every book in the house, except what his mother kept out of his reach'[6] and at a young age was exhibiting a remarkable memory. Thus, aged 8, he knew the longest Psalm in the Psalter, Psalm 119, by heart, and something approaching this level of recall seems to have also applied to large swathes of other literature, with quote and allusion bubbling unbidden into his conversation or lectures, leaving hearers sometimes bemused and at other times dumbfounded. Campbell notes that, as a child: 'At church he always sat preternaturally still, with one hand lightly resting on the other, not moving a muscle, however long the sermon might be. Days afterward he would show, by some remark, that the whole service, whether good or bad, had been, as it were, photographed upon his mind.'[7] There is no reason to doubt that this level of recall applied just as well in the school room and the lecture theatre as it did in the church pew.

Of his father it was said 'He was one of a race in whom strong individuality had occasionally verged on eccentricity'.[8] It was a remark which could almost have been said of Maxwell himself. As a student at Cambridge he tried taking exercise at 2 am, running along college corridors (a practice not popular with his fellow undergraduates),[9] and experimented with whether cats could always land on their feet. His comic verse reveals a boyish sense of fun, which it seems never left him. And as Isobel Falconer notes in passing in Chapter 4, many of his letters throughout his life have elements which are best described as whimsical. When he returned to Cambridge in the 1870s, visitors to his home in Scroope Terrace could find themselves being entertained with a demonstration of his dynamical top or other scientific toys.[10] One piece of academic gossip has it that at an evening event in Cambridge, his wife Katherine retorted 'James, you are beginning to enjoy yourself; it is time to go home!' The remark, apocryphal or not, arguably says more about Katherine's character than Maxwell's ('She was, to put it bluntly, a difficult woman', according to Mrs. J. J. Thomson), but it is hard to resist the image of a boyish professor being scolded by his matronly wife.[11]

Maxwell's earliest memory was of lying in front of his father's house in Galloway 'wondering'. As a child he was perpetually trying to find out how things worked with 'What's the go o' that? What does it do?', and if the answer did not suffice 'But what's the *particular* go of it?'[12] He would leave his aunt feeling humiliated because he asked so many questions that she could not answer. One is left with the impression that that child never quite grew up. Otto Frisch, commenting on the practical jokes and schoolboy antics that went on amongst the brilliant physicists who passed through Niels Bohr's Copenhagan Institute in the 1930s, makes a remark which could just as easily be applied to Maxwell: 'I think scientists have one thing in common with children: curiosity. To be a good scientist you must have kept this trait of childhood, and perhaps it is not easy to retain just one trait. A scientist *has* to be curious like a child; perhaps one can understand that there are other childish features he hasn't grown out of.'[13]

Poet, Renaissance man, even eccentric; but it is as scientist that Maxwell is ultimately, and rightly, remembered. When he died of cancer on the 5 November 1879, aged 48, his stock as a scientist was high. He was certainly viewed as amongst the best of his age, but not, as he is now widely viewed, as one of the best of any age. When he accepted the Cavendish Chair at Cambridge, two others – Sir William Thomson (later Lord Kelvin) and Hermann von Helmholtz – had already been sounded out for the post and declined. At the time these two men were considered to be the most important natural philosophers in their respective countries, and Maxwell was rather obviously, not. William Garnett's obituary in *Nature*, just a week after Maxwell's death may have ended with the words 'The death of James Clerk Maxwell is a loss to his University and to the world too great for words. He rests from his labours, but his works will follow him', but on the same page, in the *Notes* section of the journal, where, among other things, it is announced that The Royal Institution Christmas lectures for 1879 would be given by Prof. Tyndall, and that Miss Martha Charters Somerville – only surviving daughter of Mrs. Mary Somerville – has died in her 66th year, we find the following:

A MEMORIAL strongly recommending Lord Rayleigh's election (if he can be induced to become a candidate), to the Professorship of Experimental Physics at Cambridge, is in circulation. Lord Rayleigh's merits for such an appointment are perfectly well known to our readers. We understand that his election will be supported by many of the professoriate.[14]

To a modern reader, the haste of the announcement, and its juxtaposition with the obituary, reads as almost improper. However, only the day before, on the 17 November, Sir William Thomson had written to Rayleigh, responding to his query as to whether he, Thomson, would be willing to move from Glasgow to Cambridge take the post. He declined, ending the letter 'I need not say that it would be a great satisfaction and pleasure to myself to look forward to having you at Cambridge should you decide to take the professorship'.[15] With William Thomson declaring himself out of the running, the path was clear for Rayleigh to secure the post, to which he was duly elected on 12 December.

In February 1880, *Nature* published a longer piece, by P. G. Tait, to complement Garnett's obituary of the previous November. Tait wrote specifically on Maxwell's scientific contributions to 'let even the non-mathematical reader see how very great were his contributions to modern science'. He succeeds in under five pages in giving a picture of the breadth of Maxwell's work, but leaves the reader with the impression that his most important contribution was his work on gases: 'During the last ten years of his life he had no rival to claim concurrence with him in the whole wide domain of molecular forces, and but two or three in the still more recondite subject of electricity.'[16] However, he shrewdly notes: 'There can now be little doubt that Maxwell's theory of electrical phenomena rests upon foundations as secure as those of the undulatory theory of light. But the life-long work of its creator has left it still in its infancy, and it will probably require for its proper development the services of whole generations of mathematicians.' The anonymous author of the obituary for the Royal Society of London is more circumspect. Commenting on Maxwell's identification of light with the electromagnetic field from the inference of their almost equal velocities, he notes the differing experimental results for the velocities and states 'We cannot, therefore, affirm with certainty that Maxwell's inference is correct, but there is strong *prima facie* presumption in its favour'.[17] As Chen-Pang Yeang describes in Chapter 11, such guarded assessment of the significance of

Maxwell's electromagnetism was relatively short lived. Hertz's experimental demonstration of the existence electromagnetic waves, and the work of the Maxwellians, swiftly moved the work centre stage.

Men of the Century

The change through time in relative standing of Maxwell and Lord Kelvin makes for an interesting comparison. Lord Kelvin was seen in his lifetime as the pre-eminent natural philosopher of the Victorian age. He, like Maxwell, worked on all the major aspects of nineteenth-century physics – electricity and magnetism, thermodynamics and the nature of matter – but to this he added substantial practical technology and engineering work. In his lifetime he took out 70 patents and his heavy involvement in the laying of the transatlantic telegraph cable earned him a knighthood and made him a national figure and a wealthy man. When he died in 1907, Britain laid him to rest in Westminster Abbey beside their most famous scientist, Isaac Newton, and within a few years of his death prominent statues were unveiled in his home city of Glasgow and in Belfast, the place of his birth.

When Maxwell died, he too was buried within a church, but it was in the ruins of the Old Kirk at Parton, part of a quiet rural graveyard some seven miles from his home of Glenlair. Apart from a service attended by members of Cambridge University in Trinity College Chapel before the remains were taken to Scotland, there was no great ceremony. Indeed, Maxwell had to wait for over 100 years for Scotland to erect a statue of him in her capital.[18]

In the years following Kelvin's death, however, much of his work suffered the fate of many a great scientist's labours: it got 'ploughed in' to form anonymous components of wider scientific knowledge, or was superseded by advances in technology and engineering. Contrariwise, as the years wore on aspects of Maxwell's work, particularly electromagnetism, endured and were increasingly seen as key developments in nineteenth-century science. An example of this change can be found it the respective centenary celebrations of Maxwell and Kelvin's births.

The Kelvin centenary celebrations took place in London, at the great hall of the Institution of Civil Engineers on the 10 July 1924. There was only one lecture; given by Sir J. J. Thomson. The next day there was a banquet, and the whole event was partially combined with a conference organized by the Institution of Electrical Engineers on World Power which was taking place in Wembley during the same week.[19] The book of the centenary contains J. J. Thomson's lecture, the after dinner speeches from the banquet and a collection of 29 'Addresses Commemorative' from a range of learned societies and academies from across the globe. These latter addresses each write of Kelvin in brief and glowing terms as they variously 'venerate', 'do homage', give 'lively appreciation', 'express gratitude for his grand contributions' and show 'deep admiration' for 'one of the most illustrious sons Great Britain ever produced'. However, the over-stated tenor of some of the words, the limited scale of the event and the conspicuous absence of contributions from major contemporary figures in science, leaves the reader with the distinct sense of the celebration being a dutiful memorial to a figure who, though important in his own time, was considered to be of fading importance to the present of the 1920s.

Move forward in time to October 1931, and the centenary celebrations of Maxwell's birth are a much more substantial affair. On 30 September, Sir J. J. Thomson, now 74, unveiled memorial stones to Faraday and Maxwell in Westminster Abbey, and on the next day, 1 October, the celebrations proper began in Cambridge. After lunch at Corpus Christi College and a speech by the

Master of Corpus and University Vice Chancellor, William Spens, the delegates processed to the Senate House to hear a memorial lecture by, once again, J. J. Thomson. After Thomson's lecture the delegates were entertained at Peterhouse and St. John's Colleges. The next day there were lectures by Max Planck, Niels Bohr, Sir James Jeans, Sir Oliver Lodge and a somewhat rambling and florid contribution from the soon to retire Lucasian Professor of Mathematics, Sir Joseph Larmor. William Garnett, Maxwell's demonstrator at the Cavendish, and Sir Ambrose Fleming and Sir Richard Glazebrook, both ex-students of Maxwell, recounted memories of their time at Cambridge, with Fleming presenting the laboratory with the notes he took while a student at Maxwell's last lectures. In the afternoon, delegates were shown a reconstruction, using the original slides, of Maxwell's famous 1861 Royal Institution colour projection of a tartan ribbon. Then, to show how technology had moved on, they were shown a colour film of the previous day's procession to the Senate House. The proceedings ended with a banquet at Trinity College. It was all in all a grand affair. Cambridge, and modern physics, had made full effort to honour Maxwell as one of their most famous sons. The talks given were published in book form – *James Clerk Maxwell: A Commemoration Volume 1831–1931* (see Fig. 15.1) – to coincide with the celebrations themselves, and contained two essays which were not given in person.[20] The first was a short piece by Sir Horace Lamb, *Clerk Maxwell as Lecturer*, the second, *Maxwell's Influence on the Development of the Conception of Physical Reality*, was by Albert Einstein, who though invited to the celebrations, did not attend.[21]

The *Commemoration Volume* is, in many ways, an excellent introduction to Maxwell's life and work. It combines biography and anecdote with a perspective on Maxwell's science which is given from sufficient distance from the events to allow for a sober assessment of its value. Perhaps two contributions are worth noting above the others. The first is from Einstein, who places Maxwell's greatness firmly within the development of field theory: 'The greatest alteration in the axiomatic basis of physics – in our conception of the structure of reality – since the foundation of theoretical physics by Newton, originated in the researches of Faraday and Maxwell on electromagnetic phenomena.' In the essay, Einstein repeatedly brackets Maxwell with Newton as marking a change in epoch. He would do so again almost 30 years later in his *Autobiographical Notes*, stating:

> The most fascinating subject at the time that I was a student was Maxwell's theory. What made this theory appear revolutionary was the transition from action at a distance to fields as the fundamental variables. The incorporation of optics into the theory of electromagnetism, with its relation of the speed of light to the electric and magnetic absolute system of units as well as the relation of the index of refraction to the dielectric constant, the qualitative relation between the reflection coefficient of a body and its metallic conductivity – it was like a revelation . . . In this connection I cannot suppress the remark that the pair Faraday-Maxwell had a most remarkable inner similarity with the pair Galileo-Newton – the former of each pair grasping the relations intuitively, and the second one formulating those relations exactly and applying them quantitatively.[22]

Later in the same book, he states:

> The special theory of relativity owes its origin to Maxwell's equations of the electromagnetic field. Conversely, the latter can be grasped formally in satisfactory fashion only by the special theory of relativity.[23]

One student at Berlin in the 1920s recalls Einstein saying 'there would be no modern physics without Maxwell's electromagnetic equations; I owe more to Maxwell than to anyone'.[24]

Maxwell is, in more ways than one, a man between Newton and Einstein.

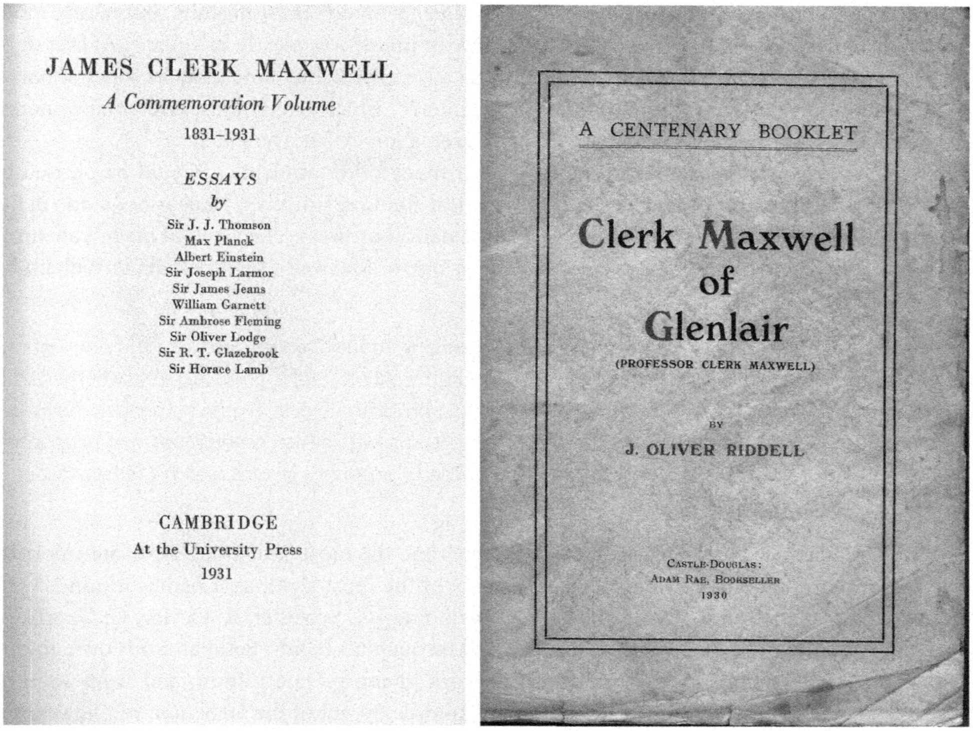
JAMES CLERK MAXWELL
A Commemoration Volume
1831–1931

ESSAYS
by
Sir J. J. Thomson
Max Planck
Albert Einstein
Sir Joseph Larmor
Sir James Jeans
William Garnett
Sir Ambrose Fleming
Sir Oliver Lodge
Sir R. T. Glazebrook
Sir Horace Lamb

CAMBRIDGE
At the University Press
1931

A CENTENARY BOOKLET

Clerk Maxwell
of
Glenlair

(PROFESSOR CLERK MAXWELL)

BY

J. OLIVER RIDDELL

CASTLE-DOUGLAS:
ADAM RAE, BOOKSELLER
1930

Fig. 15.1 Commemorating Maxwell. The 1931 Centenary of Maxwell's birth was celebrated in print not only with Cambridge University Press's *Commemoration Volume*, but also with the local tribute *Clerk Maxwell of Glenlair*. The booklet was published in Dumfries, a small town in southwest Scotland where Maxwell was still remembered as 'a richt nice mon'.

A Man of Adequate Mathematical Technique

The second essay of particular note in the *Commemoration Volume* is *Clerk Maxwell's Method* by Sir James Jeans. Discussing Maxwell's derivation of the distribution of molecular velocities in a gas he states:

> Maxwell, by a train of argument which seems to bear no relation at all to molecules, or even to the dynamics of their movements, or to logic, or even to ordinary common sense, reached a formula which according to all precedents and all the rules of scientific philosophy, ought to have been hopelessly wrong. In actual fact it was subsequently shown to be exactly right . . . It was this power of profound physical intuition, coupled with adequate, although not outstanding mathematical technique, that lay at the basis of Maxwell's greatness.[25]

Commenting on the same derivation, J. J. Thomson agrees: 'though his proof . . . has been criticized no one doubts the truth of his result – like some other men of great physical insight, his instincts were better than his reasons.'[26]

The proof, published in the *Philosophical Magazine* in 1860,[27] *is* astonishing, but hardly more so than Jeans' remark that Maxwell's mathematical technique was merely 'adequate'. At first sight Jeans' comment might be taken as hyperbole. After all, earlier in the same essay he states Maxwell did not 'show any outstanding skill as an experimenter,'[28] which given his experimental work on, for example, colour perception and gas viscosity, seems somewhat overstated.

I blush at pressing the point, as I, and I expect many other academics, would be pleased to have even a small fraction of Maxwell's mathematical abilities. However, Jeans is not alone in his less than whole-hearted praise of Maxwell's mathematical prowess. The views of his private tutor at Cambridge, William Hopkins, were recorded by one of Maxwell's fellow students William N. Lawson in his diary:

> July 15, 1853.-He (Hopkins) was talking to me this evening about Maxwell. He says he is unquestionably the most extraordinary man he has met with in the whole range of his experience; he says it appears impossible for Maxwell to think incorrectly on physical subjects; that in his analysis, however, he is far more deficient; he looks upon him as a great genius, with all its eccentricities, and prophesies that one day he will shine as a light in physical science, a prophesy in which all his fellow-students strenuously unite.[29]

Hopkins' remarks carry weight, as he was by this time the most sought after private tutor for the Tripos examinations in Cambridge, and many of the most significant names in nineteenth century mathematics and natural philosophy (including J. J. Sylvester, A. Cayley, G. G. Stokes, W. Thomson, I. Todhunter and P. G. Tait) passed through his hands. Soon after his own graduation as Seventh Wrangler in 1827, he began work giving private tuition, and from 1829 to Maxwell's own graduation year of 1854 Hopkins tutored 20 out of the 26 Senior Wranglers and 19 of the Second Wranglers. The year 1854, when Maxwell was Second Wrangler, was the most successful crop of his career, with seven of the top nine Wranglers being Hopkins' students. Thus, when Hopkins describes Maxwell as 'extraordinary' and a 'great genius', and even his analysis as 'deficient', his assessment is that of a man who had taught most of the brightest mathematical minds in the country over the previous quarter of a century.

P.G.Tait, in his *Nature* article of 1880 makes a similar point:[30]

> If the title mathematician be restricted (as it too commonly is) to those who possess a peculiarly ready mastery over symbols, whether they try to understand the significance of each step or no, Maxwell was not, and certainly never attempted to be, in the foremost rank of mathematicians. He as slow in 'writing out,' and avoided as far as he could the intricacies of analysis. He preferred always to have before him a geometrical or physical representation of the problem in which he was engaged, and to take all his steps with the aid of this: afterwards, when necessary, translating them into symbols . . . There can be no doubt that in this habit, of constructing a mental representation of every problem, lay one of the chief secrets of his wonderful success as an investigator. To this were added an extraordinary power of penetration, and an altogether unusual amount of patient determination. The clearness of his mental vision was quite on a par with that of Faraday; and in this (the true) sense of the word he was a mathematician of the highest order.

Taking Jeans, Hopkins and Tait together, it would seem that rather than damn Maxwell's mathematical capabilities with the faint praise of 'adequate', a more balanced assessment would be to note that Maxwell was first and foremost a natural philosopher. However, he was a natural philosopher who made brilliant use of mathematics when he needed to – guided by an intuition that was so deep that at times it did indeed make it 'impossible' for him to think incorrectly about physical matters. This is surely no more brilliantly exemplified than in the electromagnetic field, but it can be seen, as the contributors to this volume have shown, throughout his work across science.

Brilliant Lines of Assault

Maxwell will never be a cultural icon like Newton or Einstein. I expect he would have been bemused by the very thought. However, those within science have always honoured him well, missing no opportunities to commemorate anniversaries with books, papers and conferences down the years.[31]

When, on the eve of the new millennium, the magazine *Physics World* contacted 250 physicists around the world and asked them, amongst other things, to list the five physicists who have made the most important contributions to physics, James Clerk Maxwell came third behind Einstein and Newton.[32] Another question in the 1999 survey asked 'What have been the three most important discoveries in physics?' The authors of the article reviewing the survey state that 'Time and time again three key discoveries were singled out: quantum mechanics, Einstein's special and general theories of relativity, and Newtonian mechanics and gravitation' but go on to note that 'Maxwell's unification of electricity and magnetism was another popular choice'. When, in 2004, the same magazine asked readers to send suggestions for the greatest equations in science, Maxwell's equations topped the poll, coming joint first with Euler's relation $e^{i\pi} + 1 = 0$.[33] Neither survey can be seen as, nor would pretend to be, authoritative. In the first poll only about 130 replies were received, and in the second 120. However, both give a feel for the regard in which Maxwell is held within modern physics. They also suggest that Maxwell is chiefly remembered for electromagnetism. As Rudolf Peierls once quipped: 'If you wake up a physicist in the middle of the night and say "Maxwell", he will be sure to say "electromagnetic field."'[34] And to be remembered for the creation of electromagnetism, summed up in what must be amongst the most elegant set of equations in mathematical physics, is undoubtedly enough to make any scientist immortal. Maxwell's equations, if you will forgive the phrase, 'tick all the boxes'. Not only do they provide a unified theory of electricity and magnetism, but they describe an electromagnetic spectrum which goes from radio waves to gamma rays, and thus provide the theoretical underpinning for a vast range of topics in science and technology. Further, to any mathematician or physicist, they satisfy G. H. Hardy and P. A. M. Dirac's requirements of beauty: 'Beauty is the first test: there is no permanent place in the world for ugly mathematics' said Hardy in his elegant, if somewhat sorrowful classic *A Mathematician's Apology*.[35] Dirac, when asked during a lecture at Moscow University in 1957 to give his philosophy of physics, wrote in capitals on the blackboard 'Physical laws should have mathematical beauty'.[36] Maxwell's equations are almost the archetypal example of Dirac and Hardy's point.

However, here we must remind ourselves of one of the ironies of scientific history. As Chen-Pang Yeang points out in his chapter, the credit for the equations of electromagnetism in their modern form is not due to James Clerk Maxwell. Rather, they are due to Oliver Heaviside, a largely self-taught and hugely talented cantankerous maverick, who in the 1880s stripped down and simplified Maxwell's presentation in the *Treatise on Electricity and Magnetism* to give us the austere beauty of four concise equations. None of this, however, lessens Maxwell's brilliance. As G. F. Fitzgerald noted:

Maxwell, like every other pioneer who does not live to explore the country he opened out, had not had time to investigate the most direct means of access to the country nor the most systematic way of exploring it. This has been reserved for Oliver Heaviside to do. Maxwell's treatise is encumbered with the *débris* of his brilliant lines of assault, of his entrenched camps, of his battles. Oliver Heaviside has cleared these away, has opened up a direct route, has made a broad road, and has explored a considerable trace of country.[37]

Perhaps if Maxwell had lived beyond 48 he himself may have cleared the way and given more direct routes, but that we will never know.

The essays in this book have shown that Maxwell had brilliant lines of assault on many subjects within natural philosophy – electricity and magnetism, colour theory, molecular dynamics, Saturn's rings, and many other topics were battled with, and decisive victories were won. But they also show that he was much more than the science he has left. And perhaps in that regard it is fitting to give the last words to Maxwell's first biographers, Lewis Campbell and William Garnett. James Clerk Maxwell was, they say, a 'marvellous interpenetration of scientific industry, philosophic insight, poetic feeling and imagination, and overflowing humour'.[38] He must have been a wonderful man.

NOTES ON CONTRIBUTORS

Dr Isobel Falconer
Isobel Falconer is a historian of science, a Senior Lecturer at Glasgow Caledonian University, and former curator of the museum at the Cavendish Laboratory. She has published extensively on a number of Cavendish physicists, in particular J. J. Thomson, one of Maxwell's successors as Cavendish Professor and a prominent 'Maxwellian'.

Prof. Raymond Flood
Raymond Flood was Vice-President of Kellogg College, Oxford, and is an Emeritus Fellow of the College. His main research interest lies in the history of mathematics, and he was formerly President of the British Society for the History of Mathematics. He is currently Gresham Professor of Geometry.

Prof. Elizabeth Garber
Elizabeth Garber is Associate Professor Emerita in the Department of History at the State University of New York at Stony Brook. With Stephen Brush and C. W. F. Everitt, she has published *Maxwell on Saturn's Rings*, *Maxwell on Molecules and Gases* and *Maxwell on Heat and Thermodynamics.*

Prof. Malcolm Longair
Malcolm Longair was appointed the ninth Astronomer Royal of Scotland in 1980, as well as the Regius Professor of Astronomy, University of Edinburgh, and the director of the Royal Observatory, Edinburgh. From 1991 to 2008, he was Jacksonian Professor of Natural Philosophy at the Cavendish Laboratory and was its head from 1997 to 2005. His most recent books are *High Energy Astrophysics* and *Quantum Concepts in Physics.*

Prof. Philip L. Marston
Philip L. Marston is a Professor of Physics at Washington State University, Pullman. He is a Fellow of the Acoustical Society of America and a Senior Member of the Optical Society of America. He received advanced degrees in Physics and Electrical Engineering from Stanford University. His research interests include scattering and radiation forces.

Dr Mark McCartney
Mark McCartney is Senior Lecturer in Mathematics at the University of Ulster. His research and scholarly interests include mathematical modelling, chaotic dynamics and the history of science.

Prof. Keith Moffatt
Keith Moffatt is Emeritus Professor of Mathematical Physics at the University of Cambridge, Fellow of the Royal Society, and Fellow of Trinity College Cambridge where James Clerk Maxwell was himself a fellow. He works in fluid dynamics, and in particular on the dynamo theory of generation of cosmic magnetic fields.

Prof. Daniel M. Siegel

Daniel M. Siegel is Emeritus Professor of the History of Science at the University of Wisconsin at Madison. His interests lie in the history of physics in the nineteenth and twentieth centuries, and he has published extensively on Maxwell's work on electromagnetism, in particular publishing *Innovation in Maxwell's Electromagnetic Theory* in 1991.

Dr Stella Pratt-Smith

Stella Pratt-Smith teaches Victorian and Modern Literature at Oxford. Her forthcoming book, *Transformations of Electricity in Nineteenth-century Literature and Science* (Ashgate Publishing), reveals electricity as a site for intense and imaginative Victorian speculation, spanning a dialogue at once literary and scientific.

Dr John S. Reid

John S. Reid is Honorary Senior Lecturer in Physics and Honorary Curator of the University of Aberdeen's nationally recognized historical scientific instrument collection. He is also a Trustee of the James Clerk Maxwell Foundation. His grandfather was born in Aberdeenshire the year after Maxwell left Aberdeen, a link that helps to make Maxwell seem not so distant in time.

Sir John S. Rowlinson

John Rowlinson was Dr. Lee's Professor of Physical Chemistry at the University of Oxford and is a Fellow of Exeter College. He is a Fellow of the Royal Society and of the Royal Academy of Engineering. He received the Edelstein Award from the American Chemical Society in 2008 for his contributions to the history of chemistry.

Dr Chen-Pang Yeang

Chen-Pang Yeang is an Associate Professor at the Institute for the History and Philosophy of Science and Technology, University of Toronto. His research areas comprise the history of modern physics, engineering science and technology.

Prof. Andrew Whitaker

Andrew Whitaker is Emeritus Professor of Physics at Queen's University Belfast. His main research interest has been in the foundations of quantum theory and he also has an interest in the history of physics, recently having *The New Quantum Age* published by Oxford University Press.

NOTES AND REFERENCES

ABBREVIATIONS

CUL = Cambridge University Library

REFERENCING OF STANDARD TEXTS

A number of commonly used references are abbreviated as follows:

CG1882: Campbell, L. and Garnett, W. (1882). *Life of James Clerk Maxwell.* MacMillan & Co., London.

CG1884: Campbell, L. and Garnett, W. (1884). *Life of James Clerk Maxwell, 2/e.* MacMillan & Co., London.

SLP1: Harman, P. M. (1990). *The Scientific Letters and Papers of James Clerk Maxwell*, Vol. 1. Cambridge University Press, Cambridge.

SLP2: Harman, P. M. (1995). *The Scientific Letters and Papers of James Clerk Maxwell*, Vol. 2. Cambridge University Press, Cambridge.

SLP3: Harman, P. M. (2002). *The Scientific Letters and Papers of James Clerk Maxwell*, Vol. 3. Cambridge University Press, Cambridge.

SP1: Niven, W. D. (1890). *The Scientific Papers of James Clerk Maxwell*, Vol. 1. Cambridge University Press, Cambridge.

SP2: Niven, W. D. (1890). *The Scientific Papers of James Clerk Maxwell*, Vol. 2. Cambridge University Press, Cambridge.

CHAPTER 1

1. Einstein, A. (1931). Maxwell's influence on the development of the conception of physical reality. In *James Clerk Maxwell: A Commemorative Volume 1831–1931*, pp. 66–73. Cambridge University Press, Cambridge.
2. Feynman, R. (1964). *The Feynman Lectures on Physics*, Vol II, 1–11. Addison-Wesley.
3. An earlier version of this chapter appears in the *Bulletin of the British Society for the History of Mathematics*, 2013, 28, 2: <http://dx.doi.org/10.1080/17498430.2013.755073>. It is the text of the 2012 BSHM–Gresham Lecture, delivered at Gresham College, London, on 31 October 2012. This and other BSHM–Gresham Lectures can be seen in full on the Gresham College website: <http://www.gresham.ac.uk>.

4. Campbell, L. and Garnett, W. (1882). *The Life of James Clerk Maxwell*, p. 131. Macmillan, London. [CG1882]
5. [SLP1], p. 671.
6. [SLP1], p. 667.
7. [CG1882], p. 345.
8. Domb, C. (ed.) (1963). *Clerk Maxwell and Modern Science: Six Commemorative Lectures*, p. 43. The Athlone Press, London.
9. Mahon, B. (2004). *The Man Who Changed Everything*, p. 81. Wiley, Chichester.
10. [SLP1], p. 500.
11. [SLP1], p. 486.
12. [CG1882], p. 342.
13. Sillitto, R. M., Clerk Maxwell's Life and Work, <http://www.sillittopages.co.uk/rmsjcmlife.pdf> [Accessed July 2013].

CHAPTER 2

1. Two histories of the City of Aberdeen covering a sweep of developments are Alexander Keith (1987), *A Thousand Years of Aberdeen*, Aberdeen University Press, Aberdeen, and the compilations E. Patricia Dennison, David Ditchburn and Michael Lynch (eds.), (2002), *Aberdeen before 1800: A New History*, Tuckwell Press, East Linton, and W. Hamish Fraser and Clive H. Lee (eds.), (2000), *Aberdeen, 1800 to 2000: A New History*, Tuckwell Press, East Linton.
2. An overview of Aberdeen on either side of Maxwell's years of residence can be found in Edward Smedley, Hugh J. Rose and Henry J. Rose (eds.), (1845), *Encyclopaedia Metropolitana*; or, *Universal Dictionary of Knowledge, etc.* vol. XIV, pp. 21–2, Fellowes, Rivington et al., London, and *Encyclopaedia Britannica; A Dictionary of Arts, Science, and General Literature* (1875) 9th edn. vol. 1, pp. 39–43, Adam & Charles Black, Edinburgh.
3. A plaque was erected on the building to mark the centenary of Maxwell's residence. Details of the occupancy of all premises in Union Street are contained in the annual *Aberdeen Post Office Directories of 1856–1860*, A. King, Aberdeen.
4. A mass of factual information on Marischal College was accumulated by two Royal Commissions in the nineteenth century, investigating the possibility of a merger with neighbouring King's College. Detail quoted here is not individually referenced throughout. See (1863) *General Report of the Commissioners under the Universities (Scotland) Act 1858*, Edinburgh, HMSO. A useful summary of the Arts teaching was published by the University Librarian and former assistant to the Professor of Natural Philosophy, Peter J. Anderson (1892), *The Arts Curriculum*, Aberdeen University, Aberdeen.
5. Both for information and to attract private students, the courses at Marischal College were advertised in the press. See, for example, *The Aberdeen Journal*, 22 October 1856 for the first appearance of Professor Maxwell's name in the advertisements.
6. Murchison, Lyell and Geikie were among a strong contingent of geologists who attended the 1859 BAAS meeting, discussed later in this chapter, in which James Nicol played a significant part.
7. Ogilvie, J. (1896). *John Cruickshank, Professor in Marischal College and University of Aberdeen*, D. Wylie & Son, Aberdeen, 42–3, discusses the Marischal College reforms that took place in the late 1820s, some 60 years before similar actions became legislative requirements in Scotland.
8. Graduation cost £5, comparable to the annual class fees and equivalent to at least £500 in year 2000 currency. The reluctance of some able students to collect their degrees is understandable.

9. Parliamentary paper (1874) [C.899] Scotland. Owners of lands and heritages 17 & 18 Vict., cap. 91. 1872–3. Return I. *The name and address of every owner of one acre and upwards in extent (outside the municipal boundaries of boroughs containing more than 20,000 inhabitants)* records Glenlair as comprising only 68 acres (27.5 hectare). Glenlair had been part of a larger estate inherited by the Clerk Maxwells that had been divided and partly sold off earlier in the century. Other authors have quoted much larger figures for the size of Glenlair but Clerk Maxwell was still running the estate at the time of this return.

10. Tait remarked in his tribute to Maxwell after his death, '*But the rapidity of his thinking, which he could not control, was such as to destroy, except for the very highest class of students, the value of his lectures. His books and his written addresses (always gone over twice in MS) are models of clear and precise exposition; but his extempore lectures exhibited in a manner most aggravating to the listener the extraordinary fertility of his imagination.*' P. G. Tait, (1879) *Proceedings of the Royal Society of Edinburgh*, 10: 331–9 (p. 334); the comment being repeated in *Nature*, 19 January 1880 and in C. G. Knott, (1911) *Life and Scientific Work of Peter Guthrie Tait*, Cambridge University Press, Cambridge, 262. A similar comment by W. Garnett, (1879) *Nature*, 13: 43–6 (p. 45) '*As a professor he was wonderfully admired by those who were truly his disciples. He had not the power of making himself understood by those who listened but casually to his pithy sentences, and consequently he was not a so-called popular lecturer; nor was he a most successful teacher of careless students.*'

11. Forbes, G. (1916). *David Gill: Man and Astronomer*, John Murray, London, quotes Gill on pp. 13, 14 and 17.

12. Aberdeen Universities Commission (1858). Report of Her Majesty's Commissioners appointed to enquire into the State of the Universities of Aberdeen, with a view on their Union. Together with the Evidences and Appendices, pp. 80–1, HMSO, Edinburgh.

13. Harman, P. M. (ed.) (1990). *The Scientific Letters and Papers of James Clerk Maxwell*, vol. 1 (abbreviated [SLP1] hereafter), Cambridge University Press, Cambridge, 392–661 contain much detail that is relevant to Maxwell's Aberdeen years.

14. Maxwell's inaugural address at Aberdeen was reported at some length at the time in the *Aberdeen Journal*, 5 November 1856. Piazzi Smyth (Astronomer Royal for Scotland) was among the attendees. It is printed in full in [SLP1], pp. 419–31 and in R. V. Jones (1973), James Clerk Maxwell at Aberdeen 1856–60, *Notes & Records of the Royal Society of London*, 28: 57–81.

15. This raises questions much too lengthy to be discussed here. The philosophy in Maxwell's Natural Philosophy has been discussed in the context of his life's work at book length by P. M. Harman (1998), *The Natural Philosophy of James Clerk Maxwell*, Cambridge University Press, Cambridge.

16. The class rolls of Marischal College from 1605 to 1859 have been published by P. J. Anderson (1898), *Fasti Academiae Mariscallanae Aberdonensis*, Vol. II, 30, New Spalding Club, Aberdeen.

17. Reid, John S. (2008). James Clerk Maxwell's Scottish Chair, *Philosophical Transactions of the Royal Society A*, 366: 1661–84.

18. Gill, D. (1913). *History & Description of the Royal Observatory Cape of Good Hope*, Neill & Co., Edinburgh, xxx–xxxi.

19. Four Senior Wranglers came from Aberdeen in the decade 1858 to 1867. They were: James M. Slesser (briefly Professor of Mathematics at Belfast before succumbing to illness); James Stirling (later Sir James, Lord Justice of Appeal and Privy Councillor); Thomas Barker (Professor of Mathematics in Owen's College) and Charles Niven (at first, Professor of Mathematics at Queens College, Cork, and from 1880, David Thomson's successor as Professor of Natural Philosophy at Aberdeen).

20. Reid, John S. (1990). Eighteenth-century Scottish university instruments: The Remarkable Professor Copland, *Bulletin of the Scientific Instrument Society*, 24: 2–8.

21. Fraser, A. (1857). *Lecture Notes of Professor Maxwell Taken by Angus Fraser on Natural Philosophy*, Aberdeen University Special Libraries and Archives Ms AMCS/3/17.
22. Anon. (1893). *Alma Mater* (the student magazine of the University of Aberdeen), p. 202, 7 June contains a brief biography of Dr Angus Fraser, with photograph opposite.
23. Davidson, A. (1859). 'lecture notes' in Cambridge University Library, Ms Add 8791.
24. [SLP1], pp. 432–7.
25. Fraser, A. (1857). *Lecture Notes*, p. 65.
26. Fraser, A. (1857). *Lecture Notes*, p. 21.
27. Reid, John S. (2008). James Clerk Maxwell's Scottish Chair.
28. Not only would it be completely out of character for Maxwell to use his predecessor's notes but, by way of confirmation, an advertisement appeared in the *Aberdeen Journal*, 5 November, 1856, advertising for sale the personal apparatus and a substantial number of diagrams used by the late Professor Gray in teaching natural philosophy at the college. Given the frequency with which Maxwell mentioned teaching in his letters, he would certainly have mentioned had he purchased even some of this material.
29. James, D. Forbes own notes for his Edinburgh lectures are preserved in St. Andrews University Archives: see msdep7 items IX/9, IX/12, IX/13 for topics taught in years covering Maxwell's presence and msQ113.F88, msQ113.F8, msQA821.F8 and msQB51.F8 for lecture contents. The notes are not written to be read verbatim, being well filled with phrases and bald statements in slightly spidery, closely spaced writing that would at best constitute lecture prompts. Forbes delivered about 110 lectures per session, plus six topical exams and two general exams. His subject matter was not constant over the years. For example, in Maxwell's first year under Forbes (1847–48) the course included 37 lectures on Astronomy that were replaced in the following year by a sub-course on optics, and lectures on properties of bodies and more on heat. In subjects in common between Maxwell and Forbes, their treatment is different. Forbes gave 'extra lectures' on individual selected topics in which he introduced more advanced concepts and some detailed mathematics but he does not seem to have had formally an 'advanced class'.
30. [SLP1], pp. 558–9, 'Letter to P. G. Tait', 21 November 1857.
31. McCartney, M. (2008). William Thomson: An Introductory Biography. In Raymond Flood, Mark McCartney and Andrew Whittaker (eds.), *Kelvin: Life, Labours and Legacy*, 10, Oxford University Press, Oxford.
32. [SLP1], p. 575, 'Letter to Jane Cay', 28 November 1857.
33. [SLP1], pp. 558–9, 'Letter to P. G. Tait', 21 November 1857.
34. The formalization of the Honours degree was one result of the 1858/1860 Scottish Universities' Commission that, amongst other reforms, set up Honours in Scotland as an additional year of study beyond a three-year ordinary degree. It clearly took time to establish itself. A student at the University of Aberdeen after Maxwell left was the formidably able George Chrystal who studied under Maxwell in Cambridge after taking his Aberdeen degree. He commented: 'When I went to the University of Cambridge, I found that the course there for the ordinary degree in Arts was greatly inferior in educational quality to the Scottish one. On the other hand, the courses in honours were on a very much higher standard.' G. Chrystal (1908), Promoter's Address to Graduates of Arts, University of Edinburgh, *The Scotsman*, Edinburgh, 11 April.
35. Professor Alexander Bain of the University of Aberdeen, who had in his early days been a driving force behind the Aberdeen Mechanics' Institution, is quoted in Ogilvie (1896, p. 116) as saying that Maxwell's oral exposition suffered 'from misplacement of emphasis in his articulation'. David Gill (1913) leaves the impression that Maxwell frequently flew off at a tangent on an idea that had just occurred to him in

the lecture, abandoning the formal remit of the topic, but the two extant sets of lecture notes made by students in the class tell the story of a coherently delivered course. In truth, a student who just wants to be filled up with knowledge that can be written out in haste in an exam needs a different kind of lecturer from a student who is hoping to be inspired by the subject. Maxwell had done enough exam swatting at Cambridge to know the score on the first account but his hopes were clearly to swing the balance from exam style formalism to the inspirational. At Aberdeen he gave his students regular questions but there were no serious exams, allowing him to be discursive.

36. Flood, R. (2008). Thomson and Tait: The *Treatise on Natural Philosophy*, pp. 175–91 in Flood et al. (2008) *Kelvin: Life, Labours and Legacy*.

37. Extensive details of the various editions of Maxwell's published books are given by Edward Fenwick (2009), *A Bibliography of James Clerk Maxwell: Part I, Books by Maxwell*, privately published by Edward Fenwick, 3 Thirlestane Lane, Edinburgh EH9 1AJ.

38. James Clerk Maxwell in [SLP1], p. 557, 'From a Letter to Henry Richmond Droop', 14 November 1857.

39. Adverts for enrolment in the School of Science during Maxwell's involvement appeared in the *Aberdeen Journal* on 21 October 1857, p. 4 col. 5; 20 October 1858, p. 4, col. 4; and 19 October 1859, p. 4, col. 3.

40. This wording comes from the 1857 advert in the *Aberdeen Journal*.

41. *The Thirty-Fourth Report of the Aberdeen Mechanics' Institution; being the Third Joint-Report of the Council of the Schools of Science and Art, and of the Directors of the Institution* October 1860, printed by William Bennett, Aberdeen, 1860.

42. The link with the Aberdeen Mechanics' Institution is discussed by G. M. Fraser (1912), *Aberdeen Mechanics' Institute*, Aberdeen University Press, Aberdeen, Ch. X. See also *Aberdeen Mechanics Institution Minute Book No. 3* in Aberdeen Public Library Lo370.6 Ab 3.4 for details of the running and examinations of the School of Science and Art from 1856–9. Further details in *Minutes of the School of Science & Art Aberdeen Book no. 9* in Aberdeen Public Library Lo 3706/Ab3.4/52215.

43. James Clerk Maxwell in [SLP1], pp. 573–4, 'Letter to Cecil James Munro', 26 November, 1857.

44. *Aberdeen Journal*, Wed. 28 April, 1858. In the *Aberdeen Journal*, Wed. 27 April 1859, only the prize list was published, giving details of three prize-worthy performances in natural philosophy.

45. Reid, John S. (1987). Late eighteenth-century adult education in the sciences at Aberdeen: The natural philosophy classes of Professor Patrick Copland, pp. 168–79 in J. J. Carter and J. H. Pittock (eds.) *Aberdeen and the Enlightenment*, Aberdeen University Press, Aberdeen.

46. Coulson, C. A. (1963) in C. Domb (ed.) *Clerk Maxwell and Modern Science: Six Commemorative Lectures*, pp. 43–4, University of London, Athlone Press.

47. Niven, W. D. (ed.), (1890). *The Scientific Papers of James Clerk Maxwell*, Cambridge University Press, Cambridge, vol. 1 (abbreviated [SP1] hereafter), pp. 248–444 contain papers from Maxwell's work at Aberdeen.

48. Brush, S. G., Everitt, C. W. F. and Garber, E. (eds.), (1983). *Maxwell on Saturn's Rings*, MIT Press, Cambridge, Mass.

49. Garber, E., Brush, S. G. and Everitt, C. W. F. (eds.), (1986). *Maxwell on Molecules and Gases*, MIT Press, Cambridge, Mass.; Garber, E., Brush, S. G. and Everitt, C. W. F. (eds.), (1995), *Maxwell on Heat and Statistical Mechanics*, Associated University Presses, New Jersey.

50. Everitt, C. W. F. (1975). *James Clerk Maxwell: Physicist and Natural Philosopher*, Charles Scribner & Sons, New York.

51. Tolstoy, I. (1981). *James Clerk Maxwell: A Biography*, Canongate, Edinburgh.

52. Goldman, M. (1983). *The Demon in the Aether: The Story of James Clerk Maxwell*, Paul Harris, Edinburgh.

53. *Report of the Twenty-Sixth Meeting of the British Association for the Advancement of Science; Held at Cheltenham in August 1856*, John Murray, London, 1857, 'Communications to the Sections' pp. 12–3; also in [SP1], pp. 238–47.

54. Maxwell, J. C. (1857). On the Dynamical top, &c., *Transactions of the Royal Society of Edinburgh*, XXI, part IV: 559–70.

55. Hayward, R. B. (1856). On a Direct Method of estimating Velocities, Accelerations, &c., *Transactions of the Cambridge Philosophical Society*: 1–20.

56. Brush et al. (1983) *Maxwell on Saturn's Rings* have given a detailed overview of the context.

57. Maxwell's essay is contained within [SP1]. The quotation comes from p. 373.

58. Ogilvie, G. I. (2008), James Clerk Maxwell and the dynamics of astrophysical discs, *Philosophical Transactions of the Royal Society A*, 366: 1707–15.

59. For example, Tolstoy, I. (1981) *James Clerk Maxwell* and Goldman, M. (1983) *The Demon in the Aether* include full page photographs.

60. Maxwell, J. C. (1857), Account of Experiments on the Perception of Colour, *Philosophical Magazine*, XIV; 40–7, reproduced in [SP1], pp. 263–70.

61. For more on the longer-term context and subsequent interest in Maxwell's fish-eye lens see Beech, M. (2011), Hooke's dream lens, Maxwell's fish-eye, and Luneburg's sphere, *Bulletin of the Scientific Instrument Society*, 109: 2–8.

62. Maxwell, J. C. (1857). On the General Laws of Optical Instruments, *Quarterly Journal of Pure & Applied Maths*, Vol. II, reproduced in [SP1], pp. 271–85.

63. Westheimer, G. (1966). The Maxwellian view, *Vision Research*, 6: 669–82.

64. Maxwell, J. C. (1860). On the Theory of Compound Colours and the Relations of the Colours of the Spectrum, *Philosophical Transactions of the Royal Society*, 150: 57–78, reproduced in [SP1], pp. 410–44.

65. Garber, E. et al. (1995). *Maxwell on Heat and Statistical Mechanics* and Garber, E. et al. (1986). *Maxwell on Molecules and Gases*.

66. Garber, E. et al. (1986). *Maxwell on Molecules and Gases*.

67. Maxwell, J. C. (1860). Illustrations of the Dynamical Theory of Gases, *Philosophical Magazine* 19: 19–32 and 20: 2–37, reproduced in [SP1], pp. 377–409.

68. The inventory of John Clerk Maxwell's personal estate at his death (Ref. SC16/41/20, Kirkcudbright Sheriff Court, 1856) valued the estate at about £2500.

69. Campbell, L. and Garnett, W. (1882), *Life of James Clerk Maxwell*, [CG1882], MacMillan & Co., London.

70. Sage, D. (1899), *Memorabilia Domestica: Parish Life in the North of Scotland*, Chapter XVII, Ministry and Contemporaries in Aberdeen, Rae, Wick.

71. A lengthy account of Principal Dewar running to over 50 pages is included in James Bruce (1844), *The Aberdeen Pulpit and Universities: A Series of Sketches of the Aberdeen Clergy and of the Professors in the Aberdeen Colleges*, J. Strachan & Co., Aberdeen.

72. Dewar, D. (1826), *Elements of Moral Philosophy, and of Christian Ethics*, 2 vols., J. Duncan, Paternoster-Row, London.

73. Anderson, P. J. (1898), *Fasti Academiae Mariscallanae Aberdonensis*, Vol. II, p. 30, New Spalding Club, Aberdeen.

74. <http://www.clerkmaxwellfoundation.org/JCMLibrary.pdf> [Accessed July 2013].

75. There is no record of Maxwell mentioning the aristocratic connection of his mother-in-law in any of his published correspondence. Susan Place's father, Edward Place, had married Lady Anne Gordon,

the second daughter of George Gordon, 3rd Earl of Aberdeen, in 1787. The Place's family home was Skelton Grange, Yorkshire. Catherine was Susan's grandmother's name on the Gordon side. The 4th Earl of Aberdeen was also a grandchild of the 3rd Earl and hence a cousin of Susan. He had been Prime Minister from 1852–5 and was Chancellor of King's College, Aberdeen, from 1847–60, amongst many other offices.

76. The now sizeable city of Aberdeen on Hong Kong Island with its famous harbour was named in 1845 after George Hamilton-Gordon, 4th Earl of Aberdeen and Foreign Secretary who had over-seen Hong Kong becoming a British Crown Colony. The 4th Earl would later become Maxwell's cousin-in-law, creating a tenuous link between Maxwell and the Far East.

77. Campbell, L. and Garnett, W. [CG1884], MacMillan & Co., London, 2nd edn, p. 218, 'Letter to Miss Cay', 18 February 1858.

78. The *Aberdeen Journal* announcements included the death of Susan Place Dewar, youngest daughter, on 22 December 1844 at Clifton, York (announced 25 December, 1844); death of Edward Place Dewar, a Marischal College AM graduate of 1845, on 21 November 1855 (announced 19 December 1855); death of John Dewar, the youngest son, on 12 July 1849 (announced 25 July 1849); death of Donald Dewar, a Marischal College AM graduate of 1850 (announced 18 June 1862). Another son, William Gordon Dewar, is mentioned by P. J. Anderson (1898). Maxwell's mother-in-law died on 17 March 1876 (announced 22 March 1876).

    Donald Dewar took ill with typhus in the summer of 1861 and then with suspected cancer. He was persuaded to stay with the Maxwells in London later that year and have a leg removed by surgery. He remained with the Maxwells for several months into 1862, rallied but died that summer. By then Katherine had lost three brothers, at least, and a sister, all before they had reached middle age.

79. Anderson, P. J. (1898). *Fasti Academiae Mariscallanae Aberdonensis.*

80. [CG1884], p. 190.

81. Parliamentary papers 1874 [C.899] Scotland. Owners of lands and heritages 17 & 18 Vict., cap. 91. 1872–3. Return I. *The name and address of every owner of one acre and upwards in extent (outside the municipal boundaries of boroughs containing more than 20,000 inhabitants)* show the Trustees of the late James Dewar, Mrs Susan, as owning 637 acres at Overdurdie in Kilspindie and 5 acres at Blairgowrie.

82. Letter to the editor signed C.T., The Late Professor Clerk Maxwell, *Daily News*, London, Wed., 12 November, 1879.

83. The perception of decline was articulated forcefully by Charles Babbage (1830), *Reflections on the Decline of Science in Britain, and on Some of its Causes*, B. Fellowes, London.

84. Howarth, O. J. R. (1922). *The British Association for the Advancement of Science: A Retrospective 1831–1921*, BAAS, London.

85. A list of attendees of the 1859 BA meeting was printed for the association by W. Bennett, Aberdeen, 1859. Among the astronomers who attended were George B. Airy, Warren De la Rue (who spoke on celestial photography), Wm. Rowan Hamilton (Astronomer Royal of Ireland), Thomas McLear, T. Romney Robinson and William Lassell. Other notable natural philosophers included Neil Arnott, David Brewster, Michael Faraday, Robert Fitzroy, James D. Forbes, William Hopkins (Maxwell's former private tutor at Cambridge known academically for his contributions to geology), Moritz Jacobi (St. Petersburg, known for his pioneering work on electric motors), Fleeming Jenkin (notable electrical engineer whom Maxwell would work with in London, later to become Professor in London and Edinburgh), James P. Joule, Humphrey Lloyd, Baden Powell (Oxford mathematician and writer on natural philosophy), W. J. Macquorn Rankine (Professor of Civil Engineering and Mechanics but whose work would intersect Maxwell's in future), G. Johnstone Stoney, Balfour Stewart, John Tyndall, William Thomson and Robert Willis (Jacksonian Professor of Natural Philosophy at Cambridge).

86. Murchison was a leading geologist of his day, being by then Director-General of the British Geological Survey, who had been one of the main protagonists behind the formation and expansion of the BAAS.

87. British Association for the Advancement of Science (1860). *British Association, Aberdeen Meeting, 1859. Report of the Local Committee to the Subscribers to the Local Fund, with the Treasurers' Statement of Accounts and List of Subscribers*, D. Chalmers & Co., Aberdeen.

88. [SLP1], pp. 615–8.

89. The Mathematical and Physical Science section proceedings were reported in *The Athenæum* No. 1665, 24 September 1859, pp. 400–2; No. 1666, 1 October 1859, pp. 432–4; No. 167, 8 October 1859, pp. 465–8.

90. Millar, A. H. (1925). *James Bowman Lindsay and Other Pioneers of Invention*, Malcolm C. Macleod, Dundee.

91. The *Aberdeen Journal*, 28 September 1859, p. 6.

92. Reid, J. S. (2008). James Clerk Maxwell's Scottish Chair.

93. Scottish Universities Commission. *General report of the commissioners under the Universities (Scotland) Act, 1858. With an appendix, containing ordinances, minutes, reports on special subjects, and other documents*, p. 535 (HMSO, 1863). After Maxwell's death in November 1879 the inventory of his Estate drawn up the following month includes the item 'Balance of Compensation allowance due to deceased by H. M. Exchequer as formerly Professor of Natural Philosophy in Marischal College Aberdeen £34.6.7'. Maxwell's estate on his death has been published on the web by Scotlandspeople, <http://www.scotlandspeople.gov.uk/content/help/index.aspx?1145> [Accessed July 2013].

94. Reid, J. S. (2008). James Clerk Maxwell's Scottish Chair.

CHAPTER 3

1. (1875) *Encyclopaedia Britannica; A Dictionary of Arts, Science, and General Literature*, Adam & Charles Black, Edinburgh. 9th edn, vol. XIV, p. 821, 1882.

2. In the 34-page, double column, article on 'London', (1875) *Encyclopaedia Britannica* 'Higher Education' merits only a single modest paragraph, p. 835.

3. This issue and the reforms that came later in the nineteenth century are discussed more fully in Cantor, G. (2005), *Quakers, Jews and Science: Religious Responses to Modernity and the Sciences in Britain, 1650–1900*, Oxford University Press, Oxford.

4. Hearnshaw, F. J. C. (1929). *The Centenary History of King's College London 1828–1928*, George Harrap & Co. Ltd., London, discusses the foundation and early history of King's College. However, his brief and ill-founded comment on Maxwell (discussed at length by Cyril Domb (1980), James Clerk Maxwell in London 1860–1865) does not enhance his reputation as a historian.

5. In 1860, London's *Daily News* published the official results of the entrance exam of the Civil Service of India, competed for by young men around 20 years old who had a University Degree, or equivalent. Graduates of Oxford and Cambridge dominated the list of some 80 strong, with Dublin and Belfast making up significant numbers. The four candidates from Aberdeen all came in the top half, three in the top 20%; the four candidates from King's College, London were all in the bottom quarter. *Daily News*, London, 22 August, 1860. One would need a longer run of data (not readily available) and more background to establish a trend but the figures are suggestive that King's College was still finding its feet or, more positively as suggested in the text, aiming to provide an education with a different emphasis.

6. Lyon Playfair's lecture is reported at length in Tallis, J. (1852), *History and Description of the Crystal Palace and the Exhibition of the World's Industry in 1851*, London Printing & Publishing Co., vol. II,

pp. 194–200 in which he explicitly cites University College and King's College 'as productive of much good . . . but their instruction in science terminates just where the industrial colleges of the continent begin'.

7. King's College Council Minutes, 1860, Ref: KA/C/M, p. 231, includes the printed specification for the Professorship of Natural Philosophy; p. 232 lists the other candidates as Rev. A. Power, R. B. Clifton, G. R. Smalley and A. H. Green and the Council's decision.
8. King's College Council Minutes, 1860.
9. Fabre, C. (1889–1906). *Traité Encylopédique de Photographie*, Gauthier-Villars et Fils, Paris.
10. W. D. H. [Willem Hackmann (1998)], Thomas Sutton Panoramic Camera Lens, in Sphæra No. 8, Newsletter of the Oxford Museum of the History of Science. <http://www.mhs.ox.ac.uk/sphaera/index.htm?issue8/articl7> [Accessed July 2013].
11. Maxwell, J. C. On the Theory of Three Primary Colours, *Notices of the Proceedings at the Meetings of the Members of the Royal Institution of Great Britain*, Vol. III, 17 May, 1861, pp. 370–5; reproduced in W. D. Niven (1890), *The Scientific Papers of James Clerk Maxwell*, Cambridge University Press, Cambridge, Vol. 1 (hereafter referred to as [SP1]), pp. 445–50.
12. Evans, R. M. (1961), Maxwell's colour photograph, *Scientific American*, 205: 118–28.
13. [CG1884], p. 177.
14. *The Calendar of King's College, London for 1860–61*, p. 113. The same wording is in the Calendars of other relevant years.
15. Hosking was replaced by Prof. Robert Kerr, an Aberdonian who had been a private student of Natural Philosophy at Marischal College in 1836–37, first President of the Architectural Association who was a respected authority on domestic housing.
16. The Charles Loudon Bloxam Papers are deposited in the British Museum, additional manuscripts 81280–81289.
17. The King's College Calendars, assembled each summer, contain a report for the previous year prepared in April that includes the numbers of matriculated students in each Department.
18. King's College Applied Science Register No. 1 (1857–69), King's College archives KA/SRB/7.
19. Domb, C. (1980). James Clerk Maxwell in London 1860–1865, *Notes and Records of the Royal Society*, Vol. 35, pp. 67–103.
20. Maxwell, J. C. Manuscript notebook, Vol. 1, King's College Archives, ref: GB 0100 KCLA Maxwell.
21. Harman, P. M. (1990). *The Scientific Letters and Papers of James Clerk Maxwell*, Vol. I (hereafter referred to as [SLP1]), Cambridge University Press, Cambridge, pp. 419–31 and pp. 662–74.
22. Morton, A. Q. and Wess, J. A. (1993). *Public and Private Science: The King George III Collection*, Oxford University Press, Oxford.
23. Skeat, W. O. (1957). *King's College London Engineering Society 1847–1947*, King's College, London.
24. *The Calendar of King's College, London for 1860–61*, pp. 268–70.
25. Domb, C. (1980). James Clerk Maxwell in London 1860–1865.
26. Maxwell, J. C. (1863). Instruments Connected with Light, pp. 502–7 in *Record of the International Exhibition 1862*, William Mackenzie, London.
27. Maxwell, J. C. (1876). Special Considerations Respecting Scientific Apparatus, pp. 1–21 and Instruments Connected with Fluids, pp. 87–92 in *Handbook to the Special Loan Collection of Scientific Apparatus 1876*, South Kensington Museum Science Handbooks, Chapman and Hall, London.
28. Professor Wheatstone, Professor Williamson, Mr C. F. Varley, Professor Thomson, Mr Balfour Stewart, Mr C. W. Siemens, Dr A. Matthiessen, Professor Maxwell, Professor Miller, Dr Joule, Mr Fleeming

Jenkin, Dr Esselbach, Sir C. Bright (1864), Report of the Committee appointed by the British Association on Standards of Electrical Resistance, pp. 111–76, in *Report of the Thirty-Third Meeting of the British Association for the Advancement of Science; Held at Newcastle Upon-Tyne in August and September 1863*, John Murray, London. This and subsequent reports were the only papers published in the joint names of Wheatstone and Maxwell. Wheatstone was in name Professor of Experimental Philosophy at King's College London for 41 years. He was almost 30 years older than Maxwell but although they shared an interest in matters electrical, mechanical and optical it seems they did not collaborate on other topics. Wheatstone gave no lectures for most of his tenure, didn't attend BAAS meetings and it is likely that he wasn't paid by the college, since payment was related to number of students taught. Wheatstone had his own personal laboratory equipment and his own library (both of which he donated to King's College in his will) but his post was largely honorific and by 1860 his focus was in the commercial world of telegraphy. However, the association did bring prestige to King's College and academic credibility to Wheatstone.

29. Maxwell, J. C. On Governors, in [SP2], pp. 105–20.

30. Report of the Committee on Standards of Electrical Resistance (1865), pp. 345–67 in *Report of the Thirty-Fourth Meeting of the British Association for the Advancement of Science; Held at Bath in September 1864*, John Murray, London.

31. Report of the Committee on Standards of Electrical Resistance (1866), pp. 308–13 in *Report of the Thirty-Fifth Meeting of the British Association for the Advancement of Science; Held at Birmingham in September 1865*, John Murray, London.

32. Although Maxwell published no papers on Electricity and Magnetism while at Aberdeen, judging from some of his letters he was clearly thinking about the subject following his 1855/1856 papers 'On Faraday's Lines of Force'. His letter to Faraday in November 1857 is published [SLP1], pp. 548–52 and in a letter to his friend R. B. Litchfield, pp. 508–9, he commented: 'I believe there is a department of mind conducted independent of consciousness, where things are fermented and decocted, so that when they are run off they come clear.' His decocted thoughts on electromagnetism appeared soon after he settled in at King's College.

33. Maxwell, J. C. On Faraday's Lines of Force, in [SP1], p. 155, reprinted from the *Transactions of the Cambridge Philosophical Society*, X, part I.

34. This is oversimplifying the complex issue of Faraday's own changing conceptions on electric and magnetic fields, for Faraday did not use the term 'field' until 1845, though he had earlier talked of magnetic lines of force.

35. Darrigol, O. (2000). *Electrodynamics from Ampère to Einstein*, Oxford University Press, Oxford.

36. Maxwell, J. C. On Faraday's Lines of Force, in [SP1], p. 207, reprinted from the *Transactions of the Cambridge Philosophical Society*, X, part I.

37. Maxwell, J. C. (1861 and 1862). On Physical Lines of Force, pp. 451–513 in [SP1]; reprinted from *Philosophical Magazine* XXI: 161–75, 281–91, 338–48 (1861) and *Philosophical Magazine* XXII: 12–24, 85–95 (1862).

38. Maxwell, J. C. Dynamical Theory of the Electromagnetic Field, in [SP1], pp. 526–97.

39. Hunt, B. J. (1991). *The Maxwellians*, Cornell University Press, New York, gives an extensive account of how Maxwell's ideas influenced the next generation of electrical researchers who drew out of them their practical utility.

40. Maxwell, J. C. (1873). *A Treatise on Electricity and Magnetism*, Clarendon Press, Oxford.

41. See Chapter 11 of this book and also Hunt, B. J. (1991), *The Maxwellians* for examples of aspects of electromagnetic waves such as reflection, refraction and energy propagation not treated by Maxwell himself.

42. Everitt, C. W. F. (1975). *James Clerk Maxwell: Physicist and Natural Philosopher*, Charles Scribner & Sons, New York.

43. Garber, E., Brush, S. G. and Everitt, C. W. F. (eds.) (1986). *Maxwell on Molecules and Gases*, MIT Press, Cambridge, Mass.; Garber, E., Brush, S. G. and Everitt, C. W. F. (eds.) (1995), *Maxwell on Heat and Statistical Mechanics*, Associated University Presses, New Jersey.

44. Macquorn Rankine, W. J. (1881). Principle of the Equilibrium of Polyhedral Frames, p. 564 in *Miscellaneous Scientific Papers*, Charles Griffin & Co., London.

45. Background on the Keith Prize, courtesy of the Royal Society of Edinburgh.

46. Routh, E. J. (1896). *A Treatise on Analytical Statics with Numerous Examples*, Chapter 8, pp. 229–51, Cambridge University Press, Cambridge, 2nd edn.

47. [SLP2] reproduces four reports during his King's College years and over two dozen subsequent reports.

48. In every year from 1864 to 1879 inclusive, except 1867, Maxwell was a co-author of one or more reports to the British Association for the Advancement of Science by various committees. He was on eight committees in total, half of which were concerned with electrical issues and half with other issues.

49. King's College Council Minutes, 1860, p. 305. The rate for each matriculated student, which was essentially the salary, was advertised for Maxwell's successor as 4 guineas, in comparison with Maxwell's 5 gns.

50. The Thames was at times like an open sewer and the 'great stink' of 1858 would have been remembered by many. Poor sewerage was a contributor to infectious diseases, the cause of 20% of London's deaths in the 1860s as detailed in the *Encyclopaedia Britannica* 1875, pp. 827–8. In addition, coal smoke and city smog were responsible for the 'black wreath' – Wordsworth's phrase – of pungent and irritating poor air that contributed to respiratory diseases.

51. Cat, J. (2011). Glenlair: A Brief Architectural History, <http://www.glenlair.org.uk/glenlair_history.asp> [Accessed July 2013].

CHAPTER 4

1. Maxwell, J. C. (1890). Introductory Lecture on Experimental Physics. In W. D. Niven (ed.), *The Scientific Papers of James Clerk Maxwell*, vol. II, pp. 241–55, on p. 250. Cambridge University Press, Cambridge (hereafter [SP2]).

2. University Intelligence, *The Times*, 4 November 1873, also *Cambridge University Reporter* 4 November 1873.

3. Warwick, A. (2003). *Masters of Theory: Cambridge and the Rise of Mathematical Physics*, p. 392, University of Chicago Press; James Clerk Maxwell (1873). *Treatise on Electricity and Magnetism*, 2 vols, Clarendon Press, Oxford.

4. Report of the Physical Sciences Syndicate 27 February 1869, a copy is in *Cambridge University Reporter* (16 November 1870); Report of the Syndicate to consider the means of raising the necessary funds for establishing a Professor and Demonstrator of Experimental Physics, and for providing buildings and apparatus required for that department of science 29 March 1870. The copy in the University Registry Guard Book CUL CUR 39.33 8. has a handwritten note of the Senate discussion added by J. W. Clark, the Registrar: 'This report was discussed in the Arts School on the day fixed above. The general tone was unfavourable to the proposed increase in the capitation fee. See "Cutting book" 1870 no. 295. JWC'.

5. *Royal Commission on Scientific Instruction and the Advancement of Science, Vol. I. First, supplementary, and second reports, with minutes of evidence and appendices*, House of Commons Parliamentary Papers (1872) [C.536].

6. Cavendish, W., Duke of Devonshire, to Power, J. Vice Chancellor of Cambridge University, 10 October 1870, CUL VCCorr.1.2.1.

7. Grace to Establish in the University a Professorship of Experimental Physics 28 November 1870; the Grace was discussed and approved on 9 February 1871. A copy, with J. W. Clark's handwritten comments is at CUL CUR 39.33 12.

8. Letters from G. G. Stokes, J. W. Strutt and E. Blore to Maxwell are reproduced in footnotes in P. M. Harman (ed.) (1995). *The Scientific Letters and Papers of James Clerk Maxwell*, Vol. II, pp. 611–5n. Cambridge University Press, Cambridge (hereafter [SLP2]).

9. [CG1882], p. 348.

10. Maxwell, J. C. to Power, J., Vice Chancellor of Cambridge 23 February 1871, CUL VCCorr.VI.2. Photo reproduced by kind permission of the Syndics of Cambridge University Library.

11. Dong-Won, K. (1991). *The Emergence of the Cavendish School: an early history of the Cavendish Laboratory 1871–1900*, Harvard University PhD thesis; The voting slips are at CUL O.XIV.39. The names and colleges as given on the voting slips are: William Pike, Downing; Henry Sidgwick, Trinity; Mr Trotter, Trinity; T.G. Bonney, St John's; Sedley Taylor, Trinity; R.C. Jebb, Trinity; A. Cayley, Trinity; James Porter, Peterhouse; E.A. Swainson, Christs; Joseph Wolstenholme, Christs; R. Kalley Miller, Peterhouse; N. M. Ferrers, Gonville & Cauis; William H. Thompson, Trinity. The announcement of the election is in *Cambridge University Reporter*, 15 March 1871.

12. Maxwell, J. C. to Strutt, J. W. 15 March 1871, reprinted in [SLP2], 8, p. 615. 'Wranglers', 'Senior Optimes', and 'Junior Optimes' were (in descending order) the three classifications of honours in the Mathematical Tripos. The man who came top in the exams was the 'Senior Wrangler'.

13. Becher, H. W. (1986). Voluntary Science in Nineteenth Century Cambridge University to the 1850's, *British Journal for the History of Science*, 19: 57–87.

14. *Report of Her Majesty's Commissioners Appointed to Inquire into the State, Discipline, Studies, and Revenues of the University and Colleges of Cambridge: Together with the Evidence, and an Appendix* (1852), p. 74.

15. Lecture lists 1875–7, *Cambridge University Reporter*; Warwick (2003) *Masters of Theory*, Ch. 6.

16. *Copy of the report, dated 27th March 1876, of the syndicate appointed 27th May 1875, by the University of Cambridge, to consider the requirements of the university in different departments of study, with appendices.* House of Commons Parliamentary Papers (1876) (272).

17. Report of the Physical Sciences Syndicate 27 February 1869; *Report of Her Majesty's Commissioners Appointed to Inquire into the State, Discipline, Studies, and Revenues of the University and Colleges of Cambridge: Together with the Evidence, and an Appendix* (1852), p. 119.

18. Maxwell, J. C. to Power, J., Vice Chancellor 26 September 1871, CUL VCCorr.IV.6.

19. Maxwell, J. C. to Tait, P. G. 19 October 1871, reprinted in [SLP2], p. 836.

20. Waterston, C. D. and Macmillan, S. A. (2006). Former Fellows of the Royal Society of Edinburgh 1783–2002: Biographical Index Part One. <http://www.royalsoced.org.uk/cms/files/fellows/biographical_index/fells_indexp1.pdf> [Accessed July 2013].

21. Schuster, A. (1910) in *A History of the Cavendish Laboratory, 1871–1910*, pp. 18–9. Longmans, Green & Co., London, 1910; Venn, J. and Venn, J. A., *A Cambridge Alumni Database* <http://venn.lib.cam.ac.uk/Documents/acad/intro.html> [Accessed July 2013]; Report to Senate 2 March 1874, *Cambridge University Reporter* (3 March 1874), p. 244.

22. Shaw, W.N. A notebook of notebooks, CUL GBR/0012/MS.

23. Maxwell, J. C. to Thomson, W. 25 August 1879, reprinted in P. Harman (ed.) (2002), *The Scientific Letters and Papers of James Clerk Maxwell*, Vol. III, pp. 838–9. Cambridge University Press (hereafter [SLP3]).

24. Maxwell, J. C. (1890). On the application of Hamilton's Characteristic Function to the Theory of an Optical Instrument symmetrical about its Axis, in [SP2], pp. 439–44.

25. Maxwell, J. C. (1890). Introductory Lecture on Experimental Physics in [SP2], pp. 241–55, on pp. 244–5.

26. Maxwell, J. C. (1890). Introductory Lecture on Experimental Physics in [SP2], p. 244.

27. Maxwell, J. C. (1890). Introductory Lecture on Experimental Physics in [SP2], p. 242.

28. Maxwell, J. C. (1890). Introductory Lecture on Experimental Physics in [SP2], p. 250.

29. Maxwell, J. C. (1890). Introductory Lecture on Experimental Physics in [SP2], p. 249.

30. Todhunter, I. (1873). *The Conflict of Studies and Other Essays*, p. 19. Macmillan, London.

31. Schuster, A. (1910) in *A History of the Cavendish Laboratory.*

32. Macleod, R. and Moseley, R. (1980). The 'Naturals' and Victorian Cambridge: Reflections on the Anatomy of an Elite, 1851–1914, *Oxford Review of Education*, 6: 177–95.

33. Glazebrook, R. T. (1931). Early Days at the Cavendish Laboratory, in *James Clerk Maxwell: A Commemoration Volume, 1831–1931*, pp. 130–41, on pp. 136–7. Cambridge University Press, Cambridge.

34. Chrystal, G. to Maxwell, J. C. 5 July 1877, reprinted in [SLP3], p. 521n.

35. Schaffer, S. (1992). Victorian Metrology and its Instrumentation: A Manufactury of Ohms, in R. Bud and S. Cozzens (eds.), *Invisible Connections: Instruments, Institutions and Science*, pp. 23–56. SPIE Press, Bellingham.

36. Schuster, A. (1910) in *A History of the Cavendish Laboratory*, p. 31; Maxwell, J. C. (1877). Referees report on Royal Society grant application, University of Birmingham MSS Poynting papers 7/iv/2/1.

37. Draft of letter from Maxwell, J. C. to Cookson, H. W. Vice Chancellor 5 July 1873, reprinted in SLP2, p. 876.

38. Schaffer, S. (1992). Victorian Metrology and its Instrumentation, p. 34; Maxwell, J. C. (15 May 1877). Cavendish Laboratory Annual Report, *Cambridge University Reporter*, p. 434; Vouchers for apparatus purchase by Professor Maxwell 1876–9 CUL GBR/0265/CAV 1.

39. Chrystal, G. to Maxwell, J. C. 9 July 1878, reprinted in SLP3 p. 687n.

40. Maxwell, J. C. (2 April 1878). Cavendish Laboratory Annual Report, *Cambridge University Reporter*, p. 420; Strutt, R. J. (1968). *Life of John William Strutt, Third Baron Rayleigh*, augmented edition, p. 103. University of Wisconsin Press, London.

41. Maxwell, J. C. to Tomlinson, C. 29 May 1873, reprinted in [SLP2], p. 858; Draft of letter from J. C. Maxwell to the Duke of Devonshire, late January–early February 1873, reprinted in [SLP2], pp. 785–6; Maxwell, J. C. to Thomson, W. 25 March 1873, reprinted in [SLP2], p. 839.

42. Maxwell, J. C. to Garnett, W. 23 August 1877, reprinted in [SLP3], p. 669.

43. Maxwell, J. C. (1879). Intensity of the sensation produced by an electric discharge in J. C. Maxwell (ed.), *Electrical Researches of the Hon. Henry Cavendish*, n31 pp. 437–43. Cambridge University Press, Cambridge.

44. Schuster, A. (1910) in *A History of the Cavendish Laboratory*, p. 33.

45. Review of *The Theory of Heat*, in *The Ironmonger* (31 January 1872).

46. Lamb, H. (1931). Clerk Maxwell as a lecturer, in *James Clerk Maxwell: A Commemoration Volume, 1831–1931*, pp. 142–6, on p. 145.

47. Michel, J-B., Shen, Y. K., Aiden, A. P., Veres, A., Gray, M. K., Brockman, W., The Google Books Team, Pickett, J. P., Hoiberg, D., Clancy, D., Norvig, P., Orwant, J., Pinker, S., Nowak, M. A., and Aiden, E. L. (2011). Quantitative analysis of culture using millions of digitized books, *Science*, 331: 176–82.

48. *Report of Her Majesty's Commissioners Appointed to Inquire into the State, Discipline, Studies, and Revenues of the University and Colleges of Cambridge: Together with the Evidence, and an Appendix* (1852), p. 101.

49. Seeley, J. R. (1866). *The Student's Guide to the University of Cambridge*, pp. 183–4, p. 185. Deighton, Bell, London.

50. Report of the Physical Sciences Syndicate, 27 February 1869.

51. Shaw, W. N. autobiographical notes, Twice twenty & four: education at Cambridge supplement, CUL GBR/0012/MS.

52. For a detailed account of the history of the mathematical tripos see W. W. Rouse Ball (1889), *A History of the Study of Mathematics at Cambridge*, Cambridge University Press, Cambridge. For the role of the mathematical tripos in the rise of mathematical physics at Cambridge in the nineteenth century, see A. Warwick (2003) *Masters of Theory.*

53. Tanner, J. R. (ed.) (1917). *The Historical Register of the University of Cambridge*, pp. 348–54. Cambridge University Press, Cambridge; *Report of Her Majesty's Commissioners Appointed to Inquire into the State, Discipline, Studies, and Revenues of the University and Colleges of Cambridge: Together with the Evidence, and an Appendix* (1852), p. 21.

54. Bettany, G. T. (1874). Practical Science at Cambridge. *Nature*, 11: 132–3.

55. Tanner, J. R. (1917). *The Historical Register of the University of Cambridge.*

56. Wilson, D. B. (1982). Experimentalists among the mathematicians: physics in the Cambridge Natural Sciences Tripos, 1851–1900, *Historical Studies in the Physical Sciences*, 12: 325–71.

57. Wilson, D. B. (1982). Experimentalists among the mathematicians, p. 345; Maxwell, J. C. (1871). *Theory of Heat*. Longmans Green & Co., London.

58. Maxwell, J. C., report on physics in NST Mark Book 1873, CUL Min VIII.56.

59. Wilson, D. B. (1982). Experimentalists among the mathematicians, pp. 339–40.

60. NST Mark Book 1873, CUL Min VIII.56.; Tanner, J. R. (1917) *The Historical Register of the University of Cambridge*; Wilson, D. B. (1982), Experimentalists among the mathematicians, p. 326.

61. NST Mark Book 1873, CUL Min VIII.56.

62. Dampier, W. C. D., rev. James, F. A. J. L. (2004). Liveing, George Downing, in *Oxford Dictionary of National Biography*, online edn. Oxford University Press, <http://www.oxforddnb.com/view/article/34559> [Accessed July 2013].

63. *Cambridge University Reporter* (3 June 1879), p. 676; Rolleston, H. D. (1932). *The Cambridge Medical School: A Biographical History*, p. 26. Cambridge University Press, Cambridge.

64. Maxwell notebook, CUL Add7655/V/n/2; Venn, J. and Venn, J. A., *A Cambridge Alumni Database* <http://venn.lib.cam.ac.uk/Documents/acad/intro.html> [Accessed July 2013]; Tanner, J. R. *The Historical Register of the University of Cambridge* (1917).

65. Maxwell, J. C. (1885). Physical Sciences in *Encyclopedia Britannica* 9th edn. A&C Black, Edinburgh.

66. Maxwell, J. C. (1890). Introductory Lecture on Experimental Physics in [SP2], pp. 241–55.

67. Todhunter, I. (1873). *The Conflict of Studies*, p. 17.

68. Becher, H. W. (1986). Voluntary science.

69. *Report of Her Majesty's Commissioners Appointed to Inquire into the State, Discipline, Studies, and Revenues of the University and Colleges of Cambridge: Together with the Evidence, and an Appendix* (1852), p. 116.

70. Gooday, G. (1990). Precision measurement and the genesis of physics teaching laboratories in Victorian Britain, *British Journal for the History of Science*, 23: 25–51.

71. Trotter, C. to Maxwell, J. C. 20 April 1871, CUL Add MSS 7655/II/46.

72. Fawcett, W. to the Laboratory Building Syndicate CUL VCCorr.1.2.9; CG1882, p. 173; Fawcett edited and expanded the third edition of F. A. Paley's *Manual of Gothic Mouldings* in 1865.

73. Willis, R. and Clerk, J. W. (1886). *The Architectural History of the University of Cambridge*, Vol. 3, p. 183n. Cambridge University Press, Cambridge; *Cambridge University Reporter*, 21 May 1873, p. 73.

74. Maxwell, J. C. to Tait, P. G. 4 April 1871, reprinted in [SLP2] p. 634.

75. Schaffer, S. (1998). Physics Laboratories and the Victorian Country House, in C. Smith and J. Agar (eds.), *Making Space for Science*, pp. 149–80. Macmillan, London.

76. The plans for extending Glenlair are at the Royal Commission on the Ancient and Historic Monuments of Scotland, DPM 1880/71/1; DPM 1880/71/1/4; and DPM 1880/71/1/5.

77. Harrison, W. J., rev. Morus, I. R. (2004). Gassiot, John Peter, in *Oxford Dictionary of National Biography*, online edn. Oxford University Press, <http://www.oxforddnb.com/view/article/10439> [Accessed July 2013].

78. Maxwell, J. C. to Maxwell, K. C. 20 March 1871, reprinted in CG1882, p. 381.

79. Report of the syndicate 'to consider the question of an appropriate site of a building for the department of Experimental Physics', *Cambridge University Reporter* (18 November 1871).

80. Draft of letter from Maxwell, J. C. to Blore, E. 15 February 1871, reprinted in [SLP2], p. 612.

81. Trotter, C. to Maxwell, J. C. 20 April 1871, CUL Add MSS 7655/II/46.

82. The New Physical Laboratory at the University of Cambridge, *Nature*, 10 (1874), pp. 139–42; Schaffer, S. (1998) Physics Laboratories and the Victorian Country House.

83. The New Physical Laboratory at the University of Cambridge, *Nature*, 10 (1874), pp. 141–2.

84. The notebook is on display in the museum at the Cavendish Laboratory.

85. *Cambridge University Reporter* (27 April 1875), p. 352; *Cambridge University Reporter* (15 May 1877), p. 433.

86. Glazebrook, R. T. (1931). Early days at the Cavendish Laboratory, pp. 130–41.

87. The annual reports in the *Cambridge University Reporter* contain lists of apparatus acquired each year.

88. Smith, C. and Wise, N. (1989). *Energy and Empire: A Biographical Study of Lord Kelvin*, p. 131. Cambridge University Press, Cambridge; Becher, H. W. (1986) Voluntary science, p. 13; Maxwell, J. C. to Monro, C. J. 15 March 1871, reprinted in [SLP2], p. 619.

89. Maxwell, J. C. to Campbell, L. 25 December 1976, reprinted in [SLP3], p. 421.

90. Gordon, J. E. H. (1876). On a Repetition of Dr. Kerr's Experiments on a New Relation Between Electricity and Light, *Philosophical Magazine* Series 5, 2, 203–8.

91. *Cambridge University Reporter* (26 March 1874), p. 286; Venn, J. and Venn, J. A., *A Cambridge Alumni Database* <http://venn.lib.cam.ac.uk/Documents/acad/intro.html> [Accessed July 2013].

92. Clayden, A. W and Heycock, C. T. (1876). The Spectra of Indium, *Philosophical Magazine*, Series 5(2): 387–9; Venn, J. and Venn, J. A., *A Cambridge Alumni Database* <http://venn.lib.cam.ac.uk/Documents/acad/intro.html> [Accessed July 2013].

93. Garnett, W., quoted by Schuster, A. in *A History of the Cavendish Laboratory* (1910), pp. 35–6.

94. *Cambridge University Reporter* (15 October 1878), p. 48.

95. Tanner, J. R. (1917). *The Historical Register of the University of Cambridge.*

96. *Cambridge University Reporter* (3 June 1879) p. 676.

97. Maxwell notebook, CUL Add 7655/V/n/3.

98. Newall, H. F. (1910) in *A History of the Cavendish Laboratory*, 102–9.

99. Smith and Wise, *Energy and Empire*, p. 131; Gray, A. (1897). Lord Kelvin's Laboratory in the University of Glasgow, *Nature*, 55, 486–92.

100. Glazebrook, Early days at the Cavendish Laboratory, 130–41.

101. *Cambridge University Reporter* (26 May 1874) p. 410; *Cambridge University Reporter* (19 June 1877); Maxwell notebook, CUL Add 7655/V/j/5.

102. Maxwell, J. C., report on physics in NST Mark Book 1873, CUL Min VIII.56.

103. Newall, H. F. (1910) in *A History of the Cavendish Laboratory, 1871–1910*, pp. 109, Longmans, Green & Co., London.

104. *Cambridge University Reporter,* (17 December 1876) p. 146; Garnett, W. in *A History of the Cavendish Laboratory, 1871–1910*, p. 35, Longmans, Green & Co., London.

105. Fleming, J. A. (1931). Some Memories of Professor James Clerk Maxwell, in *James Clerk Maxwell: A Commemoration Volume, 1831–1931*, p. 117; Newall, H. F. (1910) in *A History of the Cavendish Laboratory, 1871–1910*, p. 106, Longmans, Green & Co., London.

106. *Royal Commission on Scientific Instruction and the Advancement of Science, Vol. I. First, supplementary, and second reports, with minutes of evidence and appendices*, House of Commons Parliamentary Papers (1872) [C.536], p. 165.

107. Simcock, T. (2009). Laboratories and Physics in Oxford Colleges, 1848–1947, in R. Fox and G. Gooday (eds.), *Physics in Oxford 1839–1939*, pp. 119–68. Oxford University Press, Oxford.

108. *A History of the Cavendish Laboratory, 1871–1910* (1910), Longmans, Green & Co., London.

109. Newall, H. F. (1910) in *A History of the Cavendish Laboratory, 1871–1910*, p. 108.

110. Glazebrook, R. T. (1910) in *A History of the Cavendish Laboratory, 1871–1910*, p. 44.

111. Schuster, A. (1910) in *A History of the Cavendish Laboratory, 1871–1910.*

112. Garnett, W. quoted by Schuster in *A History of the Cavendish Laboratory, 1871–1910*, p. 35.

113. *A History of the Cavendish Laboratory, 1871–1910*, p. 41.

114. Maxwell notebook CUL Add7655/V/n/2.

115. Maxwell notebook CUL Add7655/V/n/2.

116. Maxwell notebook CUL Add7655/V/n/2.

117. Inserted into Maxwell notebook, CUL MS Add 7655/V/j/5.

118. Maxwell notebook CUL Add7655/V/n/3.

119. *Cambridge University Reporter* (27 April 1875).

120. *A History of the Cavendish Laboratory, 1871–1910* (1910), Longmans, Green & Co., London.

121. Whitmell, C. T. to Adams, J. C. 27 January 1875, Cambridge, St. John's College Library, papers of John Couch Adams, 15/17/1.

122. *James Clerk Maxwell: A Commemoration Volume, 1831–1931*, (1931) Cambridge University Press, Cambridge.

123. Note that the postgraduate category includes students who had taken their final Tripos exam, and were thus not under exam pressure, even if they had not yet formally graduated.

124. Probably this Nanson. There was an Edward James Nanson at Trinity at the same time (and year), but Edward John appears to have been the more scientifically inclined.

125. Middleton was recorded as non-collegiate in Lent 1878. His formal presence in the Laboratory appears to have been dated from the time he became a fellow commoner at St. John's the following year.

CHAPTER 5

This essay is based upon a paper published in the *Philosophical Transactions of the Royal Society of London*, 366: 1685–96 (2008).

The following books and articles may be consulted for more details of the topics discussed in this essay, as well as many more detail of Maxwell's life and his contributions to the science of colour.

Campbell, L. and Garnett, W. (1882). *The Life of James Clerk Maxwell*. MacMillan, London. This book is available in electronic form at <http://www.sonnetsoftware.com/bio/maxbio.pdf [Accessed July 2013].

Dougal, R. C. and Forfar, D. O. (2012). Newsletter of the James Clerk Maxwell Foundation, No. 1, April 2012 at <http://www.clerkmaxwellfoundation.org> [Accessed July 2013].

Everitt, C. W. F. (1976). *James Clerk Maxwell: Physicist and Natural Philosopher*. Scribner, New York.

Fauvel, J., Flood, R., Shortland, M. and Wilson, R. (eds.) (1988). *Let Newton Be: A New Perspective on his Life and Works*. Oxford University Press, Oxford.

Harman, P. M. (1998). *The Natural Philosophy of James Clerk Maxwell*. Cambridge University Press, Cambridge.

Harman, P. M. (ed.) (1990, 1995, 2002). *The Scientific Letters and Papers of James Clerk Maxwell*, Volumes 1–3 (1846–62, 1862–73, 1874–9). Cambridge University Press, Cambridge.

Lamb, T. and Bourrieau, J. (eds.) (1995). *Light and Colour*. Darwin College Lectures. Cambridge University Press, Cambridge.

Mollon, J. D.(1995). Seeing Colour. In Lamb, T. and Bourriau, J. (eds.), *Colour: Art and Science*, pp. 127–50. Cambridge University Press, Cambridge.

Newton, I. (1672). *Philosophical Transactions of the Royal Society*, 7: 4002–10.

Niven, W. D. (ed.) (1890). *The Scientific Papers of James Clerk Maxwell*, (two volumes). Cambridge University Press, Cambridge. These are also available in a single volume from Dover Publications, New York (1965).

Schaffer, S.(1989). In Gooding, D., Pinch, T. and Schaffer, S. (eds.), *The Uses of Experiment: Studies in the Natural Sciences*. Cambridge University Press, Cambridge.

1. See: Newton, I. (1957–1977). *Correspondence*, 7 volumes, Turnbull, H. W., Scott, J. F., Hall, A. R. and Tilling, L. (eds.), 1, p. 95. Cambridge University Press, Cambridge.
2. Young, T. (1802). The Bakerian Lecture: On the Theory of Light and Colours, *Philosophical Transaction of the Royal Society of London*, 92: 20–1.
3. This can be downloaded from the website <http://brightbytes.com/collection/zoetrope.pdf> [Accessed July 2013].

CHAPTER 6

1. Maxwell, J. C. (1890). Scientific Papers (hereafter [SP1, SP2]). W.D. Niven, (ed.) 2 Volumes, Cambridge University Press, Cambridge. Reprinted in 1965 by Dover Publications, New York. (Page numbers here are from the Cambridge University Press edition.)
2. Campbell, L. and Garnett, W. (1882). *The Life of James Clerk Maxwell* (hereafter [CG1882]), pp. 374–5, 555–8. Macmillan, London.
3. [CG1882], pp. 554–5.
4. Mahon, B. (2003). *The Man who Changed Everything: The Life of James Clerk Maxwell*, pp. 116–9, 136–7, 199, 203. Wiley, Chichester.

5. [SP1], pp. 286–376 + Plate V.
This is reprinted from Maxwell, J. C. (1859). *On the Stability of the Motion of Saturn's Rings.* Macmillan, Cambridge.
This revised and final version is included in Brush, S. G., Everitt, C. W. F. and Garber, E. (eds.) (1983). *Maxwell on Saturn's Rings*, pp. 68–158. MIT Press, Cambridge, Mass. (Title page and advertisement are taken from the Macmillan book, and the remainder from the Dover Publications reprint of SP.)
6. [SP1], p. 291. Brush et al. (1983). *Maxwell on Saturn's Rings*, p. 73.
7. Brush et al. (1983). *Maxwell on Saturn's Rings.*
8. Brush et al. (1983). *Maxwell on Saturn's Rings*, p. ix.
9. Brush et al. (1983). *Maxwell on Saturn's Rings*, p. 25.
10. Brush et al. (1983). *Maxwell on Saturn's Rings*, p. ix.
11. Harrison, H. M. (1994). *Voyager in Time and Space: The Life of John Couch Adams, Cambridge Astronomer*. Book Guild, Lewes.
12. Eggen, O. J. (1970). George Biddell Airy in Gillipsie, C. (ed.), *Dictionary of Scientific Biography.* Scribner, New York. Now owned by University of California at Davis Libraries. Volume 1, pp. 84–7 <http://www.encyclopedia.com/doc/1G2-2830900062.html> [Accessed July 2013].
13. Eggen, O. J. (1971). In Gillispie, C. (ed.), *Dictionary of Scientific Biography*. Scribner, New York. Now owned by University of California at Davis Libraries. Volume 3, pp. 186–7.
14. Kollerstrom, N. (2006). An Hiatus in History: The British Claim for Neptune's Co-Prediction, 1845–1846. Parts 1 and 2. *History of Science* 143: 1–28 and 145: 349–71.
15. Mathematical discovery of planets. McTutor History of Mathematics. <http://www-history.mcs.st-and.ac.uk/HistTopics/Neptune_and_Pluto.html> [Accessed July 2013].
16. Sheehan, W., Kollerstrom, N. and Waff, C. B. (2004). The Case of the Pilfered Planet: Did the British steal Neptune? *Scientific American* 291(6): 92–9.
17. Urbain Jean Joseph Le Verrier. McTutor History of Mathematics <http://www-history.mcs.st-andrews.ac.uk/Biographies/Le_Verrier.html> [Accessed July 2013].
18. Dominique Francois Jean Arago. McTutor History of Mathematics <http://www-history.mcs.st-andrews.ac.uk/Biographies/Arago.html> [Accessed July 2013].
19. John Couch Adams. McTutor History of Mathematics <http://www-history.mcs.st-andrews.ac.uk/Biographies/Adams.html> [Accessed July 2013].
20. The Royal Society: Copley Medal. <http://royalsociety.org/awards/copley-medal/> [Accessed July 2013].
21. [CG1882], p. 250.
22. Airy, George Biddell (1859). Review of Maxwell's Essay on Saturn's Rings. *Monthly Notices of the Royal Astronomical Society* 19:297–304. Also Document 18 in Brush et al. (1983) *Maxwell on Saturn's Rings.*
23. Brush et al. (1983). *Maxwell on Saturn's Rings*, p. 22.
24. Brush et al. (1983). *Maxwell on Saturn's Rings*, p. 28.
25. Circular Aperture Diffraction. <http://hyperphysics.phy-astr.gsu.edu/hbase/phyopt/cirapp2.html> [Accessed July 2013].
26. Olin J. Eggen and the Eggen Archives. University of Wisconsin-Madison. <http://astronomy.library.wisc.edu/resources/eggen/eggen.html> [Accessed July 2013].
27. The Neptune Controversy: British Astronomy's Post-discovery Discovery (1992). *DIO* and *The Journal for Hysterical Astronomy* 2(3):115–42. <http://www.dioi.org/vols/w23.pdf > [Accessed July 2013].

28. Kollerstrom, N. (2001). Neptune's Prediction: The British Case for Co-Prediction, <http://www.dioi.org/kn/neptune/index.htm> [Accessed July 2013].

29. Brush et al. (1983). *Maxwell on Saturn's Rings*, pp. 1–5; [CG1882], pp. 501–4.

30. Brush et al. (1983). *Maxwell on Saturn's Rings*, pp. 6–10.

31. Papers of Stephen Parkinson. St. John's Library. <http://janus.lib.cam.ac.uk/db/node.xsp?id=EAD%2FGBR%2F0275%2FParkinson> [Accessed July 2013].

32. [CG1882], pp. 247–63.

33. [SLP1], No. 66, pp. 305–13. No. 71, pp. 319–24. No. 91, pp. 387–91. No. 94, pp. 396–7. No. 96, pp. 399–401. No. 100, pp. 406–9. No. 109, pp. 482–6. No. 110, pp. 487–91.

34. [SLP1], No. 101, pp. 410–1; also in Brush et al. (1983) *Maxwell on Saturn's Rings*, Document 1, pp. 39–40.

35. [SLP1], No. 104, pp. 416–8.

36. The Prize Essay, together with the comments of Thomson and Challis, is included in [SLP1], pp. 438–79.

37. [SLP1], pp. 438–9.

38. Brush et al. (1983). *Maxwell on Saturn's Rings*, p. 12.

39. [SLP1], No. 126, pp. 527–31. No. 128, pp. 533–7; also in Brush et al. (1983) *Maxwell on Saturn's Rings*, Documents 2 and 3, pp. 41–4 and 44–7.

40. [SLP1], No. 134, pp. 553–6. No. 137, pp. 560–3. The first letter is in Brush et al. (1983) *Maxwell on Saturn's Rings*, Document 7, pp. 54–7.

41. [SLP1], No. 138, pp. 564–7.

42. [SLP1], No. 143, pp. 578–81. Also in Brush et al. (1983) *Maxwell on Saturn's Rings*, Document 11, pp. 60–3.

43. [SLP1], No. 152, pp. 597–8. Also in Brush et al. (1983) *Maxwell on Saturn's Rings*, Document 16, pp. 67.

44. [SLP1], No. 153, p. 599.

45. Mahon, B. (2003). *The Man who Changed Everything*, p. 74.

46. [CG1882], pp. 274–5.

47. [SP1], pp. 311–2; Brush et al. (1983) *Maxwell on Saturn's Rings*, pp. 93–4.

48. Brush et al. (1983). *Maxwell on Saturn's Rings*, p. 12.

49. [SP1], p. 357; Brush et al. (1983) *Maxwell on Saturn's Rings*, p. 139.

50. Brush et al. (1983). *Maxwell on Saturn's Rings*, p. 14.

51. Maxwell, J.C. (1862). On Theories of the Constitution of Saturn's Rings. *Proceedings of the Royal Society of Edinburgh* 4:99–100. (Originally read on 19 April 1858.); [SP1], Volume 1, pp. 286–7.

52. [SP1], p. 359; Brush et al. (1983) *Maxwell on Saturn's Rings*, p. 141.

53. Brush et al. (1983). *Maxwell on Saturn's Rings*, p. 8.

54. [SLP1], No. 134.

55. [SP1], p. 333; Brush et al. (1983) *Maxwell on Saturn's Rings*, p. 115.

56. Brush et al. (1983). *Maxwell on Saturn's Rings*, p. 14.

57. Brush et al. (1983). *Maxwell on Saturn's Rings*, p. 15.

58. [SP1], p. 361; Brush et al. (1983) *Maxwell on Saturn's Rings*, p. 143.

59. [SP1], p. 362; Brush et al. (1983) *Maxwell on Saturn's Rings*, p. 144.

60. [SP1], pp. 370–1; Brush et al. (1983) *Maxwell on Saturn's Rings*, p. 153.

61. [SP1], pp. 373–4; Brush et al. (1983) *Maxwell on Saturn's Rings*, pp. 155–6.

62. Brush et al. (1983). *Maxwell on Saturn's Rings*, Documents 20–28, pp. 169–94.

63. Brush et al. (1983). *Maxwell on Saturn's Rings*, pp. 22–3.

64. Keeler, J. E. (1895). A Spectroscopic Proof of the Meteoric Constitution of Saturn's Rings, *Astrophysical Journal* 1:416–27.

65. Nicholson, P. D., Hedman, M. M., Clark, R. N. et al. (2008). A Close Look at Saturn's Rings with Cassini VIMS. *Icarus* 193(1):182–212.

CHAPTER 7

1. Tait, P. G. (1880). Maxwell's Scientific Work. *Nature*, 21: 317–21.

2. Gibbs, J. W. (1961). *Rudolf Julius Emanuel Clausius. Scientific Papers*, Vol. 2, pp. 261–7, Dover Publications, New York.

3. Clausius, R. (1857). On the Kind of Motion we Call Heat. *Philosophical Magazine*, 14: 108–27.

4. Clausius, R. (1859). On the Mean Lengths of Path Described by Separate Molecules of Gaseous Bodies. *Philosophical Magazine*, 18: 81–91.

5. Garber, E., Brush, S. and Everitt, F. (eds.) (1986). *Maxwell on Molecules and Gases.* MIT Press, Boston, Mass.

6. Maxwell, J. C. (1983). On the Stability of the Motion of Saturn's Rings. In S. Brush, F. Everitt and E. Garber (eds.) *Maxwell on Saturn's Rings*, pp. 65–158. MIT Press, Boston, Mass.

7. Larmor, J. (ed.) (1909). *Memoir and Scientific Correspondence of the Late Sir George Gabriel Stokes, Bart.* Letter from Maxwell to Stokes May 1859, Vol, 2, pp. 8–11. Cambridge University Press, Cambridge.

8. Maxwell, J. C. (1860). Illustrations of the Dynamical Theory of Gases. *Philosophical Magazine*, 19:19–32; 20: 21–37. Also in Niven, W. D. (ed.) (1890) *The Scientific Papers of James Clerk Maxwell* (hereafter [SP1]), pp. 377–409. Cambridge University Press, Cambridge.

9. Maxwell, J. C. (1866). On the Viscosity or Internal Friction of Air and Other Gases. *Philosophical Transactions of the Royal Society*, 156: 249–68 [Bakerian Lecture, 1866.] Also in [SP2], pp. 1–25.

10. Clausius, R. (1862). On the Conduction of Heat by Gases. *Philosophical Magazine*, 23: 417–35 and 512–34. Despite its title, this paper developed into a general discussion of the transport properties of gases.

11. Maxwell, J. C. (1867). On the Dynamical Theory of Gases. *Philosophical Transactions of the Royal Society*, 157: 49–88. Also in [SP2], pp. 26–78.

12. Maxwell, J. C. (1867). On the Dynamical Theory of Gases. Also in [SP2], p. 30.

13. Clausius, R. (1857). On the Kind of Motion we Call Heat. *Philosophical Magazine*, 14: 108–27, on p. 119.

14. Maxwell, J. C. (1860). Illustrations of the Dynamical Theory of Gases. Also in [SP1], p. 409.

15. Maxwell, J. C. (1860). Report of the Meeting of the British Association for the Advancement of Science, Transactions of Section A, p. 15.

16. Campbell, L. and Garnett, W. (1882). *The Life of James Clerk Maxwell.* Letter from Maxwell to H.R. Droop 28 January 1862. p. 332. Macmillan, London.

17. Maxwell, J. C. (1867). On the Dynamical Theory of Gases. Also in [SP2], p. 47.

CHAPTER 8

1. Laplace, P.-S. (1798–1805). Supplément au dixième Livre du Traité de Mécanique céleste. Sur L'Action capillaire and Supplément à la Théorie de l'Action capillaire. These two supplements of 1806 and 1807 are usually bound into later printings of volume 4 of the Traité de Mécanique Céleste, Paris.
2. Dupré, A. (1869). Théorie Mécanique de la Chaleur. Gauthier-Villars, Paris.
3. Herschel, J. F. W. (1845). President's Address. Reports of the British Association for the Advancement of Science, 15, xxvii–xliv. See p. xli.
4. Peter Guthrie. Tait to Thomas Andrews, 18/12/1861, and Tait to William Thomson, 13/12/1861, cited by Smith, Crosbie and Wise, M. Norton (1989). *Energy and Empire: A Biographical Study of Lord Kelvin*, p. 354. Cambridge University Press, Cambridge.
5. Maxwell, J. C. (1853). On the Equilibrium of Elastic Solids. *Transactions of the Royal Society of Edinburgh*, 20: 87–120.
6. Maxwell, J. C. (1871). *Theory of Heat*. Longmans Green and Co., London. (Also later editions.) Chapter 22.
7. Lamé, G. (1852). Leçons sur la Théorie Mathématique de l'Elasticité des Corps Solides, p. 332. Bachelier, Paris.
8. Thomson, W. and Tait, P. G. (1883). *Treatise on Natural Philosophy*, 2nd edn, Volume 1, p. 214. Cambridge University Press, Cambridge.
9. Rankine, W. J. M. (1851). On the Centrifugal Theory of Elasticity, as Applied to Gases and Vapours. *Philosophical Magazine*, 2: 509–42. Rankine, W. J. M. (1865). On Thermodynamic and Metamorphic Functions, Disgregation and Real Specific Heats, *Philosophical Magazine*, 30: 407–10.
10. Clausius, R. (1862). Ueber die Anwendung des Satzes von der Aequivalenz der Verwandlungen auf die innere Arbeit. *Annalen der Physik* 116: 73–112. See p. 95; Translated as On the Application of the Theorem of the Equivalence of Transformations to the Internal Work of a Mass of Matter. *Philosophical Magazine*, 24: 81–97, 201–13 (1862). See p. 201.
11. Maxwell to Tait, 13/10/1876, printed in [SLP3], No. 623. Cambridge University Press, Cambridge.
12. Maxwell, J. C. (1878). Tait's 'Thermodynamics'. *Nature*, 17: 257–9, 278–80; reprinted in Niven, W. D. (ed.) (1890). *The Scientific Papers of James Clerk Maxwell*, Volume 2, Paper XCI, pp. 660–71 (hereafter [SP2]). Cambridge University Press, Cambridge. Although this paper is nominally a review of the second edition of Tait's book it is really a short account of Maxwell's mature thoughts on the foundations of thermodynamics.
13. van der Waals, J. D. (1973). On the Continuity of the Gaseous and Liquid States. Thesis, Leiden. Available in English as Volume 14 of Studies in Statistical Mechanics. North-Holland, Amsterdam, 1988, reprinted by Dover publications, 2004. Maxwell, J. C. (1874). van der Waals on the Continuity of the Gaseous and Liquid States. *Nature*, 10:477–80; reprinted by [SP2], Ref.12, Volume 2, Paper LXIX, pp. 407–15.
14. van der Waals, J. D. (1873). On the Continuity of the Gaseous and Liquid States, §36.
15. Maxwell to Thomson, 13 July 1871, 24 July 1871 and 28 December 1874 in Thomson papers at Belfast, reprinted by [SLP1, SLP2, SLP3], No. 381, 382 and 540. Maxwell, J. C. (1875). On the Dynamical Evidence of the Molecular Constitution of Bodies. *Journal of the Chemical Society*, 13:493–508; *Nature*, 11: 357–9, 374–7; Reprinted in [SP2], Paper LXXI, pp. 418–38.
16. Watson, H. W. (1877). Treatise on the Kinetic Theory of Gases. Oxford University Press, Oxford. Reviewed by Maxwell, James Clerk Maxwell (1877). Kinetic Theory of Gases. *Nature*, 16: 242–6. Reprinted in [SLP3], No. 654 and 655. Maxwell does not name Boltzmann as the generator of these difficult proofs but the comment is clearly aimed at him.

17. Boltzmann, Ludwig (1871, 1872). Ueber das Wärmegleichgewicht zwischen mehratomigen Gasmolecülen. *Wiener Berichte* 2, 63, (1871) 397–418; Einige allgemeine Sätze über Wärmegleichgewicht. *Wiener Berichte*. 679–711; Analyscher Beweis des 2 Hauptsatzes der mechanischen Wärmetheorie aus den Sätzes über der lebendige Kraft. *Wiener Berichte*. 712–32; Weitere Studien über das Wärmegleichgewicht unter Gasmolecülen. *Wiener Berichte*. 66 (1872) 275–370. The last of these papers was translated into English by Brush, Stephen G. (1966). *Kinetic Theory*, Vol. 2, pp. 88–175. Pergamon Press, Oxford.

18. Maxwell, J. C. (1979). On Boltzmann's Theorem On the Average Distribution of Energy in a System of Material Points. *Transactions of the Cambridge Philosophical Society*, 12: 547–70. Reprinted in [SP2], Paper XCIV, pp. 713–41. Boltzmann, L. (1882) (same title), *Philosophical Magazine*, 14: 299–312.

19. Maxwell's review of Andrews's paper is in Volume 7 of the Royal Society's Referees' Reports and has been reprinted in [SLP3], No. 604.

20. Boltzmann, L. (1896). Über die Berechnung der Abweichungen der Gase vom Boyle-Charles'schen Gesetz und der Dissociation derselben. *Wiener Berichte*. 2a, 105, 695–706.

21. Jeans, J. H. (1904; also later editions). *Dynamical Theory of Gases*. Cambridge University Press, Cambridge.

22. Gibbs, J. W. (1902). *Elementary Principles in Statistical Mechanics developed with Especial Reference to the Rational Foundations of Thermodynamics*. Scribner, New York. Einstein independently arrived at Gibbs's concept of the canonical and grand canonical ensembles in 1902–4, see his (1987) *Collected Scientific Papers, Volume 2, The Swiss Years, Scientific Writings 1900–1909*. Princeton University Press, Princeton.

23. Ornstein, L. S. (1908). Toepassing der Statische Mechanica van Gibbs op molekulair-theoretische vraagstukken. Thesis, Leiden. An augmented translation was published in French, Archive Néerlandaises 4, 203–303 (1918).

24. Gibbs, J. W. (1875–6). On the Equilibrium of Heterogeneous Substances. *Transactions of the Connecticut Academy of Arts and Sciences*. 3: 108–248. Maxwell received a copy of this, the first part of the whole paper, early in February 1876. The rest of the paper did not appear until 1878, *Transactions of the Connecticut Academy of Arts and Sciences* 3: 343–524 (1877–8). The whole was reprinted in Gibbs, J. W. (1928). *Collected Works*, Volume 1, pp. 55–184, 184–353. Longmans, Green & Co., New York.

25. Postcard from Maxwell to Tait, x/3/1876, printed in [SLP3], No. 617.

26. Maxwell made some plaster casts of the $U(V, S)$ surface one of which he sent to Gibbs. There is a photograph of it in Donnan, F. G. and Haas, A. E. (eds.) (1936). *A Commentary on the Scientific Writings of J. Willard Gibbs*, Vol. 1, p. 19. Yale University Press, New Haven, CT.

27. Postscript of a letter from Maxwell to Stokes, 3/8/1875, printed in Larmor, J. (ed.) (1907). *Memoir and Scientific Correspondence of the late Sir George Gabriel Stokes*, pp. 33–4. Cambridge University Press, Cambridge; and by [SLP3], No. 570.

28. Maxwell, J. C. On the Equilibrium of Heterogeneous Substances. This work was an address delivered in May 1876 at the exhibition in South Kensington of the 'Special Loan Collection of Scientific Instruments'. It was printed in the official report, and an abstract appeared in *Proceedings of the Cambridge Philosophical Society*, 2: 427–30 (1864–76), but it was largely lost to sight until Joseph Larmor reprinted it in *Philosophical Magazine*, 33: 818–21 (1908); and [SLP3], No. 589.

29. Griffiths, R. and Wheeler, J. A. (1970). Critical points in multicomponent systems. *Physical Review* A, 2: 1047–64.

30. Maxwell, J. C. (1876). Capillary Action. *Encyclopaedia Britannica*, 9th edn, Vol. 5, pp. 56–71; Reprinted by [SP2], Paper LXXXIII, pp. 541–91.

31. Fowler, R. H. (1928; 2nd edn. 1936) *Statistical Mechanics*. Cambridge University Press, Cambridge. Fowler, R. H. and Guggenheim, E. A. (1939). *Statistical Thermodynamics*. Cambridge University Press, Cambridge.

CHAPTER 9

1. Zemansky, M. W. (1943). *Heat and Thermodynamics*. McGraw-Hill, New York, 2nd edn. Leff and Rex (see fn. 2) describe the changes in the account of Maxwell's demon given in the various editions of this influential book up to the 6th edn published in 1981, for which R. H. Dittman was an additional author.

2. Leff, H. S. and Rex, A. F. (1990). *Maxwell's Demon: Entropy, Information and Computing*. Adam Hilger, Bristol. This enormously useful book provides an overview of the whole topic; it reprints 25 of the most important accounts published up to that date, and provides a full annotated bibliography containing 250 items, including some textbook and popular accounts as well as research papers. Only 42 of these items were published before 1950.

   With the explosion of interest in Maxwell's demon in the late twentieth century, a second edition was published in 2003 by the same editors and publisher (though now styled IOP), *Maxwell's Demon 2: Entropy, Classical and Quantum Information, Computing*. For space reasons, 10 interesting papers had had to be dropped from the 1st edition, and 26 new papers were included, mostly dealing with quantum and computational aspects of the subject. The bibliography had swelled to over 550 items.

3. Morus, I. R. (2008). 'A Dynamical Form of Mechanical Effect': Thomson's Thermodynamics; and Whitaker, A. (2008). The Legacy. Both in Raymond Flood, Mark McCartney and Andrew Whitaker (eds.) *Kelvin: Life, Labours and Legacy*, pp. 160–74 and 278–306 respectively. Oxford University Press, Oxford.

4. Smith, C. and Wise, M. N. (1989). *Energy and Empire: A Biographical Study of Lord Kelvin*. Cambridge University Press, Cambridge.

5. Whitaker, A. (2013). James Thomson, Engineer and Scientist: The Path to Thermodynamics; and: James and William Thomson: The Creation of Thermodynamics. Both in M. W. Collins, C. Koenig and R. C. Dougall. *Kelvin, Thermodynamics and the Natural World*. WIT Press, Southampton, to be published.

6. Kuhn, T. (1959). Energy Conservation as an Example of Simultaneous Discovery. In Clagett, M. (ed.) *Critical Problems in the History of Science*, pp. 321–56. University of Wisconsin Press, Madison. Reprinted in Kuhn, T. (1977). *The Essential Tension*, pp. 66–104. University of Chicago Press, Chicago.

7. Smith, C. and Wise, M. N. (1989). *Energy and Empire: A Biographical Study of Lord Kelvin*, pp. 315–6. Cambridge University Press, Cambridge.

8. Smith, C. and Wise, M. N. (1989). *Energy and Empire: A Biographical Study of Lord Kelvin*, pp. 320–7. Cambridge University Press, Cambridge; Cardwell, D. S. L. (1971). *From Watt to Clausius: The Rise of Thermodynamics in the Early Industrial Age*, pp. 244–9. Heinemann, London.

9. Brush, S. G. (1976). *The Kind of Motion we Call Heat: A History of Kinetic Theory on the 19th Century*. 2 vols: Vol. 1, pp. 1–299; Vol. 2, pp. 303–769, p. 569. North-Holland, Amsterdam.

10. Brush, S. G. (1976). *The Kind of Motion we Call Heat*, p. 578.

11. Smith, C. (1976). Natural Philosophy and Thermodynamics: William Thomson and the 'Dynamical Theory of Heat'. *British Journal for the History of Science*, 9: 293–319.

12. Brush, S. G. (1976). *The Kind of Motion we Call Heat*, p. 573.

13. Brush, S. G. (1976). *The Kind of Motion we Call Heat*, p. 578.

14. Brush, S. G. (1976). *The Kind of Motion we Call Heat*, pp. 3–210.

15. Brush, S. G. (1976). *The Kind of Motion we Call Heat*, p. 157.

16. Brush, S. G. (1976). *The Kind of Motion we Call Heat*, pp. 8–9.

17. Brush, S. G. (1976). *The Kind of Motion we Call Heat*, p. 144.

18. Brush, S. G. (1976). *The Kind of Motion we Call Heat*, p. 585.

19. Garber, E. Kelvin on Atoms and Molecules. In Flood, R. et al. (2008) *Kelvin: Life, Labours and Legacy*, pp. 192–211.

20. Brush, S. G. (1976). *The Kind of Motion we Call Heat*, p. 206.

21. Brush, S. G. (1976). *The Kind of Motion we Call Heat*, p. 164.

22. Brush, S. G. (1976). *The Kind of Motion we Call Heat*, p. 186.

23. Brush, S. G. (1976). *The Kind of Motion we Call Heat*, p. 174.

24. Brush, S. G. (1976). *The Kind of Motion we Call Heat*, p. 168.

25. Garber, E. Maxwell's kinetic theory 1859–70. This volume.

26. Brush, S. G. (1976). *The Kind of Motion we Call Heat*, p. 170.

27. Brush, S. G. (1976). *The Kind of Motion we Call Heat*, p. 170.

28. Brush, S. G. (1976). *The Kind of Motion we Call Heat*, p. 3.

29. Garber, E. (1966). Maxwell, Clausius and Gibbs: Aspects of the Development of Kinetic Theory and Thermodynamics. PhD Dissertation, Case Institute of Technology; Brush, S. G. (1976). *The Kind of Motion we Call Heat*, p. 182.

30. Brush, S. G. (1976). *The Kind of Motion we Call Heat*, p. 404.

31. Klein, M. J. (1970). Maxwell, his Demon, and the Second Law of Thermodynamics. *American Scientist* 58: 84–97; also in Leff, H. S. and Rex, A. F. (1990) *Maxwell's Demon: Entropy, Information and Computing*, pp. 75–88 and (2003) *Maxwell's Demon 2: Entropy, Classical and Quantum Information, Computing*, pp. 59–72.

32. Klein, M. (1972). Mechanical Explanation at the End of the Nineteenth Century. *Centaurus*, 17: 58–82.

33. Knott, C. G. (1911). *Life and Scientific Work of Peter Guthrie Tait*, pp. 115–6. Cambridge University Press, Cambridge.

34. Harman, P. M. (1998). *The Natural Philosophy of James Clerk Maxwell*, p. 139. Cambridge University Press, Cambridge.

35. Harman, P. M. (1982). *Energy, Force, and Matter: The Conceptual Development of Nineteenth-Century Physics*, p. 8. Cambridge University Press, Cambridge.

36. Daub, E. E. (1970). Maxwell's Demon. *Studies in the History and Philosophy of Science*, 1:213–27; also in Leff, H. S. and Rex, A. F. *Maxwell's Demon: Entropy, Information and Computing*, pp. 37–51 and (2003) *Maxwell's Demon 2: Entropy, Classical and Quantum Information, Computing*, pp. 44–58.

37. Strutt, R. J., Fourth Baron Rayleigh (1924). *John William Strutt: Third Baron Rayleigh*, pp. 47–8. Edward Arnold, London.

38. Thomson, W. (1874). The Kinetic Theory of the Dissipation Of Energy. *Nature* 9: 441–4; also in Leff, H. S. and Rex, A. F. (1990) *Maxwell's Demon: Entropy, Information and Computing*, pp. 34–6 and (2003) *Maxwell's Demon 2: Entropy, Classical and Quantum Information, Computing*, pp. 41–3.

39. Brush, S. G. (1976). *The Kind of Motion we Call Heat*, pp. 240, 607–8.

40. Brush, S. G. (1976). *The Kind of Motion we Call Heat*, p. 297.

41. Knott, C. G. (1911). *Life and Scientific Work of Peter Guthrie Tait*, pp. 213–4. Cambridge University Press, Cambridge.

42. Tait, P. G. (1868). *Sketch of Thermodynamics*. Douglas, Edinburgh.

43. Brush, S. G. (1976). *The Kind of Motion we Call Heat*, p. 602.

44. Smith, C. and Wise, M. N. (1989). *Energy and Empire: A Biographical Study of Lord Kelvin*, p. 622.

45. Maxwell, J. C. (1871). *Theory of Heat*. Longmans, London.

46. Knott, C. G. (1911). *Life and Scientific Work of Peter Guthrie Tait*, pp. 214–5. Cambridge University Press, Cambridge.

47. Tolstoy, I. (1981). *James Clerk Maxwell*, p. 137. Canongate, Edinburgh.

48. Maxwell, J. C. (1878). Tait's thermodynamics. *Nature* 17: 257–9, 278–80.

49. Brush, S. G. (1976). *The Kind of Motion we Call Heat*, p. 632.

50. Jeans, J. H. (1904). *Dynamical Theory of Gases*. Cambridge University Press, Cambridge.

51. Mahon, B. (2003). *The Man who Changed Everything: The Life of James Clerk Maxwell*. Wiley, Chichester.

52. Whitaker, A. (2006). *Einstein, Bohr and the Quantum Dilemma: From Quantum Theory to Quantum Information*. Cambridge University Press, Cambridge.

53. Stanley, M. (2008). The Pointsman: Maxwell's Demon, Victorian Free Will, and the Boundaries of Science. *Journal of the History of Ideas* 69: 467–91.

54. Brush, S. G. (1976). *The Kind of Motion we Call Heat*, p. 82.

55. Clarke, B. (1966). Allegories of Victorian Thermodynamics. *Configurations* 4: 67–90.

56. Brush, S. G. (1976). *The Kind of Motion we Call Heat*, p. 593.

57. Whitaker, A. (2012). *The New Quantum Age: From Bell's Theorem to Quantum Computation and Teleportation*. Oxford University Press, Oxford.

58. Smith, C. and Wise, M. N. (1989). *Energy and Empire: A Biographical Study of Lord Kelvin*, pp. 624–5.

59. Smith, C. and Wise, M. N. (1989). *Energy and Empire: A Biographical Study of Lord Kelvin*, pp. 617–21.

60. Hayles, N. K. (1990). Self-Reflexive Metaphors in Maxwell's Demon and Shannon's Choice. In N. Katherine Hayles (1990). *Chaos Bound: Orderly Discourse in Contemporary Literature and Science*, pp. 39–47. Cornell University Press, Ithaca, New York.

61. Kragh, H. S. (2008). *Entropic Creation: Religious Contexts of Thermodynamics and Cosmology*. Ashgate, Aldershot.

62. Smith, C. and Wise, M. N. (1989). *Energy and Empire: A Biographical Study of Lord Kelvin*, p. 617.

63. Smith, C. and Wise, M. N. (1989). *Energy and Empire: A Biographical Study of Lord Kelvin*, p. 645.

64. Tyndall, J. (1874). Presidential Address. British Association for the Advancement of Science Report, 44, lxvi–xcvii.

65. Flood, R. et al. (2008). *Kelvin: Life, Labours and Legacy*, pp. 200–1.

66. Smith, C. and Wise, M. N. (1989). *Energy and Empire: A Biographical Study of Lord Kelvin*, p. 624.

67. Smith, C. and Wise, M. N. (1989). *Energy and Empire: A Biographical Study of Lord Kelvin*, p. 625.

68. Feynman, R. P., Leighton, R.B. and Sands, M. (1986). *The Feynman Lectures on Physics*, pp. 46.1–46.9. Addison-Wesley, Reading, Massachusetts.

69. Szilard, L. (1972). On the Extension of Phenomenological Thermodynamics to Fluctuation Phenomena; and On the Decrease of Entropy in a Thermodynamic System by the Intervention of Intelligent Beings. Translations of both papers are in B. T. Field and G. Weiss Szilard, (eds.) (1972) *The Collected Works of Leo Szilard: Scientific Papers*, pp. 34–102 and pp. 103–29 respectively. MIT Press, Cambridge, Mass. Another translation of the second paper by Anatol Rapoport and Mechthilde is in Wheeler, J. A. and Zurek, W. H. (1983). *Quantum Theory and Measurement*, pp. 539–48. Princeton University Press, Princeton; in Leff, H. S. and Rex, A. F. (1990) *Maxwell's Demon*, pp. 124–33; and in Leff, H. S. and Rex, A. F. (2003) *Maxwell's Demon 2*, pp. 110–9.

70. Hargittai, I. (2006). *The Martians of Science: Five Physicists who Changed the Twentieth Century*. Oxford University Press, Oxford.

71. Landauer, R. (1991). Information is Physical. *Physics Today*, 44(5): 23–9; Landauer, R. (1999). Information is Inevitably Physical. In Hey, A. J. G. (ed.) (1999) *Feynman and Computation*, pp. 77–92. Perseus, Reading, Massachusetts.

72. Leff, H. S. and Rex, A. F. (1990). *Maxwell's Demon*, pp. 16; and Leff, H. S. and Rex, A. F. (2003) *Maxwell's Demon 2*, p. 15.

73. von Neumann, J. (1955). *Mathematical Foundations of Quantum Mechanics*. Princeton University Press, Princeton. (This is a translation of the original book published in German in 1932.)

74. von Neumann, J. (1966). *Theory of Self-Reproducing Automata* (Edited and completed by A. W. Burks). University of Illinois Press, Urbana.

75. Leff, H. S. and Rex, A. F. (1990). *Maxwell's Demon*, pp. 22–5; and Leff, H. S. and Rex, A. F. (2003) *Maxwell's Demon 2*, pp. 21–3.

76. Shannon, C. E. (1948). A Mathematical Theory of Communication, *Bell Systems Technical Journal*, 27: 379–423 and 623–56.

77. Brillouin, L. (1956). *Science and Information Theory*. Academic Press, New York.

78. Brillouin, L. (1951). Maxwell's Demon Cannot Operate: Information and Entropy I. *Journal of Applied Physics* 22: 334–7; also in Leff, H. S. and Rex, A. F. *Maxwell's Demon*, pp. 134–7 and Leff, H. S. and Rex, A. F. (2003) *Maxwell's Demon 2*, pp. 120–3.

79. Landauer, R. (1961). Irreversibility and Heat Generation in the Computing Process. *IBM Journal of Research and Development*, 5: 183–91; also in Leff, H. S. and Rex, A. F. *Maxwell's Demon*, pp. 188–96 and in Leff, H. S. and Rex, A. F. (2003) *Maxwell's Demon 2*, pp. 148–56.

80. Bennett, C. E. (1973). Logical Reversibility of Computation. *IBM Journal of Research and Development*, 17: 525–32; also in Leff, H. S. and Rex, A. F. *Maxwell's Demon*, pp. 197–204.

81. Bennett, C. E. (2000). Notes on the Theory of Reversible Computation. *IBM Journal of Research and Development*, 32: 16–23; also in Leff, H. S. and Rex, A. F. *Maxwell's Demon*, pp. 281–8 and in Leff, H. S. and Rex, A. F. (2003) *Maxwell's Demon 2*, pp. 327–34.

82. Minsky, M. (1999). Richard Feynman and Cellular Vacuum. In A. J. G. Hey (ed.) *Feynman and Computation*, pp. 117–30. Perseus, Reading, Massachusetts.

83. Fredkin, E. and Toffoli, T. (1982). Conservative Logic. *International Journal of Theoretical Physics*, 21: 219–53.

84. Stolze, J. and Suter, D. (2008). *Quantum Computing: A Short Course from Theory to Experiment*, 2nd edn. Wiley-VCH, Weinheim.

85. Mermin, N. D. (2007). *Quantum Computer Science*. Cambridge University Press, Cambridge.

86. Bennett, C. H. (1985). Demons, Engines and the Second Law. *Scientific American*, 253 (November): 38–46.

87. Penrose, O. (1970). *Foundations of Statistical Mechanics*. Pergamon Press, Oxford.

88. Leff, H. S. and Rex, A. F. (2003). *Maxwell's Demon 2*, pp. 33.

CHAPTER 10

1. Maxwell, J. C. (1864). On Faraday's Lines of Force (read 10 December 1855 and 11 February 1856). *Transactions of the Cambridge Philosophical Society*, 10: 27–65; reprinted in Niven, W. D. (ed.), *The Scientific Papers of James Clerk Maxwell*, 2 vols. (1890; Dover Publications, New York, 1965 [2 Volumes in 1]), Volume 1, pp. 155–229 [SP1]. Maxwell, J. C. (1873). *A Treatise on Electricity and Magnetism*.

2 vols, Clarendon, Oxford; 2nd edn., Niven, W. D. (ed.). Clarendon, Oxford, 1881. Campbell, L. and Garnett, W. (1882) [CG1882], *The Life of James Clerk Maxwell*, 2nd edn., with a new preface and appendix with letters by Robert H. Kargon (1969), *The Sources of Science*, No. 85, Johnson Reprint, New York, pp. 197–222.

2. Maxwell, J. C. (1864). On Faraday's Lines of Force, in [SP1] pp. 155–9. For further bibliography throughout, including both primary and secondary sources, see Siegel, D. M. (1991), *Innovation in Maxwell's Electromagnetic Theory*. Cambridge University Press, Cambridge; on Faraday, pp. 5–12.
3. Siegel, D. M. (1991). *Innovation in Maxwell's Electromagnetic Theory*, pp. 5–9.
4. Siegel, D. M. (1991). *Innovation in Maxwell's Electromagnetic Theory*, pp. 9–10.
5. Siegel, D. M. (1991). *Innovation in Maxwell's Electromagnetic Theory*, pp. 12–5.
6. Siegel, D. M. (1991). *Innovation in Maxwell's Electromagnetic Theory*, pp. 6–11.
7. Siegel, D. M. (1991). *Innovation in Maxwell's Electromagnetic Theory*, pp. 8–12.
8. Maxwell's notation for field quantities and other quantities varied over the years, in particular beginning with component form for vector quantities and turning to quaternion form later. For mnemonic purposes, uniform vector notation will be employed here; associated historiographical issues are discussed in Siegel, D. M. (1991), *Innovation in Maxwell's Electromagnetic Theory*, pp. 3–4, 186 note 8.
9. Einstein, A. (1950). The Fundamentals of Theoretical Physics. In (1973), *Out of My Later Years*, revised edn., Citadel Press, Secaucus, New Jersey, pp. 98–110, also 57. Einstein, A., Autobiographical notes, in Schilpp, P. A. (ed.) (1949, 1951), *Albert Einstein: Philosopher-Scientist*, 2 vols. Volume 1, pp. 25, 33–7, 63, Library of the Living Philosophers, Evanston, Illinois; reprinted by Harper and Row, New York in 1959.
10. Siegel, D. M. (1975). Completeness as a Goal in Maxwell's Electromagnetic Theory. *Isis*, 66: 361–8.
11. Hertz, H. (1888). On Electromagnetic Waves in Air and their Reflection. In Hertz, H. *Electric Waves*, translated by Jones, D. E., pp. 124–36. MacMillan, London, 1893; Dover, New York, 1962.
12. Siegel, D. M. (1991). *Innovation in Maxwell's Electromagnetic Theory*.
13. Maxwell, J. C. (1864). On Faraday's Lines of Force. Maxwell, J. C. (1861–2), On Physical Lines of Force, Part I: The Theory of Molecular Vortices Applied to Magnetic Phenomena. *Philosophical Magazine*, 21(139, March 1861): 161–75; On Physical Lines Of Force, Part II: The Theory of Molecular Vortices Applied to Electric Currents. *Philosophical Magazine*, 21(141, April 1861): 281–91, and 21(141, May 1861): 338–45; On Physical Lines of Force, Part III: The Theory of Molecular Vortices Applied to Statical Electricity. *Philosophical Magazine*, 23(151, January 1862): 12–24; On Physical Lines of Force, Part IV: The Theory of Molecular Vortices Applied to the Action of Magnetism on Polarized Light. *Philosophical Magazine*, 23(152, February 1862): 85–95. Reprinted under the common title: On Physical Lines of Force in Niven, W. D. (ed.) (1890), *The Scientific Papers of James Clerk Maxwell*, pp. 451–513 [SP1]. Maxwell, J. C. (1865). A Dynamical Theory of the Electromagnetic Field, in [SP1] pp. 526–97.
14. Panofsky, W. K. H. and Phillips, M. (1962). *Classical Electricity and Magnetism*, 2nd edn. Addison-Wesley, Reading, Mass. A Dynamical Theory of the Electromagnetic Field, in [SP1] especially pp. 563–4. Hertz, H. (1962). *Electric Waves*, p. 21.
15. Maxwell, J. C. (1864). On Faraday's Lines of Force, in [SP1] especially pp. 160–4. On Physical Lines of Force, in [SP1] especially p. 489 and Plate VIII preceding.
16. On Physical Lines of Force, in [SP1] p. 468. Duhem, P. (1954). *The Aim and Structure of Physical Theory* [ASP], translated from 2nd French edn., Princeton University Press, Princeton, pp. 69–75, on 71, where the main examples are Oliver Lodge and William Thomson, but Maxwell is also mentioned.
17. The *locus classicus* on the inutility of Maxwell's models is Duhem, P. (1954), *The Aim and Structure of Physical Theory*, pp. 55–104, especially p. 98. See also Tolstoy, I. (1981), *James Clerk Maxwell:*

A Biography. Chicago, University of Chicago Press, quoted material on pp. 78, 124; Chalmers, A. F. (1973), Maxwell's methodology and his application of it to electromagnetism. *Studies in History and Philosophy of Science*, 4: 107–64. Siegel, D. M. (1991). *Innovation in Maxwell's Electromagnetic Theory*, pp. 2, 185–6 (n 3, 5).

18. [CG1882], pp. 530–50 (account of the molecular-vortex model presumably by Garnett). Larmor, J. (ed.) (1902). *The Scientific Writings of the Late George Francis Fitzgerald*. Hodges, Figgis, & Co., Dublin, pp. 142–62 (On a model illustrating some properties of the ether [1885] and On the structure of mechanical models Illustrating some properties of the ether [1885]); see also Hunt, Bruce J. (1991), *The Maxwellians*, Cornell University Press, Ithaca, N.Y. pp. 73–107.

19. Boltzmann, L. (ed. and tr.) (1898). *Ueber physikalische Kraftlinien*, Ostwald's Klassiker der exacten Wissenschaften, No. 102 Wilhelm Engelmann, Leipzig; Boltzmann, L., (ed. and tr.) (1895), Ueber Faraday's Kraftlinien, *Ostwald's Klassiker der exacten Wissenschaften*, No. 69, Wilhelm Engelmann, Leipzig; Boltzmann, L. (1891, 1893), *Vorlesungen ueber Maxwells Theorie der Electricitat und des Lichtes*, 2 vols, J. A. Barth, Leipzig.

20. Maxwell, J. C. (1864). On Faraday's Lines of Force, in [SP1] pp. 155–9. Siegel, D. M. (1991). *Innovation in Maxwell's Electromagnetic Theory*, pp. 30–3.

21. Siegel, D. M. (1991). *Innovation in Maxwell's Electromagnetic Theory*, pp. 12–3, 15–6, 26–8, 30.

22. Maxwell, J. C. (1864). On Faraday's Lines of Force, p. 159. Boltzmann, L. (1895). *Ueber Faraday's Kraftlinien*, Ostwald's Klassiker der exacten Wissenschaften, No. 69, p. 9. On Physical Lines of Force, in [SP1] p. 505. On Maxwell's realism, in broader context, see also Wilson, David B. (1977), Concepts of Physical Nature: John Herschel to Karl Pearson. In Knoepfelmacher, U. C. and Tennyson, G. B. (eds.), *Nature and the Victorian Imagination*, University of California Press, Berkeley, in [SP1] especially pp. 207–8.

23. Thomson, W. (1856). Dynamical Illustrations of the Magnetic and Helicoidal Rotatory Effects of Transparent Bodies on Polarized Light. *Proceedings of the Royal Society*, 8: 150–8; reprinted in *Philosophical Magazine*, 13(1857): 198–204. Thomson, W. (1884), Atmospheric Electricity (Royal Institution Friday evening lecture, 18 May, 1860). In *Reprint of Papers on Electrostatics and Magnetism*, 2nd edn. Macmillan, London, pp. 208–26, on 224 (Thomson's emphasis). Siegel, D. M. (1991), *Innovation in Maxwell's Electromagnetic Theory*, pp. 34–5.

24. On Physical Lines of Force, in [SP1] p. 505 (italics Maxwell's). Siegel, D. M. (1991). *Innovation in Maxwell's Electromagnetic Theory*, pp. 35–9.

25. On Physical Lines of Force, in [SP1] pp. 451, 467, 489, 505, and *passim*.

26. Maxwell, J. C. (1873). *A Treatise on Electricity and Magnetism*, Vol. 2, pp. 202–4, 416–7 (1st edn.), 218–20, 470 (3rd edn.). Siegel, D. M. (1991). *Innovation in Maxwell's Electromagnetic Theory*, pp. 40, 50–5.

27. On Physical Lines of Force, in [SP1] p. 486.

28. [CG1882], p. 329. Siegel, D. M. (1991). *Innovation in Maxwell's Electromagnetic Theory*, pp. 1–2, and citations.

29. On Physical Lines of Force, in [SP1] p. 462 (Equations [9]), p. 471 (Equations [33], [34]), equations in component form. Siegel, D. M. (1991). *Innovation in Maxwell's Electromagnetic Theory*, p. 68. Maxwell, James Clerk, On the Mathematical Classification of Physical Quantities, (1870), in [SP2] pp. 257–66, curl on p. 265.

30. On Physical Lines of Force, in [SP1] pp. 472–6, Equation [54] on p. 475, in component form.

31. On Physical Lines of Force, in [SP1] p. 497, Equation [115], in component form. The naming of the divergence operator stems from Maxwell's naming its negative the convergence, Maxwell, J. C. (1870). On the Mathematical Classification of Physical Quantities, in [SP2] p. 265.

32. Maxwell, J. C. (1873). *A Treatise on Electricity and Magnetism*, Vol. 2, p. 26 (1st edn.), p. 28 (3rd edn.), Equation [18], in component form. Siegel, D. M. (1991). *Innovation in Maxwell's Electromagnetic Theory*, p. 87.

33. On Physical Lines of Force, in [SP1] pp. 451–66. Siegel, D. M. (1991). *Innovation in Maxwell's Electromagnetic Theory*, pp. 57–65.

34. On Physical Lines of Force, in [SP1] pp. 467–71. Siegel, D. M. (1991). *Innovation in Maxwell's Electromagnetic Theory*, pp. 67–71. [CG1882], p. 329.

35. Siegel, D. M. (1991). *Innovation in Maxwell's Electromagnetic Theory*, pp. 71–3.

36. On Physical Lines of Force, in [SP1] p. 496 (Equation [113], in component form).

37. C.f. Siegel, D. M. (1991). *Innovation in Maxwell's Electromagnetic Theory*, pp. 75–7.

38. Maxwell, J. C. (1864). On Faraday's Lines of Force, in [SP1] p. 195.

39. On Physical Lines of Force, in [SP1] pp. 489–96 (Equations [112] on 496, in component form). Siegel, D. M. (1991). *Innovation in Maxwell's Electromagnetic Theory*, pp. 79–80, c.f. pp. 85–119. In On Physical Lines of Force, p. 491, Maxwell referred to 'the electric current due to displacement'; the abbreviation to 'displacement current' appears to be post-Maxwell.

40. On Physical Lines of Force, in [SP1] pp. 496–7.

41. Bromberg, Joan (1967). Maxwell's displacement current and his theory of light. *Archive for History of Exact Sciences*, 4: 218–34, on p. 219.

42. The locus classicus for explication of this aspect of Maxell's reasoning is [CG1882], pp. 543–4. Siegel, D. M. (1991). *Innovation in Maxwell's Electromagnetic Theory*, pp. 128–41.

43. Kohlrausch, R. and Weber, W. (1857). Elecktrodynamische maassbestimmungen insbesondere zürückführing der stromintensitäts-messungen auf mechanisches maass. In *Wilhelm Weber's Werke*, Volume 3, Heinrich Weber (ed.), Julius Springer, Berlin, pp. 609–76, on p. 651, where they evaluated a parameter equal to 155,370 x 10^6 millimeters per second, which Maxwell multiplied by 2 and converted to English units.

44. Fizeau, H. (1849). Sur une Experience Relative a la Vitesse de Propagation de la Lumiere, *Comptes Rendus Hebdomadaires des Seances de L'Academie des Sciences*, 29, 90–2, on 92; c.f. Siegel, D. M. (1991). *Innovation in Maxwell's Electromagnetic Theory*, pp. 211–2 n. 21.

45. On Physical Lines of Force, in [SP1] pp. 499–500.

46. Some scepticism concerning the closeness of the 1% agreement between V_W and V_L may be in order in view of approximations in the molecular-vortex model – see Siegel, D. M. (1991), *Innovation in Maxwell's Electromagnetic Theory*, pp. 136–41. But the historical importance of this result was in motivating Maxwell to pursue a truly electromagnetic theory of light, rather than in validating the electro-mechanical theory of On Physical Lines of Force on the basis of the claimed close agreement.

47. The wave equation appears in full generality in A Dynamical Theory of the Electromagnetic Field, in [SP1] especially pp. 577–80. The account here follows more closely the simplified and more physically suggestive derivation presented in Maxwell, J. C. On a Method Of Making A Direct Comparison Of Electrostatic With Electromagnetic Force; With A Note On The Electromagnetic Theory Of Light, in [SP2] pp. 125–43, especially pp. 139–43.

48. Siegel, D. M. (1991). *Innovation in Maxwell's Electromagnetic Theory*, pp. 8–9, 144–52.

49. A Dynamical Theory of the Electromagnetic Field, [SP1], especially pp. 577–80.

50. Hertz, H. (1962). *Electric Waves*.

51. See fn. 9.

52. Buchwald, J. Z. (1985). *From Maxwell to Microphysics: Aspects of Electromagnetic Theory in the Last Quarter of the Nineteenth Century*. University of Chicago Press, Chicago. Hunt, B. J. (1991). *The Maxwellians*.

53. Campbell, N. R. (1920). *Foundations of Science: The Philosophy of Theory and Experiment* (formerly titled *Physics: The Elements*). Cambridge University Press, Cambridge; Dover Publications, New York, 1957, pp. 119–58, especially pp. 122–9, 132–7.

CHAPTER 11

1. Hunt, B. (1991). *The Maxwellians*, pp. 122–6. Cornell University Press, Ithaca.
2. Hunt, B. (1991). *The Maxwellians*, pp. 2–4. Buchwald, J. (1985). *From Maxwell to Microphysics: Aspects of Electromagnetic Theory in the Last Quarter of the Nineteenth Century*, p. 74. University of Chicago Press, Chicago.
3. Warwick, A. (2003). *Masters of Theory: Cambridge and the Rise of Mathematical Physics*, University of Chicago Press, Chicago.
4. Buchwald, J. (1985). *From Maxwell to Microphysics*, p. 74.
5. Hunt, B. (1991). *The Maxwellians*, pp. 5–72.
6. Warwick, A. (2003). *Masters of Theory*, pp. 297–9.
7. Hunt, B. (1991). *The Maxwellians*, pp. 1–4.
8. Warwick, A. (2003). *Masters of Theory*, pp. 299–325.
9. Maxwell, J. C. On Physical Lines of Force. *Philosophical Magazine Series* 4, 21: 139 (1861), pp. 161–75; 21: 140 (1861), pp. 281–91; 23: 151 (1861), pp. 12–24; 23: 152 (1862), pp. 85–95.
10. Siegel, D. (1986). The origin of the displacement current. *Historical Studies in the Physical and Biological Sciences*, 17(1): 99–146.
11. Maxwell, J. C. (1865). A Dynamic Theory of the Electromagnetic Field. *Philosophical Transactions of the Royal Society of London*, 155: 459–512.
12. Buchwald, J. (1985). *From Maxwell to Microphysics*, pp. 30–4.
13. Hunt, B. (1991). *The Maxwellians*, p. 41.
14. Warwick, A. (2003). *Masters of Theory*, pp. 344–51.
15. Poynting, J. H. (1884). On the Transfer of Energy in the Electromagnetic Field. *Philosophical Transactions of the Royal Society of London*, 175: 343–61.
16. Poynting, J. H. (1884). On the Transfer of Energy, pp. 345–9.
17. Poynting, J. H. (1885). On the Connexion between Electric Current and the Electric and Magnetic Inductions in the Surrounding Field. *Philosophical Transactions of the Royal Society of London*, 176: 277–306; Buchwald, J. (1985), *From Maxwell to Microphysics*, p. 45.
18. Buchwald, J. (1985). *From Maxwell to Microphysics*, p. 45.
19. Nahin, P. and Heaviside, O. (1988). *The Life, Work, and Times of an Electrical Genius of the Victorian Age*, pp. 115–9. Johns Hopkins University Press, Baltimore.
20. Maxwell, J. C. (1904). *A Treatise on Electricity and Magnetism*, Vol. 2 Clarendon Press, Oxford, Art. 781–805; Hunt, B. (1991), *The Maxwellians*, pp. 28–9.
21. Hunt, B. (1991). *The Maxwellians*, p. 29.
22. Hunt, B. (1991). *The Maxwellians*, pp. 30–2.
23. Hunt, B. (1991). *The Maxwellians*, pp. 30–2; FitzGerald, G. On the Possibility of Originating Wave Disturbances in the Ether by Means of Electric Forces. In FitzGerald, G. (1902), *The Scientific Writings of the Late George Francis FitzGerald*, pp. 90–1. Dublin University Press, Dublin.

24. Hunt, B. (1991). *The Maxwellians*, pp. 30–2; FitzGerald, G. On the Possibility of Originating Wave Disturbances in the Ether by Means of Electric Forces, pp. 91–2.

25. Hunt, B. (1991). *The Maxwellians*, p. 34.

26. FitzGerald, G. (1902). On the Possibility of Originating Wave Disturbances in the Ether by Means of Electric Forces, Part II, pp. 93–8.

27. FitzGerald, G. (1902). On the Possibility of Originating Wave Disturbances in the Ether by Means of Electric Forces – Corrections and Additions, pp. 99–101.

28. FitzGerald, G. (1902). On the Quantity of Energy Transferred to the Ether by a Variable Current, pp. 122–6.

29. Hunt, B. (1991). *The Maxwellians*, pp. 45–6.

30. Warwick, A. (2003). *Masters of Theory*, pp. 340–3; Thomson, J. J. (1884), On Electrical Oscillations and the Effects Produced by the Motion of an Electrified Sphere. *Proceedings of the London Mathematical Society*, 15: 197–218.

31. Hertz, H. (1888). On the Finite Velocity of Propagation of Electromagnetic Action. In *Electric Waves*, pp. 107–23. MacMillan, London, For the details of Hertz's experiment, see Buchwald, J. (1994), *The Creation of Scientific Effects: Heinrich Hertz and Electric Waves* University of Chicago Press, Chicago.

32. Buchwald, J. (1985). *From Maxwell to Microphysics*, p. 335.

33. Buchwald, J. (1985). *From Maxwell to Microphysics*, pp. 78–9.

34. Buchwald, J. (1985). *From Maxwell to Microphysics*, pp. 79–83.

35. Hall, E. (1879). On a New Action of the Magnet on Electric Currents. *American Journal of Mathematics*, 2: 225–30.

36. Buchwald, J. (1985). *From Maxwell to Microphysics*, pp. 102–3.

37. Buchwald, J. (1985). *From Maxwell to Microphysics*, pp. 103–5; Rowland, H. (1880), On the General Equation of Electro-Magnetic Action, with Application to a New Theory of Magnetic Attractions, and to the Theory of the Magnetic Rotation of the Plane of Polarization of Light. *American Journal of Mathematics*, 3: 89–113.

38. Buchwald, J. (1985). *From Maxwell to Microphysics*, pp. 111–7; Glazebrook, R. (1881), On the Molecular Vortex Theory of Electromagnetic Action. *Philosophical Magazine*, 11: 397–413.

39. Buchwald, J. (1985). *From Maxwell to Microphysics*, pp. 117–29.

40. Buchwald, J. (1985). *From Maxwell to Microphysics*, pp. 65–6.

41. Hunt, B. (1991). *The Maxwellians*, pp. 15–21; Buchwald, J. (1985), *From Maxwell to Microphysics*, pp. 287–8.

42. Buchwald, J. (1985). *From Maxwell to Microphysics*, pp. 66–8.

43. Hunt, B. (1991). *The Maxwellians*, pp. 209–14; Buchwald, J. (1985), *From Maxwell to Microphysics*, pp. 141–3.

44. Hunt, B. (1991). *The Maxwellians*, pp. 215–28; Buchwald, J. (1985), *From Maxwell to Microphysics*, pp. 144–72.

45. Warwick, A. (2003). *Masters of Theory*, pp. 367–95.

46. Arabatzis, T. (2006). *Representing Electrons: A Biographical Approach to Theoretical Entities*, pp. 74–85. University of Chicago Press, Chicago.

47. Arabatzis, T. (2006). *Representing Electrons*, pp. 95–103; Buchwald, J. and Warwick, A. (eds.) (2001), *Histories of the Electron: The Birth of Microphysics*, pp. 19–168. MIT Press, Cambridge, Mass.

CHAPTER 12

* This question is number 101 in a collection of problems in Maxwell's notebook 2, which may be found in the King's College London Archives; for a set of solutions to these problems, see Fuller (fn. 2).

** I am grateful to David Forfar who drew my attention to Maxwell's ingenious treatment of this problem.

1. Harman, P. M. (1990, 1995, 2003). *The Scientific Letters and Papers of James Clerk Maxwell.* Vol. I (1990), Vol. II (1995), Vol. III (2003). Cambridge University Press, Cambridge (hereafter [SLP1], [SLP2] and [SLP3]).
2. Fuller, A. T. (1986). *International Journal of Control*, 43: 1135–68.
3. Rayleigh, Lord (1917). On the Dynamics of Revolving Fluids. *Proceedings of the Royal Society of London A*, 93: 148–54.
4. Kelvin, Lord (W. Thomson) (1869). On Vortex Motion. *Transactions of the Royal Society of Edinburgh*, 25: 217–60.
5. Taylor, G. I. (1923). Stability of a Viscous Fluid Contained Between Two Rotating Cylinders. *Philosophical Transactions of the Royal Society A*, 223: 289–343.
6. Stuart, J. T. (1958). On the Non-Linear Mechanics of Hydrodynamic Stability. *Journal of Fluid Mechanics*, 4: 1–21.
7. Lin, C. C. (1966). *The Theory of Hydrodynamic Stability.* Cambridge University Press, Cambridge.
8. Drazin, P. G. and Reid, W. H. (1981). *Hydrodynamic Stability.* Cambridge University Press, Cambridge.
9. Koschmieder, E. L. (1993). *Bénard Cells and Taylor Vortices.* Cambridge University Press, Cambridge.
10. Iooss, G. (1994). *The Couette-Taylor Problem.* Springer-Verlag, Berlin.
11. Stokes, G. G. (1842). On the Steady Motion of Incompressible Fluids. *Transactions of the Cambridge Philosophical Society*, 7: 439–54.
12. Arnold, V. I. (1966). Sur un Principe Variationelle pour les Écoulements Stationnaires des Liquides Parfaits et ces Applications aux Problèmes de Stabilité Non-Linéaires. *Journal de Mécanique*, 5: 29–43.
13. Moffatt, H. K. (1986). Magnetostatic Equilibria and Analogous Euler Flows of Arbitrarily Complex Topology. Part II. Stability Considerations. *Journal of Fluid Mechanics*, 166: 359–78.
14. Helmholtz, H. (1858). On Integrals of the Hydrodynamical Equations, which Express Vortex Motion. *Crelle's Journal*, 55 (translated from German by Tait, P.G. (1867)). *Philosophical Magazine Supplement* 33.
15. Maxwell, J. C. (1856). On Faraday's Lines of Force. In [SP1]. Cambridge University Press, Cambridge, 1890.
16. Maxwell, J. C. (1859). On the Stability of Motion of Saturn's Rings. In [SP1], pp. 286–377.
17. Jeans, J. H. (1902). The Stability of a Spherical Nebula. *Philosophical Transactions of the Royal Society A*, 312: 1–53.
18. Stokes, G. G. (1847). On the Theory of Oscillatory Waves. *Transactions of the Cambridge Philosophical Society*, 8: 441–55.
19. Kelvin, Lord (W. Thomson) (1867). On Vortex Atoms. *Philosophical Magazine*, 34: 15–24.
20. Tait, P. G. (1898). *On Knots. I, II and III. Scientific Papers*, Vol. 1, pp. 273–347. Cambridge University Press, Cambridge.
21. Silver, D. S. (2008). The Last Poem of James Clerk Maxwell. *Notices of the American Mathematical Society*, 55: 1266–70.
22. Carr, B. (ed.) (2007). *Universe or Multiverse?* Cambridge University Press, Cambridge.

23. Moffatt, H. K. (1985). Magnetostatic Equilibria and Analogous Euler Flows of Arbitrarily Complex Topology. Part I. Fundamentals. *Journal of Fluid Mechanics*, 159: 359–78.

24. Alfvén, H. (1942). Existence of Electromagnetic-Hydrodynamic Waves. *Nature*, 150: 405–6.

25. Woltjer, L. (1958). A Theorem on Force-free Magnetic Fields. *Proceedings of the National Academy of Science*, 43: 489–91.

26. Moffatt, H. K. (1969). The Degree of Knottedness of Tangled Vortex Lines. *Journal of Fluid Mechanics*, 35: 117–29.

27. Maxwell, J. C. (1870). On the Displacement in a Case of Fluid Motion. In [SP2], pp. 208–14.

28. Maxwell, J. C. (1873). *Treatise on Electricity and Magnetism*. 2 vols. Clarendon Press, Oxford.

29. Bullen, K. E. and Bolt, B. A. (1985). *An Introduction to the Theory of Seismology*, 4th edn. Cambridge University Press, Cambridge.

30. Moffatt, H. K. (1979). *Magnetic Field Generation in Electrically Conducting Fluids*. Cambridge University Press, Cambridge.

CHAPTER 13

1. Campbell, L. and Garnett, W. (1882). *The Life of James Clerk Maxwell, With a Selection from his Correspondence and Occasional Writings and a Sketch of his Contributions to Science* Macmillan, London. The biography has been reprinted by Cambridge University Press (2010); an online version is also accessible at <http://www.sonnetsoftware.com/bio/maxbio.pdf> [Accessed July 2013] © copyright 1997 Sonnet Software, Inc., all rights reserved. [CG1882]

2. Beer, G. (1983). *Darwin's Plots: Evolutionary Narrative in Darwin, George Eliot and Nineteenth-Century Fiction*, p. 7. Routledge, London.

3. See Willis, M. (2006). *Mesmerists, Monsters, and Machines: Science Fiction and the Cultures of Science in the Nineteenth Century*, Kent State University Press, Ohio; King, A. (2007). Searching out science and literature: hybrid narratives, new methodological directions, and Mary Russell Mitford's Our Village. *Literature Compass*, 4(5): 1485–1503; and Lightman, B. (2007). *Victorian Popularizers of Science*, University of Chicago Press, Chicago.

4. The poem's origin is John Polidori's novella, *The Vampyre* (1819).

5. [CG1882], p. 190, pp. 386–7.

6. Gold, B. J. (2010). *Thermopoetics: Energy in Victorian Literature and Science*, p. 7. MIT Press, Cambridge, Mass.

7. Scott-Moncrieff, W. G. (1883). James Clerk Maxwell. *Scottish Review*, 1(2): 316.

8. Glazebrook, R. T. (1896). *James Clerk Maxwell and Modern Physics*, p. 27. Cassell, London.

9. Wordsworth, W., Coleridge, S. T., in D. Porter and M. Gamer (eds.) (2008). *Lyrical Ballads: 1798 and 1800*, p. 423. Broadview Press, Peterborough, Ont. Peacock, T. L. (1820) The Four Ages of Poetry. In *The Works of Thomas Love Peacock* (1875), Vol. 3, p. 336.Richard Bentley and Son, London Spencer, H. and Cavenagh, F. A. (eds.) (1861; reprinted 1932), *On Education*, p. 50. Cambridge University Press, Cambridge.

10. Maxwell, J. C. General Considerations Concerning Scientific Apparatus. In W. D. Niven (ed.) (1965), *The Scientific Papers of James Clerk Maxwell*, p. 505. Dover Publications, New York.

11. Cantor, G., Gooding, D. and James, F. A. J. L. (1991). *Faraday*, p. 99. Macmillan, Basingstoke.

12. Turner, F. (1993). *Contesting Cultural Authority: Essays in Victorian Intellectual Life*, p. 38. Cambridge University Press, Cambridge.

13. [CG1882], pp. 612–7, pp. 594–6, pp. 622–5. The poem titles are abbreviated hereafter to 'A Vision', 'Hymn', and 'Lines'.
14. Tolstoy, I. (1981). *James Clerk Maxwell: A Biography*, p. 158. Canongate, Edinburgh.
15. Line numbers from poems are given in brackets throughout, in the body of the text.
16. Wordsworth, in Porter and Gamer (2008), *Lyrical Ballads*, p. 423.
17. Carlyle, T. (1829). Signs of the Times. In *Critical and Miscellaneous Essays* (2003), p. 189. Kessinger Publishing, Montana.
18. For further discussion of women and consumerism in the period, see R. Beachy, B. Craig and A. Owens (2006). *Women, Business and Finance in Nineteenth-century Europe: Rethinking Separate Spheres*. Berg, Oxford.
19. Willis, M. (2006). *Mesmerists, Monsters, and Machines*, p. 96.
20. Lightman (2007). *Victorian Popularizers of Science*, p. 653.
21. [CG1882], p. 484.
22. 'A Vision' is dated 10 November 1852, while 'Lines' is dated 10 November, 1853.
23. William Cowper, *The Task: The Garden* (1785).
24. Tolstoy, I. (1981). *James Clerk Maxwell: A Biography*, p. 158.
25. Harman, P. M. (1998). *The Natural Philosophy of James Clerk Maxwell*, p. 11. Cambridge University Press, Cambridge; Turner, F. (1993), *Contesting Cultural Authority*, p. 178.
26. Maxwell's title is also reminiscent of Samuel Taylor Coleridge's *Religious Musings* (1794); beyond this, however, there appears to be little correlation between the two works.
27. Further discussion of Faraday's and Maxwell's fluid analogies is offered in Pratt-Smith, S. (2012). The poetic science of nineteenth-century electricity. *Nineteenth-Century Contexts: An Interdisciplinary Journal*, 34(1): 51–62.
28. Tennyson, A. L. (1850). *In Memoriam*, 3, pp. 85–8: 'And all the phantom, Nature, stands— / With all the music in her tone, / A hollow echo of my own,— / A hollow form with empty hands.'
29. An additional presence, preceding comment or question is also implied, in the style of the seventeenth-century metaphysical dialogues, such as sonnet 25 of George Herbert's *The Bishop* (1633), and John Donne's *Canonisation* (1633).
30. [CG1882], p. 625.
31. Whitworth, M. (2001). *Einstein's Wake: Relativity, Metaphor and Modernist Literature*, p. 3. Oxford University Press, Oxford.
32. Lloyd, J. T. (1970). Background to the Joule-Mayer Controversy. *Notes and Records of the Royal Society of London*, 25(2): 211.
33. Gooding, D. 'Magnetic Curves' and the Magnetic Field: Experimentation and Representation in the History of a Theory. In D. Gooding, T. Pinch and S. Schaffer, (eds.) (1989). *The Uses of Experiment: Studies in the Natural Sciences*, pp. 183–298. Cambridge University Press, Cambridge.
34. Campbell and Garnett do not say whether the accompanying graph is by Maxwell.
35. Rauch, A. (2002). Poetry and Science. In R. Cronin, A. Chapman, and A. H. Harrison, (eds.) *A Companion to Victorian Poetry*, p. 475. Blackwell, Oxford.
36. Scott-Moncrieff, W. G. (1883). James Clerk Maxwell. *Scottish Review*, 1(2): 310.
37. Carpenter, W. B. (1851). *Elements of Physiology: Including Physiological Anatomy*, p. 526. Blanchard and Lea, Philadelphia.
38. Symonds, J. A. (1851). *Sleep and Dreams: Two Lectures*, p. 46. John Murray, London.

39. Thomas Nashe, quoted in D. Pick and L. Roper (2004). *Dreams and History: The Interpretation of Dreams from Ancient Greece to Modern Psychoanalysis*, p. 6. Routledge, London.

40. Carlyle, T. (1866). *On Heroes, Hero-Worship and the Heroic in History*, p. 27. J. Wiley, New York.

41. Brown, D. (2000). Victorian Poetry and Science. In J. Bristow (ed.), *The Cambridge Companion to Victorian Poetry*, p. 141. Cambridge University Press, Cambridge.

42. Daston, L. and Galison, P. (2007). *Objectivity*. Zone Books, New York.

43. Brown, D. (2000). Victorian Poetry and Science. In J. Bristow (ed.) (2000), *The Cambridge Companion to Victorian Poetry*, p. 141.

44. [CG1882], p. 601, p. 606. The title (tr. 'The Secret of Happiness') is the name of a drinking song from Gaetano Donizetti's opera Lucrezia Borgia, first performed on 26 December 1833 at La Scala in Rome. Hereafter, the abbreviated titles 'Tune' is used.

45. [CG1882], p. 603, p. 603, p. 604, p. 608.

46. [CG1882], p. 276.

47. [CG1882], p. 299.

48. Fletcher, P. (2002). Landscape and Cityscape. In R. Cronin, et al. (eds.) (2002), *A Companion to Victorian Poetry*, p. 500.

49. Marvell, A. and Aitken, G. A. (eds.) (1892). *The Poems of Andrew Marvell*, pp. 39–40. Lawrence & Bullen, London.

50. [CG1882], p. 603.

51. Young, M. A. (2005). *Negotiating the Good Life: Aristotle and the Civil Society*, p. 17. Ashgate Publishing Ltd., Aldershot.

52. [CG1882], p. 297.

53. [CG1882], p. 298.

54. Somerville, M. (1834). *On the Connexion of the Physical Sciences*, p. 2. John Murray, London.

55. Forfar, D. O. (2002). James Clerk Maxwell: His Qualities of Mind and Personality as Judged by his Contemporaries. *Mathematics Today*, 38(3): 83.

56. Everitt, C. F. W. (1975). *James Clerk Maxwell: Physicist and Philosopher*, p. 55. Charles Schribner, New York.

57. [CG1882], p. 631, p. 632.

58. Anonymous (1868). A College for Women. *John Bull*, 24 October, 498.

59. Aldrich Kidwell, P. (1984). Women astronomers in Britain, 1780–1930. *Isis*, 75(3): 536.

60. Gould, P. (1997). Women and the culture of university physics in late nineteenth-century Cambridge. *British Journal for the History of Science*, 30(2): 145.

61. For further discussion of women's participation in nineteenth-century science, see G. Kirkup and L. Smith Keller (eds.) (1992), *Inventing Women: Science, Technology, and Gender*, Open University Press, Cambridge; Tavris, C. (1992). *The Mismeasure of Woman*, Simon and Schuster, New York; and Shteir, A.B. (2004). Green-Stocking or Blue? Science in Three Women's Magazines, 1800–50. In L. Henson, G. Cantor, G. Dawson, R. Noakes, S. Shuttleworth, and J. R. Topham, (eds.) *Culture and Science in the Nineteenth-Century Media*. pp. 3–29. Ashgate, Aldershot.

62. Tennyson, A. L. (1847). 'The Princess: Splendour Falls on Castle Walls'.

63. Gould, P. (1997). Women and the culture of university physics in late nineteenth-century Cambridge, p. 146.

64. [CG1882], p. 632.

65. Keen, L. (1987). *The Legacy of Sonya Kovalevskaya: Proceedings of a Symposium*, Association for Women in Mathematics (US), Radcliffe College Mary Ingraham Bunting Institute, Contemporary Mathematics, 64, 6.

66. Kovalevskaya visited England in 1869, but it is not established whether she ever met Maxwell (see Peter C. Pozefsky (1999). Love, science, and politics in the fiction of Shestidesiatnitsy N. P. Suslova and S. V. Kovalevskaia. *Russian Review*, 58(3): 376). For further discussion of Maxwell's research during this period, see P. Harman (ed.) (2002), *The Scientific Letters and Papers of James Clerk Maxwell*, p. 27.

67. Not only was Kovalevskaya's thesis based on work by Laplace that Maxwell had earlier proved to be unworkable but much of it had been covered previously by Cauchy; hence her theorem is known as the Cauchy-Kovalevskaya Theorem. (See Cooke (1984), p. 66; P. Harman (ed.) (2002), *The Scientific Letters and Papers of James Clerk Maxwell*, p. 27.)

68. Egorov, Yu. V. and Shubin, M. A., tr. R. Cooke (1998). *Foundations of the Classical Theory of Partial Differential Equations*, p. 29. Springer, New York.

69. Harman, P. (2004). Maxwell, James Clerk (1831–1879), *Oxford Dictionary of National Biography*, Oxford University Press, Oxford. <http://www.oxforddnb.com/view/article/5624> [accessed 20 March 2013].

70. [CG1882], p. 634.

71. [CG1882], p. 634.

72. Dawson, G. (2007). *Darwin, Literature and Victorian Respectability*, p. 106. Cambridge University Press, Cambridge.

73. Dawson, G. (2007). *Darwin, Literature and Victorian Respectability.*

74. Schaffer, S. (1997). Metrology, Metrication, and Victorian Values, in B. Lightman (ed.) (1997), *Victorian Science in Context*, p. 465. University of Chicago Press, Chicago.

75. Schaffer, S. (1997). Metrology, Metrication, and Victorian Values.

76. Schaffer, S. (1997). Metrology, Metrication, and Victorian Values.

77. Huxley, T. H. (1860). The Origin of Species. *Westminster Review*, 17: 59.

78. Lewes, G. H. (1874). *Problems of Life and Mind*. Osgood, Boston, Vol. 1, p. 2. For further contemporary representations of religion and science as a battle, see J. W. Draper (1874). *History of the Conflict between Religion and Science*. D. Appleton, New York; A. D. White (1876). *The Warfare of Science*. Henry King, London; E. B. Pusey (1878). *Un-Science, Not Science, Adverse to Faith*. J. Parker, London.

79. [CG1882], p. 359.

80. Tyndall, J. (1874). Belfast Address. *The Times*, Thursday, 20 August; pg. 4; Issue 28086; col A.

81. [CG1882], p. 639.

82. Huxley, T. H. (1868; reprinted 2004). *A Liberal Education*, p. 51. Montana: Kessinger Publishing.

83. Paley, W. and Paxton, J. (1802; reprinted 2003). *Natural Theology: Or, Evidences of the Existence and Attributes of the Deity Collected from the Appearances of Nature*. Kessinger Publishing, Montana; P. L. Marston (2007). Maxwell and Creation: Acceptance, Criticism, and his Anonymous Publication. *American Journal of Physics*, 75(8): 731–40.

84. Marston (2007) Maxwell and Creation. The Eranus Club (f. 1872) was a discussion group for senior, former members of Cambridge University's Apostles' Club.

85. Bowler, P. J. (2001). *Reconciling Science and Religion*, p. 30. University of Chicago Press, Chicago.

86. Maxwell, J. C. 'Molecular Evolution'. In [CG1882], p. 637.

87. Irwin, J. (1996). *George Eliot's Daniel Deronda Notebooks*, p. 17. Cambridge University Press, Cambridge.

88. Feuer, L. S. (1992). *The Scientific Intellectual: The Psychological and Sociological Origins of Modern Science*, p. xxxi. Transaction Publishers, New York.

89. [CG1882], p. 638. British Association members were known as 'Red Lions'; hence, the reference to younger members as 'cubs'.

90. Huxley (2004), *A Liberal Education*, p. 217, p. 97.

91. Shelley, P. B. Prometheus Unbound (1820), Act IV, ll. 153–8. In J. A. V. Chapple (1986). *Science and Literature in the Nineteenth Century*, p. 46. Macmillan, London.

92. An amphicheiral knot is one that can be deformed into its mirror image. Daniel Silver notes that Maxwell's 'liquid vortex wrought' is a reference to William Thomson's vortex atom theory (Daniel S. Silver (2007). The Last Poem of James Clerk Maxwell. *Notices of the American Mathematical Society*, 55(10): 1266).

93. Stableford, B. M. (2006). *Science Fact and Science Fiction: An Encyclopedia*. Routledge, New York, p. 354; G. Beer (1992). Helmholtz, Tyndall, Gerard Manley Hopkins: Leaps of the Prepared Imagination. In E. Schaffer, (ed.) *Comparative Criticism*, Cambridge University Press, Cambridge, p. 127.

94. Maxwell refers to William Clifford (1845–79); Felix Klein (1849–1925).

95. Spencer, H. (1867). *Principles of Biology*, 2, p. 48.Williams and Norgate, London.

96. [CG1882], p. 359.

97. Gerard Manley Hopkins (1844–89).

98. Bowler, P. J. (2001). *Reconciling Science and Religion*, p. 31.

99. [CG1882], p. 458.

100. Barton, R. (1988). Review of Herbert McLeod, Chemistry and Theology in Mid-Victorian London: The Diary of Herbert McLeod, 1860–1870. *Isis*, 79(2): 350–1.

101. [CG1882], p. 336.

102. [CG1882], p. 359.

103. Smith, J. (1994). *Fact and Feeling: Baconian Science and the Nineteenth-Century Literary Imagination*, p. 35. University of Wisconsin Press, Madison.

104. Flint, K. (2000). *The Victorians and the Visual Imagination*, p. 2. Cambridge University Press, Cambridge.

105. Tyndall, J. (1871). *Fragments of Science for Unscientific People: A Series of Detached Essays, Lectures, and Reviews*, p. 83. D. Appleton and Company, New York.

106. Harman, P. (2004). Maxwell, James Clerk (1831–1879), *Oxford Dictionary of National Biography*.

107. Levine, G. (2004). *Dying to Know: Scientific Epistemology and Narrative in Victorian England*, p. 8. University of Chicago Press, Chicago

108. Hunt, L. (1844; reprinted, 1891). *What is Poetry? Imagination and Fancy*, pp. 3–4. Smith, Eder. London.

109. Bergland, R. (2008). *Maria Mitchell and the Sexing of Science: an Astronomer Among the American Romantics*, p. 37.Beacon Press, Boston, Mass.

CHAPTER 14

1. Garnett, W. (1879). James Clerk Maxwell, F.R.S. *Nature*, 21: 43–6. Maxwell's phrase 'know in whom we have believed' appears to be an allusion to 2 Timothy 1:12.

2. Campbell, L. and Garnett, W. (1882). *Life of James Clerk Maxwell.* Macmillan & Co., London. [CG1882]. See p. 32. The 176 verses of the 119th Psalm give an exposition of the benefits of trusting God's precepts.

3. Hilts, V. L. (1975). A Guide to Francis Galton's English Men of Science. *Transactions of the American Philosophical Society* 65(5): 3–85, p. 59.

4. Everitt, C. W. F. (1983). Maxwell's Scientific Creativity. In R. Aris, H.T. Davis, and R.H. Steuwer, (eds). *Springs of Scientific Creativity*, pp. 71–141. University of Minnesota Press, Minneapolis.

5. [CG1882], p. 32. The quotation was widely known from Lord Thomas Erskine's 1797 courtroom speech against the publication of Thomas Paine's *Age of Reason*, in contrasting Paine with Boyle and Newton. It also appears in at least one volume in Maxwell's library: A. Barnes on the *Book of Revelation* in the discussion of Rev. 4:11. This observation is based on a Free Library of Cambridge List from 1887 of books bequeathed by Maxwell's widow and will be designated with {FLCL}.

6. Concerning Knox, a recent church historian wrote, '. . . it is an indisputable fact that his ideas and ideals have coloured the main stream of Scottish religious and ecclesiastical history since his time'. See: Burleigh, J. H. S. (1960). *A Church History of Scotland*. p. 176. Oxford University Press, London.

7. Dewar, D. (1845). *The Church: or a comprehensive view of the doctrines, constitution, government, and ordinances of the church and of the leading denominations into which it is divided*, pp. 667–70. Glasgow {FLCL}.

8. Knox's concern over the latter is evident in his 1571 warning against suffering 'unworthy men to be thrust into the ministry of the church under whatsoever pretence it shall be.' as quoted in Dewar, D. (1845). *The Church*, p. 667.

9. Burleigh, J. H. S. (1960). *A Church History of Scotland*, Chapter 4.

10. Campbell, F. P. A. and Campbell, L. (1914). *Memorials in Verse and Prose of Lewis Campbell*, p. 384 from a letter written by Lewis in 1901. Clowes, London.

11. Macdonald, J. H. A. (1915). *Life Jottings of an Old Edinburgh Citizen*, pp. 63–6. Foulis, London. Maxwell's cousin Jemima Wedderburn Blackburn also noted years later that she had observed the procession: Blackburn, J. (1988). *Jemima: The Paintings and Memoirs of a Victorian Lady*, pp. 124–5. With an Introduction by R. Fairley, (ed.) Trafalgar Square Publishing, North Pomfret, VT.

12. [CG1882], p. 420. For other contemporary criticism of the Free Church movement, see for example: Simpson, P. C. (1909). *The Life of Principal Rainy*, p. 73 of the combined edition (no date). Hodder and Stoughton, London.Ross, J. (1887). *W. Lindsay Alexander: his life and work*, pp. 116–9, Nisbet, London, 1887.

13. T. J. Crawford became a minister at St. Andrew's in 1844 shortly after publishing his *Reasons of Adherence to the Church of Scotland*. Coincidentally, it was near Crawford's Church that a statue of Maxwell was erected early in the twenty-first century.

14. Crawford's clarity of teaching on free agency and God's desire for the salvation of all people still attracted attention near the end of the twentieth century: Murray, I. H. (1995). *Spurgeon Vs Hyper-Calvinism: The Battle for Gospel Preaching*, pp. 115–7, 143–8. Banner of Truth Trust, Carlisle, PA.

15. Crawford, T. J. (1871). *The Doctrine of Holy Scripture Respecting the Atonement*, Blackwood, Edinburgh. A copy in my possession is marked by several prior users.

16. Crawford, T. J. (1874). *The Mysteries of Christianity: Being the Baird Lecture for 1874.* Blackwood, Edinburgh.

17. [CG1882], pp. 116 and 127. Maxwell's library included several volumes by Reid, including the *Works* edited by Hamilton {FLCL}. It also included several supporting volumes by the common sense Scottish philosopher Dugald Stewart.

18. [CG1882], pp. 107–9. Hamilton's emphasis on the relations between physical objects were likely important for Maxwell's discussion of relativity and causality in his 1877 book *Matter and Motion*.

19. Hamilton, W. (1859). *Lectures on Metaphysics and Logic, Vol. 1*, H. L. Mansel and J. Veitch (eds.), p. 30. Gould, Boston. Hamilton's position was summarized 30 years later as 'I know *that* I am free, but I cannot conceive *how* I am free.': Bowen, F. (1877). *Modern Philosophy*, p. 297. Scribner, New York.

20. Hamilton, W. (1859). *Lectures on Metaphysics and Logic*, p. 12. On p. 10, Hamilton wrote 'Strictly speaking, every one must educate himself', a quality evident early in Maxwell. He lectured that 'Consciousness is to the philosopher, what the Bible is to the theologian.' Hamilton, W. (1859). *Lectures on Metaphysics and Logic*, p. 222. Hamilton noted in 1856 in a testimonial concerning Maxwell that in his class ' . . . he laboured zealously, and received by general suffrage one of the honours.' [SLP1], p. 5, where [SLP1], [SLP2] and [SLP3], denote: Harman, P. M. (1990, 1995, 2002). *The Scientific Letters and Papers of James Clerk Maxwell*, Vol. 1, 2 & 3. Cambridge University Press, Cambridge.

21. Ross, J. (1887). *W. Lindsay Alexander*, p. 95 and McCosh, J. (1875). *Scottish Philosophy*, p. 454–5, Carter, New York. Hamilton's discussions pertaining to religious thought were extended by H. L. Mansel. Maxwell classified Mansel as a philosopher in the exclusive sense of Kant and Hegel, [CG1882], p. 335. Hamilton at times referred to Christ as 'our Saviour' in his published lectures.

22. Following usage developed by Hamilton and Mansel, in the various editions (1875–1880) of Stewart and Tait's *Unseen Universe*, the 'unapproachable creator' is referred to as 'the unconditioned' in Chapter 7. Maxwell's unfavourable response is evident in a critical remark (pertaining to Tait) in his 1879 review of Thomson and Tait's *Natural Philosophy*, [SP2], p. 778 where [SP1] and [SP2] denote: Niven, W.D. (1890). *The Scientific Papers of James Clerk Maxwell*, Vol. 1 & 2. Cambridge University Press, Cambridge. Hamilton and Mansel were widely criticized for related views: Hodge, C. (1873). *Systematic Theology, Vol. I*, Part 1, Chap. 4. Scribner, New York.

23. [CG1882], p. 203.

24. Young Maxwell may have met Wilson when Maxwell's father (John) brought him to meetings of the Royal Society of Edinburgh beginning in 1845 since Wilson and John were both members. Wilson was also a friend of a clock and instrument maker, Alexander Bryson, frequently consulted by John. The likely interval of Maxwell's early interaction with Wilson correlates with his early interest in chemistry, [CG1882], p. 85. Wilson was affiliated with an independent Congregational Chapel in Edinburgh. He is not to be confused with Maxwell's moral philosophy professor, John Wilson.

25. Cairns, J. (1860). The Late Dr. George Wilson of Edinburgh. *Macmillan's Magazine* 1, 199–203. See also Ross, J. (1887). *W. Lindsay Alexander*, p. 215.

26. Wilson, G. (1862). *Counsels of an Invalid: Letters on Religious Subjects*, pp. 1–87. Macmillan, London.

27. The scientific topics mentioned in Wilson, G. (1862). *Counsels of an Invalid*, pp. 1–87: crystals, optics, chemistry and galvanic batteries, were topics of interest to Maxwell at that time. Periods of poor health of the recipient overlap with known intervals of poor health for Maxwell with the exception of some days where from Maxwell's notes, he attended William Hamilton's lectures. Years later Maxwell described his health as 'often laid up' prior to 1850 (Hilts, V. L. (1975). *A Guide To Francis Galton's English Men Of Science*, p. 59). The description (by an editor in 1862, Wilson, G. (1862). *Counsels of an Invalid*, p. 1) of the recipient as 'settled in life, and actively engaged in its pursuits' also fits Maxwell. Books mentioned in the correspondence overlap items in Maxwell's library {FLCL} and Maxwell's reading prior to 1860. Notice also that Wilson mentored from 1844 to mid-1846, David Forbes (1828–1876), the mineralogist. The tone of letters in Wilson, G. (1862) *Counsels of an Invalid* differs from that of a *known letter* from Wilson to Forbes in 1850 (RD162 E. 2001.6 Edinburgh University Library Special Collections). The recipient in Wilson, G. (1862) *Counsels of an Invalid* remains unidentified.

28. Wilson, J. A. (1860). *Memoir of George Wilson*, p. 351. Macmillan, London.

29. [CG1882], p. 125. Parker's essay is reprinted in (1864). *The Collected Works of Theodore Parker*, Vol. 9. Maxwell's library included: Beard J. R. et al., *Voices of the Church, in Reply to Strauss* {FLCL}. Maxwell probably read an essay concerning Strauss and Beard in *Edinburgh Review* (October 1849).

30. Larsen, T. (2006). *Crisis of Doubt, Honest Faith in Nineteenth-Century England.* Oxford University Press, Oxford.

31. [CG1882], p. 165.

32. [CG1882], p. 173. Sir Thomas Browne's *Religio Medici* [The Religion of a Doctor] (1643). It may be significant that *Religio Medici* was also admired by George Wilson.

33. Sir Thomas Browne's *Religio Medici* (1643), Part I, Section 16.

34. [CG1882], p. 158; [SLP1], p. 208. The expression 'Man's chief end . . . ' appears in the *Westminster Shorter Catechism* of 1648 that remained in use in the *Church of Scotland* for several centuries. See for example: *The confessions of faith, catechisms . . .* (Glasgow, 1764) {FLCL}. The Westminster Divines mentioned by Maxwell included representation from Scotland through a treaty with England, the 'Solemn League and Covenant.' Dewar, D. (1845). *The Church*, pp. 757–77, Burleigh, J. H. S. (1960). *A Church History of Scotland*, pp. 224–32.

35. [CG1882], p. 345.

36. [CG1882], p. 596. Maxwell, however, had reservations on these stanzas.

37. [CG1882], p. 345. Tayler, who himself had attended Cambridge in 1815–19, was especially sensitive to the stress on Cambridge students; a central character in one of the earliest of his numerous evangelical novels succumbed to illness while a student: C. B. Tayler, C. B. (1833). *Montague; or, Is This Religion?*, pp. 52–143. Smith & Elder, London. For more information on C. Tayler see: Marston, P. L. (2007) Maxwell and Creation. *American Journal of Physics*, 75(8): 731–40.

38. [CG1882], p. 170.

39. [CG1882], pp. 188–9.

40. Theerman, P. (1986). James Clerk Maxwell and Religion. *American Journal of Physics*, 54: 312.

41. H. M. Butler was friend of Monro's from Harrow School, who delivered a memorial sermon in Cambridge following Maxwell's death in 1879. See p. 284.

42. Maxwell's dear friend Pomeroy died from overwork during the mutiny in India of 1857. Maxwell had another loss from the mutiny: two relatives of his Aunt Wedderburn died.

43. The nature and intensity of the discussions during meetings of the Apostles is suggested by an incident during Maxwell's term: a member Henry Roby declined to participate for reasons judged by some members to be inadequate and was dismissed from the society. See: Levy, P. (1981). *Moore: G.E. Moore and the Cambridge Apostles.* Oxford University Press; Deacon, R. (1986). *The Cambridge Apostles.* Farrar. Note also: F. D. Maurice, a famous early 'Apostle', on one occasion said he thought some of the Apostles ought to 'thank God for having passed through a debating Society with any part of their souls undestroyed'. See: Brookfield, F. M. (1906). *The Cambridge 'Apostles'*, p. 221. Scribner, New York. Of Maxwell's surviving Apostle essays, those on *Nature of Evidence of Design* and *Analogies* are especially revealing of his Christian perspective; [CG1882], pp. 223–46.

44. London Metropolitan Archives Acc. 1063; [SLP1], p. 280.

45. Farrar, F. W. (1897). *Men I Have Known*, pp. 136–7. Crowell.

46. [CG1882], p. 194. Maxwell's library included Thomas Erskine's *The Unconditional Freeness of the Gospel* (1828) {FLCL}.

47. [CG1882], p. 170: 'At the same time, it is not to be supposed that Maxwell was ever completely identified with any particular school of religious opinion.' Concerning Campbell's perspective, by the mid

1850s Campbell had developed an affectionate life-long relationship with Jowett of Oxford, a leader of the liberal wing of the Broad Church party. Faber, G. C. (1957). *Jowett: A Portrait with Background*, pp. 425–32. Harvard University Press, Boston. Jowett, and it would appear also Campbell, respected the 'early phase' of the 'Evangelical school.' See: Campbell, L. (1897). *The Life and Letters of Benjamin Jowett*, Vol. 1, pp. 10–2. John Murray, London.

48. [CG1882], p. 184. It is plausible that Maxwell had read J. C. Hare's popular *The Victory of Faith, and Other Sermons* delivered in Cambridge about 1840. It may be significant, however, that while Maxwell's library contained numerous works of Owen and Edwards, there is no evidence that he owned the books of Hare and Maurice. See {FLCL} and a list of books (denoted here by {FLEL}) transferred in 1887 from the Free Library of Cambridge to the (then planned) Free Library of Edinburgh. Both {FLCL} and {FLEL} are in the Cambridgeshire Public Library, Central Library Archives.
49. [CG1882], p. 176.
50. [CG1882], p. 254.
51. [SLP1], p. 426.
52. [SLP1], p. 427.
53. Walker, G. (May 1883). James Clerk Maxwell. *Life & Work, a Parish Magazine*, 5: 75–9.
54. [CG1882], p. 367.
55. Peterhouse Library Maxwell Archives.
56. [CG1882], p. 264. The italics are Maxwell's. Martin resided on Union Street near Maxwell.
57. Martin, W. (1855). *On the nature and objects of the argument of Butler in his 'Analogy of Religion'*. Brown; [CG1882], pp. 294–6. Maxwell's library contained multiple copies of Bishop Butler's *Works* {FLCL}.
58. [CG1882], p. 304.
59. For example Dewar, in a speech in Aberdeen some years prior to the Disruption, indicated that in a prior period of unrestricted patronage: '*In many churches* of the Establishment, the truths of the gospel were not preached; and *you might have travelled a long summer's day* without hearing the glad tidings of *salvation* proclaimed from any pulpit.' Emphasis as quoted in: Anon. [An Observer], (1843). *A friendly expostulation with the ministers and members of the new secession of Free Presbyterians*, p. 26. William Inne, Edinburgh. Dewar's concerns were consistent with those of Chalmers.
60. Faculty in areas of religious studies were generally required to remain with the *Established Church*: Chalmers for example lost his position in Edinburgh, and there was even an unsuccessful attempt to remove Principal David Brewster of St. Andrews University (a natural philosopher) for his shift to the *Free Church*.
61. The Disruption is noted only briefly on p. 824 of Dewar, D. (1845) *The Church* in a footnote though in 1857 Dewar described the event as 'The great succession'. Correspondence suggests that Dewer retained cordial relations with Chalmers at least up to the time of the Disruption: CHA 4.303.11 and CHA 4.303.13, New College Library, Edinburgh.
62. Peter Bayne, who described himself as a 'schoolfellow' in Aberdeen at the time of the Disruption, in an account 50 years afterwards remained critical of Daniel Dewar for taking the position 'don't outrun Providence.' See: Bayne, P. (1893) *The Free Church of Scotland: her origin, founders and testimony*, p. 255. T. & T. Clark, Edinburgh. For additional information on Dewar and his family, see: Reid, Chapter 2, this volume.
63. [CG1882], p. 309.

64. [CG1882], pp. 309–12.

65. For a study of the revival see: Jeffrey, K. S. (2002). *When The Lord Walked The Land, The 1858–62 Revival in the North East of Scotland.* Paternoster. Grayfriars was located near the entrance to Marischal college and not far from Dewar's residence and from Maxwell's residence prior to his wedding. Early in his career Dewar had ministered at Greyfriars, prior to being called to the Tron Church in Glasgow.

66. [CG1882], p. 260. The term Antinomianism, invented by Luther in defense of the doctrine of 'Justification by Faith in Christ Alone,' suggests that Christians are released by grace from observing the moral law. Antinomianism is in contrast to earning salvation from legalism that Maxwell was suspicious of.

67. For Maxwell on the Parable of the Sower, Mark 4:1–20, see CG1882, p. 618.

68. Omicron, (1859). *Five Letters on the Religious Movement in Aberdeen*, p. 55. Crown.

69. [SLP1], 606–11.

70. [CG1882], p. 328–9. From the context the chapters are in 1 Cor., not in 2 Cor. as printed in [CG1882].

71. 1861 census record. The address is confirmed by a receipt in Cambridge University Library MS.Add.7655 (Maxwell archive).

72. [SLP1], pp. 653–88.

73. [SP2], pp. 786–93. Faraday's quotation is of Romans 1:20.

74. Jones, B. (1870). *The Life and Letters of Faraday*, 2nd edn. p. 479. Longmans {FLCL}.

75. Concerning Stewart, see p. 275 and 277.

76. Anon. [Jenkin, F.] (1868). *The North British Review*, 48: 111–28. Jenkin ended by noting that Lucretius 'failed in his bolder attempts to abolish the gods, and dispense with creation, or even to reconcile universal causation with free-will.' He requested comments from Maxwell prior to publication: [SLP2], p. 363. Jenkin's anonymous critique of Darwin, (1867). *The North British Review*, 46: 277–318, was more widely studied and is generally presumed to have been written years earlier.

77. Greater London Council (1974). *Survey of London Vol. 37: Northern Kensington.* Athlone Press.

78. [CG1882], p. 345.

79. John Offord's brother Robert, a coach-maker, helped finance the building of Palace Gardens Chapel [PGC]. He had ties with Rev. William Brock of Bloomsbury Chapel.

80. He had ministered for 14 years (sometimes with a famous elder, Samuel Prideaux Tregelles) at the Compton Street Chapel in Plymouth that had separated from the Plymouth Brethren. Previously, John had served various Brethren and Baptist fellowships in southwest England. For more information on John Offord and PGC see Marston (2007). Maxwell and Creation.

81. [CG1882], p. 338–40.

82. Noel, like Offord, was highly regarded by London's evangelical pastors, though he had left the Anglican communion in 1848. For more information see Marston (2007). Maxwell and Creation.

83. Cotterill, H. (1883). *Does science aid faith in regard to Creation?*. Cotterill was the Episcopal Bishop of Edinburgh.

84. Anon. (1884). *The Priest's Prayer Book*, ed. by Two Clergymen [R.F. Littledale and J.E. Vaux], p. 37, #160. Masters, London. Maxwell's prayer does not appear to be from a prior edition. If it had been in prior editions, Bishop Cotterill would likely have noted it.

85. [CG1882], p. 323.

86. From an account of Maxwell's work: Tait, P. G. (5 February 1880). *Nature*, 21: 317–21.

87. Kragh has traced Caro's interest in the heat death of the universe back to discussions of thermodynamics by Dupre and Clausius: Kragh, H. (2007). Cosmology and the entropic creation argument. *Historical Studies in the Physical and Biological Sciences*, 37: 369–82.

88. [SLP2], pp. 362–8. Also viewed by the present author at Bodleian Library, Oxford.

89. See Whitaker, Chapter 9 this volume.

90. A Biblical commentary *Revelation* by Albert Barnes, in Maxwell's possession by the 1870s and popular in London and Scotland in the 1860s, in the discussion of this verse suggested the appearance of a 'renovated universe' {FLCL}. It may be significant that Barnes' commentary was endorsed by Rev. William Brock, a Baptist with ties to PGC.

91. [SLP2], p. 306.

92. [SP2], p. 225.

93. The *New King James Version* reads 'things which are visible' at the end. The centrality of Hebrews 11:3 to Christian perspectives on origins was noted in 1857 by the German Calvinist Lange; see: Lange, J. P. (1915). *Genesis*, 6th edn., p. 71. Scribner's, New York. Hebrews 11:3 was referenced in the *Westminster Catechism* and was the opening phrase for the chapter on 'Physical Axioms' in several editions of Stewart and Tait's *Unseen Universe*.

94. Murphy, J. J. (1873). *The Scientific Basis of Faith*, pp. 203–4. Macmillan, London.

95. [SP2], p. 226.

96. Maxwell, J. C. (1871). *Theory of Heat*. Longmans, London. Section on the 'Nature and Origin of Molecules.'

97. Maxwell, J. C. (1871). *Theory of Heat* section on the 'Nature and Origin of Molecules.'

98. Hobbes, T. (1651). *Leviathan*, Sec. 15. Maxwell had studied *Leviathan* in the late 1840s.

99. Cambridge University Library, CUL MS.Add.7655 (Maxwell archive).

100. Sidgwick nevertheless remained affiliated with the University. Concerning the intellectual climate, see: Balfour, A. J. (1930). *Retrospect, an Unfinished Autobiography, 1848–1886*. Houghton Mifflin.

101. [SP2], pp. 241–55.

102. [SP2], p. 399.

103. Maxwell, J. C. (26 July 1877). [Review of Watson's] Kinetic Theory of Gases, *Nature* 16: 242–6.

104. Clausius, R. (1872). A Contribution to the History of the Mechanical Theory of Heat, *Philosophical Magazine*, 43: 106–15.

105. Maxwell, J. C. (1873). *Treatise on Electricity and Magnetism*: Article 791. Oxford University Press, Oxford.

106. Maxwell, J. C. (1873). *A Treatise on Electricity and Magnetism*: Article 81, also 55, and 128. Maxwell was less concerned with avoiding theological singularities (that concerned Hamilton and Mansel) than avoiding field singularities.

107. Guillemard is remembered for: Guillemard, W. H. (1879). *The Greek Testament, Hebraistic Edition*. Deighton, Bell & Co. and for bringing the influence of the Oxford Movement to Cambridge. Maxwell and George Gabriel Stokes (1819–1903), the Lucasian Professor of Mathematics, were known to receive communion at Guillemard's church. See: Walker, R. (1916). James Clerk Maxwell. *Aberdeen University Review*, 3 (9): 13–21.

108. Maxwell's library contained: Westcott, B. F. (1873). *On Some Points in the Religious Office of the Universities*. Macmillan, London. {FLCL}.

109. Benson, A. C. (1911). *Leaves of the Tree*, p. 45. Putnams. Concerning Westcott, Hort, and Lightfoot, see: Reardon, B. M. G. (1971). *From Coleridge to Gore*, pp. 346–55. Longman, London.

110. Maxwell corresponded with Lightfoot at various times in the late 1850s concerning the affairs of the University and his wedding plans: Lightfoot Correspondence, Durham Cathedral Library {DCL}. Maxwell also wrote Lightfoot at least three times in 1875. See also [CG1882], p. 301.

111. Westcott, A. (1903). *Life and Letters of Brooke Foss Westcott*, Vol. 1, p. 385. Macmillan, London.

112. Marston, P. L. (2007) Maxwell and Creation. See also: Hort, A. F. (1896). *Life and Letters of Fenton John Anthony Hort*, Vol. 2, pp. 184, 201–2. Macmillan, London. It is implied that during the Maxwell era the orientalist Edward Byles Cowell was also a member and that in 1877 Arthur Balfour attended and presented a paper as a guest. Cowell's participation is also noted in Lightfoot Correspondence {DCL}.

113. Henry Sidgwick in Hort, A. F. (1896). *Life and Letters of Fenton John Anthony Hort*, Vol. 2, p. 185. Stuart emphasized Sidgwick's leadership: Stuart, J. (1911). *Reminescences*, pp. 195–6. Chiswick, London.

114. CUL MS.Add.7655 (Maxwell archive) contains the announcement of V. H. Stanton's Eranus discussion for 28 May 1878 and one for a discussion by 'C.T.' (Trotter) on 26 Nov. 1878.

115. [SP2], p. 322.

116. Maxwell's attention during the summer of 1873 had been given partly to the determination of molecular properties from macroscopic observables, such as transport coefficients and optical scattering and refractive properties: [SP2], p. 341; [SLP2], p. 890, and [SLP2], p. 919.

117. [SP2], p. 376; Maxwell, J. C. (25 September 1873). Molecules, *Nature*, 8, 437–41. Of this lecture *The Guardian* (1 October 1873), p. 1255, noted 'A course of reasoning which leads strait up to a Personal Creator is, unhappily, so rare a phenomenon in the records of the British Association, that it deserves a very special attention . . . '. Campbell and Garnett ([CG1882], p. 358) indicated Molecules 'has been more often quoted than, perhaps any other of his writings.'

118. Herschel, J. F. D. (1830). *A Preliminary Discourse on the Study of Natural Philosophy*, Article 28. London. Reprinted in *Edinburgh Cabinet Library* {FLCL}.

119. Whewell, W. (1833). *Bridgewater Treatise on Astronomy and General Physics*. Pickering. In Darwin's *Origin* (1859) an epigraph from this book was printed facing the title page.

120. Maxwell had, however, previously remarked on 'manufactured articles' in the context of 'natural selection' and 'permanence of species' in a humorous account of the history of electrometers in his anonymous review (January 1873, *Nature*) of William Thomson's *Reprint of Papers on Electrostatics and Magnetism*. Maxwell may have noticed an addendum by Thomson (p. 180) requesting the publication of the electrical manuscripts of Cavendish, a task completed by Maxwell in 1879.

121. [SP2], pp. 376–7.

122. Chalmers, T. (1833). *Bridgewater Treatise on the Power, Wisdom, and Goodness of God*, p. 20. Pickering. {possibly in FLCL in one of various works by Chalmers}.

123. [SP2], p. 377. P. Theerman (1986). James Clerk Maxwell and Religion has noted that 'perfect in number and measure and weight' is an allusion to Wisdom of Solomon 11:20 from the old testament apocrypha which reads ' . . . thou hast ordered all things in measure and number and weight'.

124. Lodge, O. (1931). *Past Years*, p. 137. Hodder & Stoughton, London.

125. CUL Manuscripts. See Darwin Correspondence Project: <http://www.darwinproject.ac.uk>.

126. CUL Manuscripts. See Darwin Correspondence Project: <http://www.darwinproject.ac.uk>. It is highly likely that Darwin had by this time received a favourable opinion of Maxwell from his son, George, who was returning to Cambridge to study physics.

127. Westcott, B. F. (1892). *The Gospel of Life: Thoughts Introductory to the Study of Christian Doctrine*, pp. 22–4. Macmillan, London.

128. Marston (2007). Maxwell and Creation.

129. Anon. [Maxwell, J. C.] (15 January 1874). [Review of Stewart's] The Conservation of Energy, *Nature*, 9: 198–200. During the 1870s Maxwell is known to have published other unsigned book reviews in Nature. *The Conservation of Energy* was in Maxwell's library. It is included in a list of many of Maxwell's scientific books (denoted here by {Fenwick}) in: Fenwick, E. (2011). *James Clerk Maxwell and Nineteenth Century Physics, A Bibliographical Catalogue and Biographical Guide*. Martino Pub., Mansfield Center, CT.

130. Clifford, W. K. The First and the Last Catastrophe, reprinted in: Clifford, W. K. (1879). *Lectures and Essays*, Vol. 1, pp. 191–227. Macmillan, London.

131. To put Clifford's perspective in a broader context, consider a recollection of the 1870s (likely written in the 1890s) by his contemporary F. W. H. Meyers (1843–1901): 'It must be remembered that this was the very flood-tide of materialism, agnosticism,—the mechanical theory of the Universe, the reduction of all spiritual facts to physiological phenomena. It was a time when not the intellect only but the moral ideals of men seemed to have passed into the camp of negation. We were all in the first flush of triumphant Darwinism, when terrene evolution had explained so much that men hardly cared to look beyond. Among my own, W. K. Clifford was putting forth his series of triumphant proclamations of the nothingness of God, the divinity of man.' Meyers, F. W. H. (1904). *Fragments of Prose & Poetry*. Longmans, London. Maxwell's surviving remarks concerning Clifford were considerably softer.

132. This translation is Psalm 111:2, RSV. Psalm 111:2 is an epigraph of a book by J. Pye Smith (1840) in Maxwell's library.

133. Lodge, O. (1931). *Past Years*, p. 138.

134. Lodge, O. (1931). *Past Years*, p. 139.

135. Tyndall, J. (20 August 1874). Inaugural Address. *Nature*, 10: 309–19. An early reprint of Tyndall's *Belfast Address* in my possession has an opposing commentary by the American Joseph Cook (October 1874) pasted on the covers. The commentary (with a favorable epigraph from Maxwell's 1873 lecture on 'Molecules') was later reprinted in: Cook, J. (1877). *Boston Monday Lectures, Biology*, Osgood, Boston.

136. Kim, S. S. (1996). *John Tyndall's Transcendental Materialism and the Conflict between Religion and Science in Victorian England*. Mellen University Press.

137. Macloskie, G. (27 August 1874). *Northern Whig*, p. 8 in a letter, accused Tyndall of saying in the Address 'sin is a mere defectibility of matter'.

138. Robertson Smith, W. (22 August 1874). *Northern Whig*, p. 8 in a letter. This broad but sometimes controversial scholar had worked with P. G. Tait. As a cautionary measure, Tyndall, who had composed his address while in the Alps during the weeks prior to the meeting, had sent an advance copy to Charles Darwin for his impressions. CUL Manuscripts. See also Darwin Correspondence Project: <http://www.darwinproject.ac.uk>.

139. Several examples are listed in Marston (2007). Maxwell and Creation. Some of the important examples referring to favourably Maxwell omitted there include: Hutton, R. H. (1874). The Materialists' Stronghold. *Spectator*, reprinted in Hutton, R. H. (1901). *Aspects of religious and scientific thought: selected from the Spectator*. Macmillan, London; McCosh, J. (1875). Ideas in Nature Overlooked by Dr. Tyndall. *International Review*, 2: 88–111; Martineau, J. (1876). Modern Materialism: Its Attitude Towards Theology. *Contemporary Review*, 27: 323–46, 522–48; and Hodge, A. A. (1878). *Outlines of Theology, Re-Written and Enlarged*, pp. 34, 35, 241. Carter, NY. Maxwell may have been aware of Hutton's interest since his library contained several copies of Hutton's journal *Spectator* which followed the controversy {FLCL}. Hodge also quotes Maxwell's later article on 'Atoms' that 'matter . . . must have been created'.

140. [CG1882], pp. 639–41.

141. [CG1882], pp. 638. See also [SLP3], p. 100.

142. Peterhouse Library Maxwell Archives. See also [SLP3], pp. 100, 131–2.

143. Wurtz, M. (27 August 1874). The Theory of Atoms in the General Conception of the Universe. *Nature*, 10: 345–50. Wurtz concludes: 'Such is the order of nature, and as science penetrates it farther, she brings to light both the simplicity of the means set at work and the infinite variety of the results. Thus, through the corner of the veil we have been permitted to raise, she enables us to see both the harmony and the profundity of the plan of the universe. Then we enter on another domain which the human spirit will be always impelled to enter and explore. It is that, and you cannot change it. It is in vain that science has revealed to it the structure of the world and the order of all the phenomena; it wishes to mount higher, and in the conviction that things have not in themselves their own *raison d'être*, their support and their origin, it is led to subject them to a first cause—unique, universal God.'

144. Monro, C. J. (15 October 1874). Manufactured articles. *Nature*, 10: 481. The same issue contained Maxwell's review of Van der Waals' dissertation *On the Continuity of the Gaseous and Liquid States*.

145. Pye-Smith, P. H. (12 November 1874). *Nature*, 11: 22–4. It would have been appreciated in 1874 that by reputation, Virchow was no proponent of Christianity.

146. [SP2], p. 461.

147. Darwin, C. (1868). *The Variation of Animals and Plants Under Domestication*, Vol. 2. J. Murray, London.

148. [CG1882], pp. 390–1.

149. Darwin appreciated the nature of Maxwell's criticism and, with the help of his son George, included a discussion of molecular size in his subsequent publication on pangenesis.

150. Chambers, W. F. D. (1910). *Radio-activity and Vortex Theories*. Williams, London. Pamphlet archived at Bodleian Library, Oxford.

151. Anon. [Stewart, B. and Tait, P. G.] (1875). *The Unseen Universe or Physical Speculations on a Future State*, Articles 213, 214, p. 167 of 1st edn. Macmillan, London. A subsequent edition also in my possession is heavily marked with margin notes from the 1880s.

152. Maxwell's criticism is evident from his 1879 review of a later book by Stewart and Tait discussed subsequently.

153. Anon. [Stewart, B. and Tait, P. G.] (1875). *The Unseen Universe*, Art. 89, p. 67.

154. For example, Robert Watts, a Presbyterian Belfast theologian (who had previously been highly critical of Tyndall's Belfast Address) repeatedly during the following decade took the authors of the *Unseen Universe* to task for their poor understanding of Scripture. His criticism, originally published in journal essays, was eventually reprinted in a single volume, Watts, R. (1888). *The Reign of Causality*. Clark, Edinburgh. Watts had studied under Charles Hodge at Princeton.

155. Tyndall, J. (1875). Materialism and its Opponents. *Fortnightly Review*, 24: 579–99.

156. [CG1882], pp. 405–6.

157. CUL MS.Add.7655 (Maxwell archive).

158. [CG1882], p. 394. In his reply Maxwell also noted that the latter part of his encyclopedia article 'Atoms' was intended to meet C. J. Monro's published criticism.

159. [CG1882], pp. 394–5.

160. Lightfoot, J. B. (1875). *St. Paul's Epistles to the Colossians and to Philemon*. Macmillan, London. The relevant scripture includes Col. 1:12–20. In Maxwell's note of appreciation to Lightfoot for a copy of this book, he bestows a blessing on Lightfoot from Ephesians 3:8–12 (11 May 1875) Lightfoot Correspondence {DCL}.

161. FLEL includes: Henry Cotterill, *On the True Relations of Scientific Thought and Religious Belief* (reprint from: (1878). *Journal of the Transactions of the Victoria Institute*, 12: 312–41).

162. [CG1882], p. 371.
163. [CG1882], p. 387.
164. Riddell, J. Oliver [Aitken, Jean R.] (1930). *Clerk Maxwell of Glenlair: A Centenary Booklet*. Adam Rae, Castle Douglas.
165. Notice of the Parochial Board of Parton for 2 November 1877 addressed to Maxwell. CUL MS.Add.7655 (Maxwell archive). It is plausible that the region was affected by the British agricultural recession of the 1870s.
166. Greenhill, A. G. to Maxwell, 27 February 1878, CUL MS.Add.7655 (Maxwell archive).
167. Morley, J. (1917). *Recollections*, Vol. 1, p. 88. Macmillan, London. Morley became a well-known British statesman.
168. [CG1882], pp. 452–63.
169. Anon. [Stewart, B. and Tait, P. G.] (1878). *Paradoxical Philosophy*. Macmillan, London. The preface is dated October 1878.
170. Milton, *Paradise Lost* VIII: Lines 277–9.
171. The partial parallel between the Paradoxical Society and the Eranus Club could not have escaped Maxwell, and he deemed it appropriate to inject material from his latest Eranus essay, which was at the time unpublished
172. Maxwell likely read *New Republic* as it was released in the periodical *Belgravia*, present in his collection {FLCL}.
173. Stoffkraft as a conjunction of the German for material and force.
174. [SP2], p. 762. See also Maxwell's 1858 poem 'To His Wife' which notes 'Love only, cleansed from sins and cares, Shall live beyond the grave'. [CG1882], pp. 608–9.
175. Silver, D. S. (2008). The Last Poem of James Clerk Maxwell, *Notices of the AMS*, 55: 1266–70. From Maxwell's statement to Hort it is unlikely that he knew he was dying when he composed this poem, contrary to what is implied by Silver.
176. Haeckel neglected to mention that the term 'plastidule' originated in an 1874 address by an American physician Elsberg as 'the smallest partition-particles, or molecules, of a plastid,' plastid being Haeckel's early term for the formative matter of a cell. See: Elsberg, L. (1874). Regeneration; or, The Preservation of Organic Molecules; a Contribution to the Doctrine of Evolution. *Proceedings of the AAAS* 23.
177. As quoted by Maxwell, [CG1882], pp. 456–7. In this essay and in his review of *Paradoxical Philosophy*, Maxwell was critical of Nägeli's 'levelling up' of the distinction between organic and inorganic by designating qualities of life to inorganic objects. Concerning 'levelling up' see: Watson, J. (1895). *Comte, Mill, and Spencer*, pp. 101–3. Maclehose, Glasgow.
178. Drysdale, J. J. (1877). Presidential Address 1877. *Literary and Philosophical Society of Liverpool*. According to Drysdale, Haeckel's disciple (Posner) says he dismisses a critic because he 'is a believer in a God and Creator'. Drysdale adds that according to Posner '. . . you must deny God and trample the Cross under foot before you can become even a scholar, far less a master in Natural Science'. Assuming Maxwell read that part of Drysdale's essay, he didn't feel it appropriate to report his reaction. Maxwell had in his library Drysdale's *Protoplasmic Theory of Life* (1874) {Fenwick}.
179. The letter, quoted in part in M. Chisholm's *Such Silver Currents: The Story of William and Lucy Clifford, 1845–1929* is presently archived in Trinity College Library GBR/0016/CLIF/A7/1.
180. Perhaps Clifford's best known Metaphysical Society essay, 'The Ethics of Belief', appeared in *The Contemporary Review* in 1877. This essay, arguing the immorality of religious belief not considered scientifically provable, was reprinted in Clifford, W. K. (1879). *Lectures and Essays*.

181. [CG1882], p. 420. Clifford and Maxwell shared ideas in May 1871: TCL GBR/0016/CLIF/A/13.

182. [CG1882], p. 420. By 1880 a lengthy anonymous review of *Lectures and Essays* appeared in *Edinburgh Review*, within a few years known as written by W. H. Mallock.

183. Maxwell, J. C. (1879). [Report on] Cavendish Laboratory, *Cambridge University Reporter*, 676–7. The Attendant mentioned was likely William Pullin: [SLP3], p. 795.

184. Lent Term (Electrostatics) 9 students and Easter Term (Electro-magnetism) 3 students.

185. Walker, G. (May 1883). James Clerk Maxwell, *Life & Work, a Parish Magazine* 5, pp. 75–9. The text is from 1 Corinthians 2:9 and Isaiah 64:4. See also 2 Corinthians 4:18.

186. [CG1882], p. 414 from letter of May 1881. It is noteworthy that George Herbert was also a favourite author of Maxwell's early mentor, George Wilson (according to a memorial service for Wilson in 1859 archived in the National Library of Scotland).

187. [CG1882], p. 415–6. See 1 John 4:18 on 'love had cast out fear.'

188. [CG1882], p. 421.

189. [CG1882], p. 421. This agrees with what Maxwell said of himself in Galton's survey Hilts, V. L. (1975). *A Guide To Francis Galton's English Men Of Science.*

190. Rupp, G. (1981). *The Historical Journal* 24: 412–28; Hort, A. F. (1896). *Life and Letters of Fenton John Anthony Hort*, Vol. 2, p. 231.

191. [CG1882], p. 419.

192. Hort, A. F. (1896). *Life and Letters of Fenton John Anthony Hort*, Vol. 2, p. 276.

193. C. T. [probably Eranus member, Coutts Trotter], (Wednesday, 12 November 1879). The Late Professor Clerk Maxwell, *Daily News*, London.

194. [CG1882], p. 423. H. M. Butler became the Master of Trinity College in 1886.

195. For example Maxwell's fellow Eranus member, V. H. Stanton, contributed to a session of the Anglican Church Congress of 1875 in which Maxwell's views were noted by ArchDeacon Reichel. Stanton cited related work in his lectures on *Bishop Butler's Analogy in relation to his times & our own* about 25 years later. Stanton's undated lecture notes are in a storage room of the Westcott House Library. He referenced the published 1896–98 Gifford Lectures of James Ward that examined the controversy surrounding Maxwell's 1873 lecture. Stanton also cited Robert Flint's, *Theism* (Blackwood, Edinburgh, 1877), whose favourable mention of Maxwell incited criticism by Romanes, as discussed in the 'Remembering Maxwell as a Man of Faith' section. He had earlier cited Flint's *Theism* in: Stanton, V. H. (1891). *The place of authority in matters of religious belief,* pp. 212–3. Longmans, London. Lightfoot mentions Tyndall's Belfast Address in his surviving Eranus essay {DCL}.

196. Stanley, M. (2008). The Pointsman: Maxwell's demon, Victorian free will, and the boundaries of science. *Journal of the History of Ideas*, 69: 467–91; Nye, M. J. (1976). The moral freedom of man and the determinism of nature: The Catholic synthesis of science and history in the Revue des questions scientiques. *British Journal for the History of Science*, 9: 274–92; Jeans, J. (1943). *Physics and Philosophy.* Macmillan, London; Watts, R. (1888). *Reign of Causality.* Clark, Edinburgh; Harris, S. (1883). *The Philosophical Basis of Theism*, p. 368. Scribner's, New York; Flint, R. (1880). *Anti-Theistic Theories.* Blackwood, Edinburgh.

197. [CG1882], p. 438.

198. [CG1882], p. 336.

199. [CG1882], p. 444.

200. Gladstone archives, British Library.

201. Edwards, as quoted in Watts, R. (1880). *Reign of Causality.* Clark, Edinburgh. Maxwell's library contained the works of Edwards {FLCL}.

202. Boussinesq, J. (1878). *Conciliation Du Veritable Determinisme Mecanique: Avec L'Existence de La Vie Et de La Liberte Morale*. Gauthier-Villars, Paris. {Fenwick}. Boussinesq was affiliated with the Scientific Society of Brussels.

203. [SLP3], pp. 756–8. The obscure term Fixt Fate is used by Milton, one of Maxwell's favourite authors, *Paradise Lost* (II; Line 560), in a discussion of free will.

204. Hunt, B. R. and Yorke, J. A. (1993). Maxwell on chaos. *Nonlinear Science Today*, 3: 1–4.

205. Anon. [Maxwell, J. C.] (15 January 1874). [Review of Stewart's] The conservation of energy. *Nature*, 9: 198–200.

206. Westcott, B. F. (11 February 1882) as quoted in *The Christian Life* 8, 68. Through Westcott's 1892 treatise *Gospel of Life* we know of Maxwell's anonymous 1874 book review.

207. Physicus [Romanes, G. J.] (1878). *A Candid Examination of Theism*, pp. 152–6. London.

208. Since Flint was aware of the criticism by Physicus, and eventually of his identity as Romanes, it is plausible he informed P. G. Tait of this in the 1890s for they knew each other on the faculty of the University of Edinburgh. Flint also knew Lewis Campbell and Westcott.

209. Romanes, G. J. (1902). *Thoughts on Religion*, pp. 137–8. Longmans, London. Romanes resided for a period in the 1870s in Cambridge, though it is not known if he had personal knowledge of Maxwell.

210. CUL Darwin Collection, p. 112 of Physicus [Romanes, G. J.] (1878). *A Candid Examination of Theism*. The cursive revision was compared with Romanes' handwriting. See also: Romanes, E. (1896). *The Life and Letters of George John Romanes*, pp. 81–91. Longmans, London.

211. Cox, J. (1913). *Beyond the Atom*, pp. 12–4. Cambridge University Press, Cambridge. Cox (McGill Univ.) when a student had known Maxwell in the 1870s: CUL MS.Add.7655 (Maxwell archive).

212. Thomson, J. Arthur (1920). *The System of Animate Nature: The Gifford Lectures in the Years 1915 and 1916*, Vol. 1, pp. 65–6. Holt, New York.

213. Jeans, J. (1930). *Astronomy and Cosmology*, p. 1. Cambridge University Press, Cambridge.

214. Casimir, H. (1983). *Haphazard Reality—Half a Century of Science*, p. 50. Harper & Row.

215. LMA Acc. 1063. Litchfield had not previously understood aspects of Maxwell.

216. Direct remarks by Maxwell on the nebular hypothesis are not known.

217. Anon. (April, 1883). *London Quarterly Review*, pp. 1–29; reprinted in the American periodical: *The Living Age*. It also refers to recent developments in optics and reveals a direct knowledge of Maxwell's publications.

218. Anon. (April, 1883). *London Quarterly Review*, p. 29.

219. Gladstone archives, British Library. As mentioned earlier, Gladstone also noted Maxwell's discussion of free will.

220. Glazebrook, R. T. (1901). *James Clerk Maxwell and Modern Physics*, Cassell, London.

221. Pearson, K. (February, 1936). Old Tripos Days at Cambridge, as Seen from Another Viewpoint. *The Mathematical Gazette*, 20: 27–36. The comparison is likely an allusion to: Gladstone, W. E. (1890). *The Impregnable Rock of Holy Scripture*, Isbister, London.

222. [CG1882], p. 322.

223. Pearson had not been attracted to Maxwell's lecture style, having withdrawn from his class on heat; he commented also that Maxwell's lectures were not good preparation for the Mathematical Tripos exam. He was also likely unaware of Maxwell's deteriorating physical condition at the time of his visit to Maxwell's home, and Katherine would have likely also been present to hear his disrespect for the Bible. Pearson made his perspective clear when he castigated G. G. Stokes for contributing to 'the mire of natural theology'. See: Pearson, K. (1888). *Ethic of Freethought: A Selection of Essays and Lectures*, pp. 33–53. Unwin, London.

224. Pearson, K. (1978). *The History of Statistics in the 17th and 18th Centuries: Against the Changing Background of Intellectual, Scientific, and Religious Thought/Lectures by Karl Pearson Given at University College, London, During the Academic Sessions, 1921–1933*, E. S. Pearson, (ed.) p. 360. Macmillan, New York. Pearson's remark was evidently made when he had recently discovered Maxwell's 1879 letter to Galton concerning free will.

225. Darwin, G. H. (1911). *Scientific Papers*, Vol. 4, p. 306. Cambridge, Cambridge University Press.

226. Tyndall, J. (1882). Atoms, Molecules & Ether Waves. *Longman's Magazine*, 1: 29–40. Maxwell's expression 'foundation stones' remained of interest as late as 1925: Larmor, J. (1925). The Permanence of Substance. *Nature*, 115: 231–2.

227. [CG1882], p. 431.

228. Huxley, T. H. (1866). On the Advisableness of Improving Natural Knowledge. Reprinted in: Huxley, T. H. (1871). *Lay Sermons*, p. 18. Appleton, New York.

229. Clifford, W. K. (1877) Ethics of Religion. *Fortnightly Review*. Reprinted in Clifford, W. K. (1879). *Lectures and Essays*.

230. Lightfoot's Cambridge lectures in the 1860s and 1870s refuted the new views: Lightfoot, J. B. (1893). *Biblical Essays*, Macmillan, London. See also: Westcott, B. F. (1881). *Gospel of St.John*, Murray, London. It was not until the twentieth century that papyri were discovered directly supporting the traditional early dating of John's Gospel: Metzger, B. M. (1964). *The Text of the New Testement*, pp. 36–42, Oxford University Press. Maxwell's library contained a detailed essay by George Campbell defending the traditional authorship of John's gospel printed with Campbell's translation *The Four Gospels* {FLCL}. Lightfoot's Essays on Supernatural Religion (1874–1877) are also relevant.

231. Luther, M. (1961). *Lectures on Romans*. Westminster Press, Philadelphia.

232. For additional discussion of Hebrews see Westcott's *Hebrews* or the various commentaries in Maxwell's library noted in FLCL.

233. See e.g.: Stokes, G. G. (1907). Letter to A. H. Tabrum in 1895. In *Memoir and Scientific Correspondence Volume 1*, pp. 76–8. Cambridge University Press, Cambridge.; Lincoln, J. L. (1894). James Clerk Maxwell. In *In Memorium, John Larkin Lincoln 1817–1891*, pp. 544–67. Houghton Mifflin, Boston; Strutt, R. J. (1968). *Life of John William Strutt, Third Baron Rayleigh*, p. 360. University of Wisconsin, Madison.

234. Spencer Jones, H. (1956). In L. H. Ahrens, K. Rankama, and S. K. Runcorn (eds.) *Physics and Chemistry of the Earth*, Vol. 1, p. 1. Pergamon, London.

235. CUL MS.Add.7655 (Maxwell archive).

236. Abercrombie, J. (1841). *Think on These Things*. A copy was in Maxwell's and Katherine's library {FLEL}.

237. Blackburn, J. (1988). *Jemima*, p. 100. Katherine's family also knew John Abercrombie. Daniel Dewar's *Evidences of Divine Revelation* (1838) {FLCL} was dedicated to him.

238. Some of these comments were previously quoted. Maxwell's sense of reverence for his Creator is also evident from an undated religious sketch attributed to him and archived in the Peterhouse Library with his family correspondence. Henry J. S. Smith, an associate of Maxwell's from the London Mathematical Society and the British Association for the Advancement of Science, mentions Maxwell's 'personal love of Christ' in: Smith, H. J. S. (1883). [Review of] Life of James Clerk Maxwell. *The Academy*, 558 (13 January), p. 19, 20; 559, (20 January), p. 35, 36. The quotation is on p. 36.

239. Stokes, G. G. (1907). Letter to A. H. Tabrum in 1895. Stokes, G. G. (1907). *Memoir and Scientific Correspondence*, pp. 76–8. Lady Stokes described Maxwell as 'like the shadow of a rock' near the day of his death, CUL MS.Add.7656/S1437 (Stokes archive).

240. See Campbell's poetic tribute to Maxwell prior to the Preface of [CG1882]. The poem opens with an allusion to the depiction of Christ in Hebrews 5:9. See also Hort, A. F. (1896). *Life and Letters of Fenton John Anthony Hort*, Vol. 2, pp. 230 and 276.
241. [CG1882], p. 426.

CHAPTER 15

1. Campbell and Garnett date the poem as 'June ? 1856'. Campbell, L. and Garnett, W. (1882). *The Life of James Clerk Maxwell*, p. 599. Macmillan, London. [CG1882]
2. [CG1882], p. 205.
3. [CG1882], p. 114.
4. Obituary Notices of Fellows Deceased (1881). James Clerk Maxwell. *Proceedings of the Royal Society*, 33: i–xvi.
5. Everitt, C.W.F. (1975). *James Clerk Maxell: Physicist and Natural Philosopher*, p. 102. Charles Scribner's Sons, New York.
6. [CG1882], p. 35.
7. [CG1882], p. 83.
8. [CG1882], p. 5.
9. [CG1882], p. 153.
10. Lamb, H. (1931). Clerk Maxwell as Lecturer, in *James Clerk Maxwell: A Commemoration Volume 1831–1931*, p. 146. Cambridge University Press, Cambridge.
11. Tolstoy, I. (1981). *James Clerk Maxwell: A Biography*, pp. 154–5. Canongate, Edinburgh.
12. [CG1882], p. 28.
13. Frisch, O. R. (1979). *What Little I Remember*, p. 86. Cambridge University Press, Cambridge.
14. Garnett, W. (1879). James Clerk Maxwell, FRS. *Nature*, 21: 43–4.
15. Thompson, S. P. (1910). *Life of Lord Kelvin*, Vol. 2, pp. 694–5. Macmillan & Co, London.
16. Tait, P. G. (1880). Clerk-Maxwell's Scientific Work, *Nature*, 21: 317–21, p. 319.
17. Obituary Notices of Fellows Deceased (1881). James Clerk Maxwell. *Proceedings of the Royal Society*, 33: i–xvi.
18. For the story of the creation and siting of Edinburgh's Maxwell statue, see Stoddart, A (2008), Notes on the Maxwell Monument, George Street, Edinburgh, in Forfar, D. (ed.), *Celebrating the Achievements and Legacy of James Clerk Maxwell*, Royal Society of Edinburgh, Edinburgh.
19. The Kelvin Centenary (1924). *Nature*, 114(2855): 103–5.
20. *James Clerk Maxwell: A Commemoration Volume 1831–1931*, Cambridge University Press 1931. Bohr's lecture did not appear in the *Commemoration Volume*, possibly because as a fastidious author he missed the deadline for copy. It was published shortly after the event: Bohr, N. (1931). Maxwell and modern theoretical physics. *Nature*, 128(3234): 691–2.
21. Letters of invitation to Einstein from Patrick Blackett can be found at 1–99, 1–100, 1–101, Albert Einstein Archives, Hebrew University of Jerusalem.
22. Einstein, A. (1996). *Autobiographical Notes*, pp. 31–3. Open Court, Illinois.
23. Einstein, A. (1996). *Autobiographical Notes*, p. 59. Open Court, Illinois.
24. Salaman, E. (1955). A Talk with Einstein. *The Listener*, 8 September, 370–1.

25. Jeans, J. (1931). Clerk Maxwell's Method in *James Clerk Maxwell: A Commemoration Volume 1831–1931*, pp. 97–8. Cambridge University Press, Cambridge.

26. Thomson, J. J. (1931). James Clerk Maxwell, in *James Clerk Maxwell: A Commemoration Volume 1831–1931*, p. 14. Cambridge University Press, Cambridge.

27. Maxwell, J. C. (1890). Illustrations of the Dynamical Theory of Gases, in [SP1].

28. Jeans, J. (1931). Clerk Maxwell's Method in *James Clerk Maxwell: A Commemoration Volume 1831–1931*, p. 93. Cambridge University Press, Cambridge.

29. [CG1882], p. 133, footnote 2.

30. Tait, P. G. (1879). James Clerk-Maxwell, *Proceeding of the Royal Society of Edinburgh*, 10: 331–9, p. 317.

31. See, for example, Domb, C. (ed.) (1963). *Clerk Maxwell and Modern Science*. The Athlone Press, London.

    Domb, C. (1979). James Clerk Maxwell – 100 years later. *Nature*, 282: 235–9; Berger, M. S. (ed.) (1984), *J.C. Maxwell, The Sesquicentennial Symposium*, Elsevier, Amsterdam; *James Clerk Maxwell Commemorative Booklet*, (1999). Fourth International Congress on Industrial and Applied Mathematics. Pub. James Clerk Maxwell Foundation, Edinburgh; James Clerk Maxwell 150 years on. *Philosophical Transactions of the Royal Society A*. (Themed issue), (2008) 366(1871): 1649–1874; Forfar, D. ed., (2008) *Celebrating the Achievements and Legacy of James Clerk Maxwell*, Royal Society of Edinburgh, Edinburgh.

    <http://www.nature.com/news/specials/maxwell/index.html> [accessed December 2012].

32. Durrani, M. and Rodgers, P. (1999). Physics: past, present, future. *Physics World*, 12(12): 7–13. IOP, Bristol.

33. Crease, R. P. (2004). The greatest equations ever. *Physics World*, 17(10): 14–5. IOP, Bristol.

34. Peierls, R. E. (1963). Field Theory Since Maxwell; in Dombe, C. (ed.) (1963) *Clerk Maxwell and Modern Science*, p. 26. The Athlone Press, London.

35. Hardy, G. H. (1967). *A Mathematician's Apology*, p. 85. Cambridge University Press, Cambridge.

36. Farmelo, G. (2009). *The Strangest Man*, p. 359. Faber & Faber, London.

37. Fitzgerald, G. F. (1893). Electrical papers by Oliver Heaviside (review). *The Electrician*, 31: 389–90.

38. [CG1882], p. 433.

INDEX

Note: Page numbers in bold relate to illustrations.

C

D

E

F

G

H

I

J

K

L

M

N

O

P

Q

R

S

T

U

V

W

X

Y

Z

The manufacturer's authorised representative in the EU for product safety is Oxford University Press España S.A. of El Parque Empresarial San Fernando de Henares, Avenida de Castilla, 2 - 28830 Madrid (www.oup.es/en or product.safety@oup.com). OUP España S.A. also acts as importer into Spain of products made by the manufacturer.
Printed and bound by CPI Group (UK) Ltd, Croydon, CR0 4YY
14/09/2026
02216284-0017